HYDROCYCLONES

FLUID MECHANICS AND ITS APPLICATIONS
Volume 12

Series Editor: R. MOREAU
MADYLAM
Ecole Nationale Supérieure d'Hydraulique de Grenoble
Boîte Postale 95
38402 Saint Martin d'Hères Cedex, France

Aims and Scope of the Series

The purpose of this series is to focus on subjects in which fluid mechanics plays a fundamental role.

As well as the more traditional applications of aeronautics, hydraulics, heat and mass transfer etc., books will be published dealing with topics which are currently in a state of rapid development, such as turbulence, suspensions and multiphase fluids, super and hypersonic flows and numerical modelling techniques.

It is a widely held view that it is the interdisciplinary subjects that will receive intense scientific attention, bringing them to the forefront of technological advancement. Fluids have the ability to transport matter and its properties as well as transmit force, therefore fluid mechanics is a subject that is particulary open to cross fertilisation with other sciences and disciplines of engineering. The subject of fluid mechanics will be highly relevant in domains such as chemical, metallurgical, biological and ecological engineering. This series is particularly open to such new multidisciplinary domains.

The median level of presentation is the first year graduate student. Some texts are monographs defining the current state of a field; others are accessible to final year undergraduates; but essentially the emphasis is on readability and clarity.

For a list of related mechanics titles, see final pages.

Hydrocyclones

Analysis and Applications

edited by

L. SVAROVSKY
University of Bradford, U.K.

and

M. T. THEW
University of Southampton, U.K.

KLUWER ACADEMIC PUBLISHERS
DORDRECHT / BOSTON / LONDON

ISBN 0-7923-1876-5

Published by Kluwer Academic Publishers,
P.O. Box 17, 3300 AA Dordrecht, The Netherlands.

Kluwer Academic Publishers incorporates
the publishing programmes of
D. Reidel, Martinus Nijhoff, Dr W. Junk and MTP Press.

Sold and distributed in the U.S.A. and Canada
by Kluwer Academic Publishers
101 Philip Drive, Norwell, MA 02061, U.S.A.

In all other countries, sold and distributed
by Kluwer Academic Publishers Group,
P.O. Box 322, 3300 AH Dordrecht, The Netherlands.

Front cover courtesy of A. B. Sinker and M. T. Thew,
University of Southampton, UK.

All Rights Reserved
© 1992 Kluwer Academic Publishers
No part of the material protected by this copyright notice may be reproduced or
utilized in any form or by any means, electronic or mechanical,
including photocopying, recording or by any information storage and
retrieval system, without written permission from the copyright owner.

Produced by Technical Communications (Publishing) Ltd., Letchworth, England.

Printed by Information Press Ltd., Oxford, England.

This volume consists of papers presented at the 4th International Conference on Hydrocyclones, 23-25 September 1992 in Southampton U.K.

ACKNOWLEDGEMENTS

The valuable assistance of the Technical Advisory Committee and panel of referees is gratefully acknowledged.

TECHNICAL ADVISORY COMMITTEE

Dr L Svarovsky (Chairman)	University of Bradford
Dr E G Arato	BHR Group Ltd
Mr K Bassiti	BHR Group Ltd
Prof M T Thew	University of Southampton
Mr H J Walker	Vortoil Separation Systems Ltd

CORRESPONDING MEMBERS

Mr L Chu	Northeast University of Technology, China
Prof J Listewnik	Maritime University of Szczecin, Poland
Mr P G Michaluk	Serck Baker Ltd, UK
Dr V Milisic	IFTS, France
Mr R Mozley	Richard Mozley Ltd, UK
Dr T J Napier-Munn	J K Mineral Research Centre, Australia
Prof H F Trawinski	Amberger Kaolin-Werke GmbH, Germany (retired)

Organised and sponsored by BHR Group Limited
Co-sponsored by the Institution of Mechanical Engineers
Endorsed by the Department of Trade and Industry

Contents

4th International Conference on
HYDROCYCLONES
Southampton, UK : 23-25 September 1992

Foreword
L Svarovsky : University of Bradford; M T Thew : University of Southampton, UK

Opening Address : The Hydrocyclone - A Story of Continuous Innovation
J M Stinson : Chairman and Managing Director, Conoco (UK) Ltd, UK

FUNDAMENTALS AND NEW AREAS

Bradley Hydrocyclones : Design and Performance Analysis   3
M Antunes : Institut de Recherches sur la Catalyse, France; R A Medronho : Federal University of Rio de Janiero, Brazil

The Characteristics of Hydraulic Interaction Between Adjacent Hydrocyclones   15
A B Sinker, M T Thew : University of Southampton, UK

An Evaluation of Experimental Design Procedures for Hydrocyclone Modelling   31
J J Cilliers, R C Austin, J P Tucker : University of Cape Town, South Africa

The Effect of the Air Core on the Flow Field within Hydrocyclones   51
Q Luo, J R Xu : Northeast University of Technology, China

An Evaluation of the Plitt and Lynch & Rao Models for the Hydrocyclones   63
M A Z Coelho, R A Medronho : Federal University of Rio de Janiero, Brazil

Determination of Phase Behaviour in De-watering Hydrocyclones using Resistivity Measurements : Feasibility Study   73
M T Thew, M Becker : University of Southampton, UK

Fluid Flow Model of the Hydrocyclone for Concentrated Slurry Classification   95
R K Rajamani, L Milin : University of Utah, USA

Prospects for the use of Hydrocyclones for Biological Separations   109
D Rickwood : Essex University; J Onions, B Bendixen, I Smyth : University of Southampton, UK

A Parametric Evaluation of the Hydrocyclone Separation of Drilling Mud from Drilled Rock Chippings   121
K J Walker, T J Veasey, I P T Moore : University of Birmingham, UK

DESIGN AND OPERATION

A New Method of Testing Hydrocyclone Grade Efficiencies 135
L Svarovsky : University of Bradford; J Svarovsky : Bradford Grammar School, UK

Effect of Solids Feed Grade on the Separation of Slurries in Hydrocyclones 147
E Ortega-Rivas : University of Chihuahua, Mexico; L Svarovsky : University of Bradford, UK

A Cylindrical Hydrocyclone 177
H Yuan : Wuxi Institute of Light Industry, China

Prediction of Underflow Medium Density in Dense Medium Cyclones 191
T J Napier-Munn, I A Scott, R Tuteja, J J Davis, T Kojovic : Julius Kruttschnitt Mineral Research Centre, Australia

The Effect of Rheology on the Performance of Hydrocyclones 215
R R Horsley, Q K Tran : Curtin University; J A Reizes : University of New South Wales, Australia

Development of a Cyclonic De-Gassing Separator for use in a Roadside Fuel Dispenser 229
P D G Massingberd-Mundy, K B Snooks, J G Gulliver : University of Southampton, UK

PRACTICAL APPLICATIONS

Revolutionary Metal/Ceramic and Polymer Composite Technologies to form Net Shape Erosion/Corrosion Resistant Hydrocyclone Apexes and Liners 245
J Weinstein, M Schreiner, R Webb : Alanx Products L.P., USA

LARCODEMS Separator - Development of Three-Product Unit 263
C L Shah : British Coal Corporation TSRE, UK

Liquid Hydrocyclone Separation Systems 275
M F Schubert : Conoco Inc, USA; F Skilbeck, H J Walker : Vortoil Separation Systems Limited, UK

Testing of the Vortoil Deoiling Hydrocyclone using Canadian Offshore Crude Oil 295
K M Simms, S A Zaidi : Wastewater Technology Centre; K A Hashmi : Energy, Mines and Resources, Canada; M T Thew, I C Smyth : University of Southampton, UK

A New Method of Starch Production from Potatoes 309
S Bednarski : Technical University of Cracow, Poland

DEVELOPMENT

Separation of Liquid-Liquid-Solid Mixtures in a Hydrocyclone-Coalescer System   329
S Bednarski : Technical University of Cracow; J Listewnik : Maritime University of Szczecin, Poland

LDV Measurements in the Vortex Flow Created by a Rotating Wall Dewatering Cyclone   359
P Schummer, P Noe : ACB/CERG, France; M Baker : B P Engineering, UK

Application of a Novel In-Line Free-Vortex Separator in a Two Phase Pumping System   377
E Arato, N Barnes : BHR Group Ltd, UK

Three-Product Hydrocyclone for Simultaneous Separation of Solids both Heavier and Lighter than Liquid Medium   397
S Bednarski : Technical University of Cracow, Poland

Operational Parameters of Hydrocyclones in the Urea Crystals Thickening   405
S Bednarski : Technical University of Cracow, Poland

FOREWORD

It is with great pleasure and satisfaction that we introduce this volume which comprises the papers accepted for the 4th International Conference on Hydrocyclones held in Southampton from 23rd to 25th September 1992. As the name implies, this is the fourth Conference in the series, with the previous ones held in Cambridge in 1980, Bath in 1984 and Oxford in 1987.

The papers cover a wide span of activities, from fundamental research to advances in industrial practice and, as in the earlier volumes, make a significant contribution of lasting value to the technical literature on hydrocyclones.

Hydrocyclones continue to widen their appeal to engineers; besides their traditional role in mineral processing they now attract a lot of attention in chemical engineering, the oil and gas industry, power generation, the food industry, textiles, metal working, waste water treatment, pharmaceuticals, biotechnology and other industries. The reason for this continuously increasing attention is, as David Parkinson (General Manager of Conoco (UK)) said recently, that "...a hydrocyclone is an engineering dream, a machine with no moving parts." Yet as this Volume clearly shows, the hydrocyclone can do so many things and do them well, whether the application is in solid-liquid, liquid-liquid or liquid-gas separation.

It remains for us to thank the members of the Technical Advisory Committee and the referees for their time and efforts and, of course the sponsors (BHR Group Limited) for their support. In particular, our thanks should go to Carl Welch and Kit Stones in the BHR Group Conference office who have shown a great deal of patience and limitless energy in collecting the papers for this volume and organising this Conference. We are also grateful to the authors who so gallantly faced the criticisms of the referees and contributed to the success of both the Conference and this publication.

So we commend this volume to you and hope to see everyone again, at the next hydrocyclone meeting.

Lado Svarovsky

Martin Thew

"THE HYDROCYCLONE - A STORY OF CONTINUOUS INNOVATION"

OPENING ADDRESS BY

**J. MICHAEL STINSON
CHAIRMAN & MANAGING DIRECTOR
CONOCO (U.K.) LIMITED**

AT

**THE 4TH INTERNATIONAL CONFERENCE ON HYDROCYCLONES
SOUTHAMPTON, 23 SEPTEMBER 1992**

Thank you very much for inviting me to address you this morning. Although I am an engineer by original training, I'm not much of one now. But I maintain a respect for research and engineering which is deeply seated and represent a company which both relies on and creates technical innovation . Conoco has been deeply involved in hydrocyclone technology as many of you perhaps know.

Technical conferences, like this one, are ideal occasions for industry and the research community to communicate. Industry can learn about new and emerging technologies and provide direction for future R&D, by highlighting its needs. The research community has an opportunity to see the impact of their previous work and identify potential applications for new R&D.

I would like to address the importance of identifying or defining needs and/or applications as critical factors and driving forces in the innovation process.

It has often been said that "Necessity is the mother of Invention". I would like to recommend that we change that saying to "Necessity is the mother of Innovation".

Invention, to me, is restrictive. It seems to limit the importance of clear needs and applications to just the act of initial discovery. Invention is usually only the first step in the driven innovation process.

What do I mean by the term "Innovation Process"? The Innovation Process to me is best described as: Invention, Research, Development and Engineering.

> Invention - the idea,
> Research - establish the credibility of the idea,
> Development - proving the practicality of the idea and
> Engineering - proving the applicability of the idea.
> Manufacturing and Marketing - commercializing the idea.

The Innovation Process is not composed of separate, discrete steps, nor is it a once through system, but a continuous cycle of discovery, application and re-discovery.

Hydrocyclones are an excellent example of an invention that has gone through many innovation cycles. In the late 1800's, the mining and mineral processing industry started to use a new separation technology. The separation of solids from liquids using hydrocyclones was a major step forward and one of the first innovation cycles for this technology.

The oil and drilling industries have long used simple, rugged cyclone devices for sand separation from oil, drilling muds and other fluids.

Today, more than one hundred years later, this same technology has been applied to liquid-liquid separation.

The trigger and driving force for this amazing success story was the identification of a critical need for an improved means to separate oil and water.

As a result of the "Torrey Canyon" accident in 1967, academic and industry leaders called for increased research into oil spill clean-up. This event triggered Martin Thew and his team of fluid mechanics specialists at the University of Southampton to test the possibility of using hydrocyclones to separate oil-water mixtures.

In 1968, their initial tests showed that, contrary to previously held beliefs, hydrocyclones might provide adequate separation if certain geometry and design parameters could be established. By 1970, the Southampton team began an eight-year research and development programme funded by industry and the Science and Engineering Research Council.

In 1978, Southampton's first hydrocyclone patents were filed by the British Technology Group. Ten years after the initial tests, Martin Thew and his team had completed three of the four steps in the innovation process: Invention, Research and Development. However, the fourth step, Engineering, required focusing on an application, defining the potential market and field testing. This final step would take a further five years.

In 1983, the Southampton liquid-liquid hydrocyclone separation technology was licensed to two companies for commercial development. That same year, the first commercial high-pressure Vortoil hydrocyclone was successfully tested aboard a platform in the Bass Strait, Australia.

The Bass Strait field test marked the completion of the first cycle through the innovation process. However, several concurrent innovation cycles were begun by Southampton and others during this period. Most highly successful innovations share this continuous discovery characteristic.

Conoco's experience with liquid/liquid hydrocyclones started in 1985. We were among the first North Sea operators to install hydrocyclones for handling produced water. Although the Murchison platform had been commissioned only four years earlier, the level of water production was about to exceed the designed treatment capacity of the existing conventional facilities.

Due to restrictions of platform space and weight, Conoco was faced with the prospect of restricted oil production and possible reduction of recoverable oil, both of them major economic penalties. Hydrocyclones provided a unique solution to the space and weight limitations as well as being able to meet all of the present and future

Murchison water handling needs. The equipment had the added benefit of being simple with very low maintenance costs.

In effect, produced water was no longer a consideration in determining the Murchison field production profiles. Very soon, the entire produced water handling duty was entrusted to hydrocyclones.

In 1986, Conoco's Hutton field suffered a similar produced water problem. The Hutton Tension Leg Platform had the added complication of being subjected to motion from waves and wind. Motion inhibits oil-water separation in conventional systems. However, when hydrocyclones were installed on Hutton they did not exhibit any reduced performance due to platform motion.

What started out as a search for an improved oil spill clean-up method finished as the premier technology for safe disposal of produced water from offshore production operations.

Today, nearly 8,000,000 barrels per day of produced water can be treated with Vortoil hydrocyclones. They have been installed on production operations in 18 countries worldwide.

Liquid-liquid hydrocyclones have proven to be so successful in offshore production operations because they meet the industry's need for mechanically simple, light-weight, compact, efficient oil-water separation devices.

I would be remiss if I didn't discuss another important driving force in the search for elegant, low-capital cost and low maintenance equipment such as hydrocyclones - economics. In real terms today's oil is about one-third what it was ten years ago. Capital costs have continued to rise and probably not much better than they were then. Operating expenses have been driven by inflation. Therefore, the oil industry, like many other industries, is in a heavy cost squeeze. The only way we can respond is to simplify, simplify, simplify. Hydrocyclones, so wonderfully using centrifugal force, are an example of what many industries need to maintain their costs in a competitive range. It isn't an option for us to continue to use expensive, outmoded technology. I salute you all for being on the forefront of a vital effort yielding effective technology with economies in mind.

This is not the end of the liquid-liquid hydrocyclone story, only the completion of the first chapter.

Separation of lighter and heavier phases of liquid mixtures by hydrocyclone is not limited to oil industry applications. The next cycle of the innovation process has already begun with hydrocyclones planned for the nuclear, shipping, pharmaceutical and food processing industries.

Today, we are living in a rapidly changing world with great challenges. These challenges (financial, environmental, safety, etc.) should not be viewed as threats but opportunities. In order to achieve the greatest benefit from these opportunities, we all must learn to focus our efforts on those activities which will yield the highest overall value.

The offshore petroleum production business must make major changes in the way it does its business if it is to prosper. Embracing innovation in every facet of our business and the way we go about our business, is the way ahead.

I look forward to these challenges as I know you do.

FUNDAMENTALS AND NEW AREAS

BRADLEY HYDROCYCLONES: DESIGN AND PERFORMANCE ANALYSIS

M. ANTUNES and R.A. MEDRONHO*
*Escola de Química/UFRJ, CT, Bloco E
Federal University of Rio de Janeiro
21941 - Rio de Janeiro - RJ, Brazil

ABSTRACT

Hydrocyclones are an important class of solid/liquid separation equipments. In this work, three Bradley hydrocyclones with 15,30 and 60 mm diameters were used. Aqueous suspensions of calcium carbonate, with volumetric concentration ranging from 0% to 10% were tested. Based on the experimental results, it was possible to obtain correlations for the product between Stokes number and Euler number, $Stk_{50}Eu$, for the Euler number, Eu, and for the fraction of feed liquid reporting to underflow, R_w. Based on these equations it is possible either to predict Bradley hydrocyclones performance or to design them.

NOMENCLATURE

C_v	Feed solids concentration by volume
C_{vu}	Underflow concentration by volume
d	Particle size
d'_{50}	Reduced cut size
D_c	Hydrocyclone body diameter
D_i	Inlet diameter
D_o	Overflow diameter
D_u	Underflow diameter
Eu	Euler number, equation (2)
G'	Reduced grade efficiency
K	Constant
ℓ	Vortex finder length
L	Total length of cyclone
L_1	Length of the cylindrical part of the hydrocyclone
n	Parameter in Plitt's equation (12)
ΔP	Static Pressure Drop

Q Feed volumetric flow rate
Q_u Volumetric flow rate to underflow
R_e Reynolds number, equation (3)
R_f Underflow-to-throughput ratio (flow ratio)
R_w Recovery of feed liquid to underflow
Stk_{50} Stokes number, equation (1)
T Temperature
v_c Superficial velocity in the cyclone body
α Parameter in Lynch and Rao's equation (11)
μ Liquid viscosity
ρ Liquid density
ρ_s Solids density

INTRODUCTION

A hydrocyclone is a very simple equipment to make. It has moving parts and consists of a conical section joined to a cylindrical portion, which is fitted with a tangential inlet and closed by an end plate with an axially mounted overflow pipe. The end of the cone terminates in a circular apex opening.

Despite its simplicity, hydrocyclones are very efficient when promoting solid-liquid separations. Actually, they are also used for solid-solid [1], liquid-liquid [2,3], and gas-liquid [4] separations.

As pointed out in an earlier work [5], most manufacturers produce only a limited range of cyclone diameter and, in order to be able to cover a wide range of flow rates and cut sizes, each hydrocyclone can have its design proportions altered. This approach requires an accurate knowledge of how geometric proportions affects the equipment performance.

An alternative approach to this, is to use a custom-made hydrocyclone based on a geometrically similar family. In this case, all the cyclone proportions are related to body diameter and the only design variable is the cyclone size (the underflow orifice diameter is being considered here to be an operating variable). Naturally, this approach is only open to users who are in a position to build a hydrocyclone of any diameter.

Contrasting with gas cyclones, for which there are several families, there are only two well-known families of geometrically similar hydrocyclones. These are due to Rietema [6] and Bradley [7]. Table 1 gives the geometric proportions for these two families.

TABLE 1

Families of geometrically similar hydrocyclones

Cyclone	D_i/D_c	D_o/D_c	ℓ/D_c	L_1/D_c	L/D_c	θ
Rietema [6]	0.28	0.34	0.40	–	5.0	10° – 20°
Bradley [7]	1/7	1/5	1/3	1/2	–	9°

According to Svarovsky [8], the dimensionless groups involved when studing hydrocyclones are the Stokes, Euler, and Reynolds numbers, defined as:

$$Stk_{50} = (\rho_s - \rho) v_c {d'_{50}}^2 / (18 \mu D_c) \qquad (1)$$

$$Eu = \Delta P / (\rho v_c^2 / 2) \qquad (2)$$

$$Re = D_c v_c \rho / \mu \qquad (3)$$

where:

$$v_c = 4Q / (\pi D_c^2) \qquad (4)$$

Svarovsky [8] and Medronho [9] have shown that most of the available theories of separation lead to the conclusion that the product between the Stokes number and the Euler number is constant for the geometrically similar hydrocyclones. They have also shown that this product depends on the cyclone design, but is not affected by the relative size of the inlet orifice D_i/D_c [5].

Based on a massive experimental work, it has been shown [9] that the product Stk_{50} Eu is, as a matter of fact, a function of the underflow-to-throughput ratio and feed concentration. In that work, three geometrically similar hydrocyclones, which followed the Rietema's optimum design [6], were used. Based on the available theories of separation and on test data, Medronho [9] has proposed a model composed by three dimensionless equations, which fully describe the operation of Rietema's hydrocyclones. This model is given by the following correlations:

$$Stk_{50} Eu = 0.0474 [\ell n(1/R_w)]^{0.742} \exp(8.96 C_v) \qquad (5)$$

$$Eu = 371.5 \, Re^{0.116} \exp(-2.12 \, C_v) \qquad (6)$$

$$R_f = 1218 \, (D_u/D_c)^{4.75} \, Eu^{-0.30} \qquad (7)$$

Based on these equations, it is possible either to predict Rietema's hydrocyclone performance or to design it (see references 9, 10 and 11 for examples). It is important to mention that, when designing hydrocyclones, this model is capable to give not only the cyclone body diameter, bul also the underflow orifice size needed to meet a specific performance requirement.

The objetive of the present work is to derive a similar model valid for hydrocyclones which follow the recommeded proportions proposed by Bradley [7].

MATERIALS AND METHODS

The test hydrocyclones were manufactured from brass in three body diameters: 15, 30 and 60 mm. All design proportions were according to Bradley's optimum design as per Table 1. Each cyclone had a set of three pieces with alternative sizes of the underflow orifice. These sizes were 1, 1.5 and 2 for the small cyclone, 2, 3 and 4 for the medium cyclone and 4, 6 and 8 for the large one. The pressure gauge was placed as close as possible to the hydrocyclone body. This care assures that measured pressure drop is due only to the hydrocyclone itself, excluding any pressure losses in the inlet pipe.

The test rig uses a centrifugal pump as a prime mover. The holding tank has a volume of 40 liters and is equipped with a variable speed agitator operated in an inclined plane to avoid vortex action in the slurry. A cooling coil is placed in the tank to keep constant the temperature of the circulating suspension during an experiment. Both the underflow and the overflow discharged into atmospheric pressure. The test rig is also provided with a by-pass line which takes the pumped suspension back into the feed vessel, to facilitate gradual start-up and easy control of the circuit.

The test material was calcium carbonate of density 2.45 g/cm^3, and its size distribution can be found in the Figure 1.

The feed concentration was varied from 1 to 10% by volume.

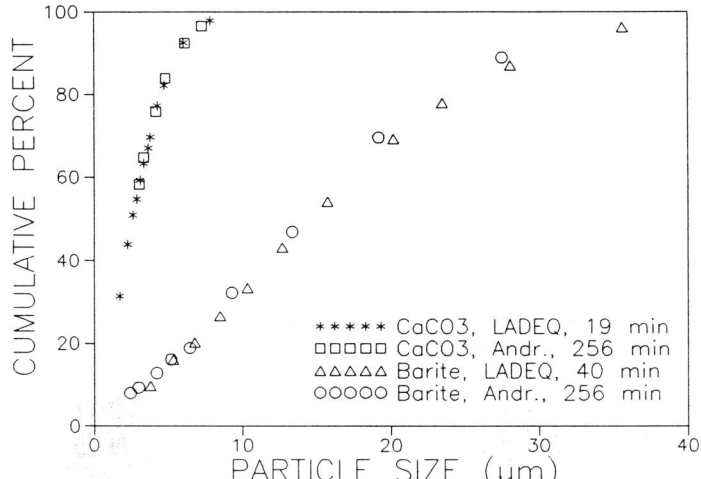

Figure 1. Comparison of size analyses obtained with LADEQ equipment (12) and Andreasen pipette (15) for CaCO3 used in this work and barite used in another work (16).

The particle size analyses were carried out using a LADEQ equipment [12]. This equipment determines the cumulative distribution (percentage undersize) by mass of the equivalent Stokes' diameters. The principle of the method is gravity sedimentation and the incremental concentrations changes are determined by taking samples from the settling suspension. This equipment was developed from a Svarovsky's original idea [13,14]. It consists of a graduated sedimentation vessel (0 to 35 cm) which holds 1000 ml when filled to the 35 cm mark. A small diameter stainless steel pipe axially inserted at the bottom of the vessel goes up to the zero mark. The suspension placed into the sedimentation vessel is allowed to flow through the pipe by regulation of a tap. This method gives the same size distributions as the ones obtained with Andreasen pipette [15], as shown in Figure 1. The advantage is to finish the analises quickly: 18 minutes for coarse powders up to 40 minutes for fine ones. With Andreasen pipette more than four hours are required.

The underflow and overflow rates were measured by the bucket and stopwatch method. The concentration of solids in the underflow, overflow, and size analyses samples were measured gravimetrically by evaporation and weighing.

The suspending liquid both in the hydrocyclone tests and in the size analyses was water with 0.1% Calgon as a dispersing agent.

The reduced grade efficiency, G', and the reduced cut size, d'_{50}, were evaluated according to the procedure recommended by Svarovsky [10].

RESULTS AND DISCUSSION

An amount of 60 experiments have been obtained using feed concentrations ranging from 1 to 10%. Another set of 12 experiments were obtained when running the hydrocyclone with water only.

Through multiple linear regression, the following equations were derived.

$$Stk_{50} Eu = 0.055 \left[\ln(1/R_w) \right]^{0.66} \exp(12\ C_v) \qquad (8)$$

$$Eu = 258\ Re^{0.37} \qquad (9)$$

$$R_w = 1.21 \times 10^6 (D_u/D_c)^{2.63} Eu^{-1.12} \qquad (10)$$

The correlation coefficient obtained for equations (8), (9) and (10) were 0.91, 0.96 and 0.97, respectively.

Figure 2 shows a comparison of calculated and observed values for $Stk_{50} Eu$, Eu and R_w.

Figure 3 shows the experimental points for the reduced grade efficiency. This curve can be adjusted by the following equations:

Lynch and Rao [17]:

$$G' = \frac{\exp(\alpha\ d/d'_{50}) - 1}{\exp(\alpha\ d/d'_{50}) + \exp(\alpha) - 2} \qquad (11)$$

Plitt [18]:

$$G' = 1 - \exp[-0.693(d/d'_{50})^n] \qquad (12)$$

Based on 60 experimental curves obtained in this work for the reduced grade efficiency, G', it was possible to calculate $\alpha = 5.1$ and $n = 3.12$.

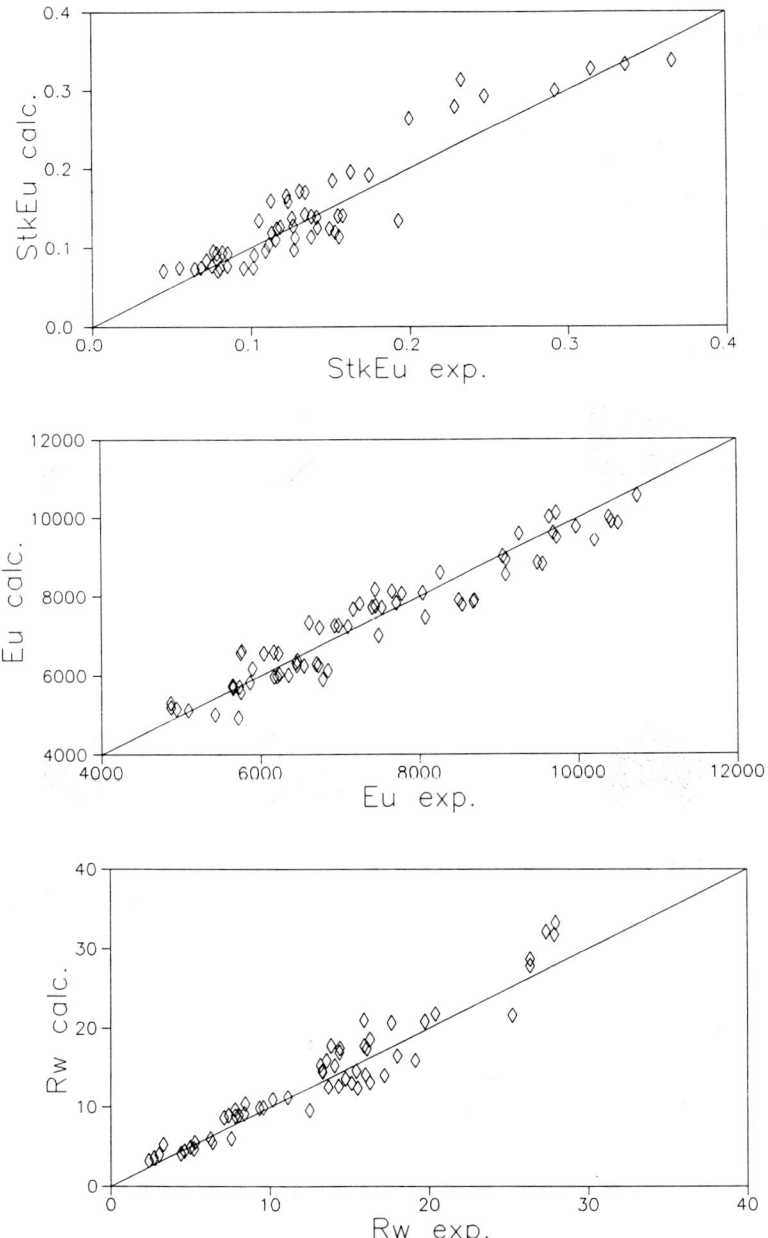

Figure 2. Comparison of observed and calculated values of StkEu, Eu and Rw, using equations (8), (9) and (10), respectively.

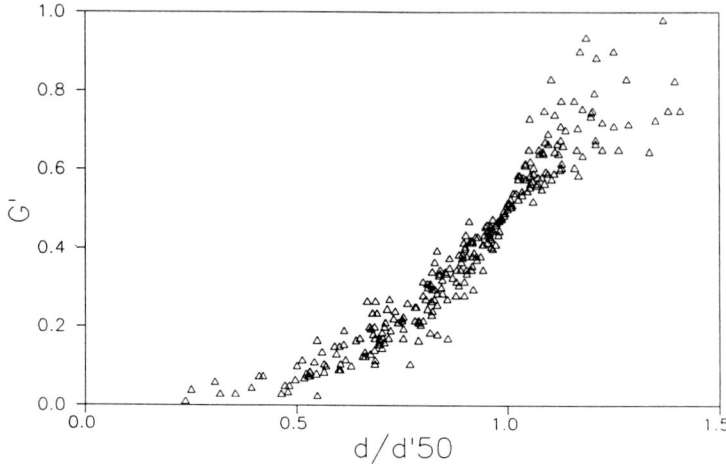

Figure 3. Experimental points for the reduced grade efficiency.

Bradley and Pulling [19] presented an experimental curve for the reduced grade efficiency, when working with a Bradley's hydrocyclone. Figure 4 shows a comparison of their experimental values and the curve calculated through equation (11) with $\alpha = 5.1$. As can be seen, there is a reasonable fit.

Figure 5 shows a comparison of Bradley and Pulling's [19] experimental points for the flow rate and the curve given by equation (9), assuming $T = 35°$ C. As can be seen, there is a very good fit.

With equations (8), (9) and (10) it is possible either to predict Bradley's hydrocyclone performance or to design it. For instance, if it is desired to design a Bradley's hydrocyclone, it can be shown that replacing equations (2) and (3) in equation (9) results:

$$D_c = 3.5 \frac{\rho^{0.31} Q^{0.54}}{\mu^{0.085} \Delta P^{0.23}} \qquad (13)$$

Equation (13) is dimensionally consistent.

An interesting comparison can be made based on the equations (5), (6) and (7) for Rietema's hydrocyclones and equations (8), (9) and (10) for Bradley's ones. For hydrocyclones of same diameter working at the same operating conditions, the latter gives a flow rate around 2.8 times lower than the former. However, in that situation, Bradley's hydrocyclone will

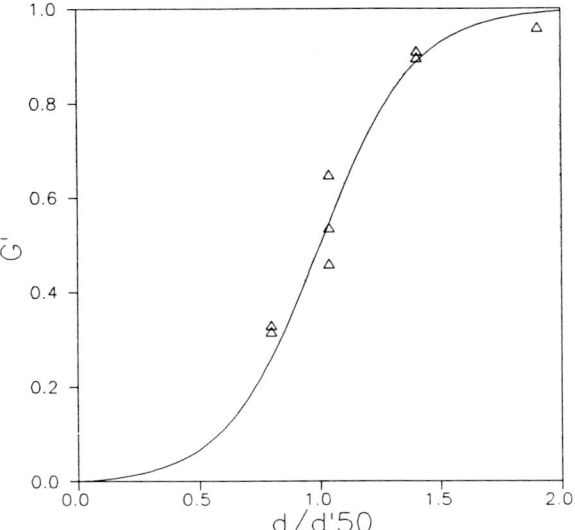

Figure 4. A comparison of Bradley and Pulling's (19) experimental points and the curve given by equation (11) with alpha = 5.1.

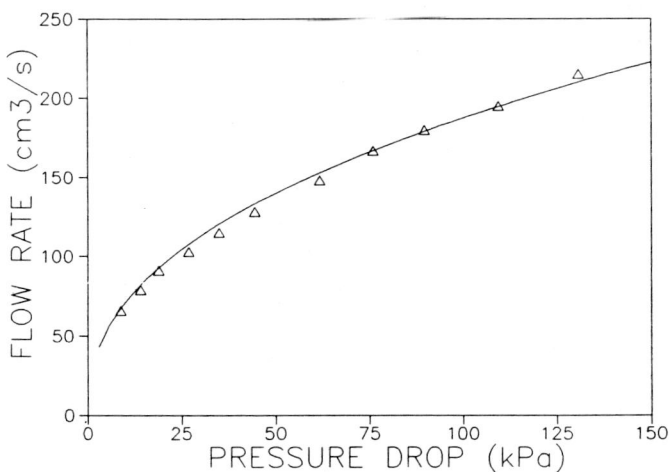

Figure 5. A comparison of Bradley and Pulling's (19) experimental points and the curve given by equation (9).

give a lower reduced cut size and larger underflow-to-throughput ratio. Therefore, they show lower capacity than Rietema's hydrocyclones but are more efficient.

Despite the fact that hydrocyclones obeying Rietema's design show larger capacity than Bradley's, it is important to note that when designing cyclones to perform a given duty, it does not necessarily mean that the required number of Rietema's hydrocyclones will be lower than the Bradley's one.

For instance, if a minimum total efficiency coupled with a minimum underflow concentration are required, the number of Rietema's hydrocyclones will not be necessarily lower than Bradley's one, since the latter is more efficient than the former.

CONCLUSIONS

The conclusions of this work are summarized as follow:
1 - The model given by equations (8), (9) and (10) is capable to predict Bradley's hydrocyclone performance and serves to design these hydrocyclones for a given duty.
2 - Bradley's hydrocyclone show a lower capacity than the Rietema's hydrocyclones, but are more efficient.
3 - With experimental data coming from a given hydrocyclone of any geometry, the constants in equations (8), (9) and (10) may be appropriately adjusted to serve as an operational model or to be used for scale-up purposes.

REFERENCES

1. Davies, J.J. and Napier-Munn, T.J., The influence of medium viscosity on the performance of dense medium cyclones in coal preparation. Proc. 3rd Int. Conference on Hydrocyclones, Oxford, Elsevier for BHRA, 1987, 155-65.

2. Nezhati, K. and Thew, M.T., Aspects of the performance and scaling of hydrocyclones for use with light dispersions. Proc. 3rd Int. Conference on Hydrocyclones, Oxford, Elsevier for BHRA, 1987, 167-80.

3. Medronho. R.A., Russo, C. and Olteanu, D., Application of hydrocyclones to the treatment of oil wastes. Proc. Int. Symposium on Environmental Problems, Istanbul/Turkey, 1991, 147-59.

4. Nebrensky, J.R., Morgan, G.E. and Oswald, B.J., Cyclone for gas/oil separation, Proc. 1st Int. Conference on Hydrocyclones, BHRA, Cambridge, 1980, 167-78.

5. Medronho, R.A. and Svarovsky, L., Tests to verify hydrocyclones scale-up procedure. *Proc. 2nd Int. Conference on Hydrocyclones*, BHRA, 1984, 1-14.

6. Rietema, K., Performance and design of hydrocyclones - Parts I to IV, *Chem. Eng. Science*, 1961, **15**, 298-325.

7. Bradley, D., *The Hydrocyclone*, Pergamon Press, Oxford, 1965, pp. 129.

8. Svarovsky, L., *Hydrocyclones*, Holt, Rineart & Winston, Eastbourne, 1984, pp. 1-11 and 44-57.

9. Medronho, R.A., Scale-up of hydrocyclones at low feed concentrations. Ph.D. Thesis, University of Bradford, 1984.

10. Svarovsky, L., Efficiency of separation of particle from fluids (Chapter 3) and Hydrocyclones (Chapter 6). In *Solid-Liquid Separation*, ed. L. Svarovsky, Butterworths, London, 1990, pp. 43-73 and 202-50.

11. Svarovsky, L., Selection of hydrocyclone design and operation using dimensionless group, *3rd Int. Conference on Hydrocyclones*, Oxford, Elsevier for BHRA, 1987, 1-5.

12. Silva, M.A.P. and Medronho, R.A., An easy method for particle size analysis, *Anais do XIV Encontro sobre Escoamento em Meios Porosos*, Campinas/Brazil, 1986, 267-75 (in Portuguese)

13. Svarovsky, L., personal communication, Bradford, 1980.

14. Medronho, R.A. and Svarovsky, L., Particle size analysis and slurry concentration with the aid of a digital density meter. *Anais do XIII Encontro sobre Escoamento em Meios Porosos - Vol. II*, São Paulo/Brazil, 1985, 496-506 (in Portuguese)

15. Allen, T., *Particle Measurement*, Chapman and Hall, London, 1981, pp. 270-6.

16. Coelho, M.A.Z. and Medronho, R.A., An evaluation of the Plitt and Lynch and Rao models for the hydrocyclone, *4th Int. Conference on Hydrocyclones*, Southampton, Elsevier for BHRA, 1992.

17. Lynch, A.J. and Rao, T.C., Modelling and scale-up of hydrocyclone classifiers. *Proc. 11th Int. Mineral Processing Congress*, Cagliari, 1975, 1-25.

18. Plitt, R.A., A mathematical model of the hydrocyclone. *CIM Bulletin*, Dez. 1976.

19. Bradley, D. and Pulling, D.J., Flow patterns in the hydraulic cyclone and their interpretation in terms of performance. *Trans. Instn. Chem. Engrs*, 1959, **37**, 34-44. Discussion, **ibid**, 44-5.

THE CHARACTERISTICS OF HYDRAULIC INTERACTION BETWEEN ADJACENT HYDROCYCLONES

A.B.SINKER & M.T.THEW

Department of Mechanical Engineering
University of Southampton
SOUTHAMPTON S09 5NH (UK)

Summary

This paper describes a preliminary experimental investigation into a claim made by a Finnish patent which stated that there were marked performance improvements to be gained by hydraulically interacting the vortices of two or more neighbouring hydrocyclones by removing the common walls between them. The design and manufacture of an interacted pair of 28.5mm hydrocyclone units is described which allowed for the effect of a varying inter-axial spacing (keeping the axes parallel) to be investigated. A parameter is suggested for qualifying the degree of interaction.

Separation performance was evaluated using a dilute dispersion in fresh water of fine nylon powder with a mass mean diameter (MMD) of 31μ and a density of 1140 kgm^{-1}. The nylon powder was chosen to provide a high degree of sensitivity to separation fluctuations since its MMD was very close to the d_{50} of the 28.5mm hydrocyclone used, for a readily attainable range of flows. The experiments were conducted with an inlet flowrate range of 10 to 18 lmin^{-1}.

Although not wholly conclusive, there is evidence that an improvement in separation, accompanied by a reduction in pressure drop, can be obtained over a narrow range of interaction.

INTRODUCTION

The need to improve hydrocyclone performance

The traditional duties for hydrocyclones lay in the mineral processing industries where patents for these go as far back as the late nineteenth century. The specialised needs demanded from this industry focused the research efforts down a narrow path and thus it took over sixty years for other industries to realise the benefits that hydrocyclones could offer. Today, hydrocyclone technology is as specialised as it is diverse. It now ranges from the separation of starch cells employing very small scale hydrocyclones (dia. = 10mm) to the oily water separators, used extensively in the offshore oil industry, where the units are of unusually high length:diameter ratio (typical installations having a capacity in excess of 20-30 m^3min^{-1}).

It is clear that in a number of industries, the use of cyclones and hydrocyclones would be more widespread if their separation performance was improved still further to encompass the need for effective separation of smaller particles within a low differential density system and for an improved sharpness of classification. The price paid for these improvements in separation performance is normally an increased pressure drop combined with an additional manufacturing complexity when compared to conventional designs. In most circumstances, these two factors alone militate against the use of hydrocyclones.

In this paper, a small scale experiment, carried out within the very limited resources of an undergraduate project, is described that explores some claims made by a Finnish patent which suggest that judicious removal of part of the wall between adjacent hydrocyclones can be beneficial to both separation and classification. These claims are detailed more fully in the next section.

The Finnish patent concept

The close coupling of cyclones and hydrocyclones to produce parallel multi unit systems is not particularly new; there have been all manner of these types of systems proposed and developed over the years. Indeed, as an example of this, Bradley[1] described such a design that was in existence as early as 1950 in which a number of small hydrocyclone units were arranged in parallel with a common inlet, underflow and overflow header, in an effort to reduce critical system cutsize. At the 2nd International Conference on Hydrocyclones in 1984, J D Boadway[2] described a novel arrangement for a set of multiple hydrocyclones where the vortex flow was set up partially in the inlet manifold, the units being linked (dictated by the inlet manifold) in the form of a Karman vortex trail.

On December 4th 1985, a patent was published by the Finnish company Nobar Ky[3] for a very radical change in hydrocyclone design for solid-liquid separation. The concept central to the patent was the beneficial hydraulic interaction of the vortices within adjacent swirl chambers. Fig.1 shows a cross-section taken from the patent which illustrates a square grid of sixteen hydrocyclones, all with parts of their wall removed. The flow dividers have the additional task of injecting the feed flow into each unit.

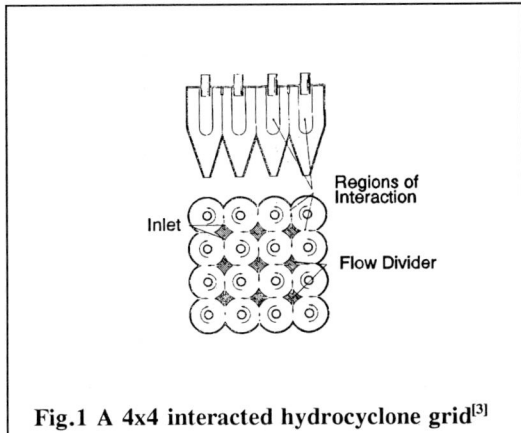

Fig.1 A 4x4 interacted hydrocyclone grid[3]

According to the patent, the advantages of an hydraulic interaction are twofold. Firstly, it argues that a significant proportion of the pressure drop across a hydrocyclone is due to the energy loss caused by the shear stress between the inner wall and the primary vortex resulting in the retardation of the spin. By interacting the swirl chambers to form 'vortex collision zones', the area of wetted wall is decreased and thus, assuming that the flow velocities at entry are matched and the angle of impact between the two adjacent vortices is small enough, there will be very limited shear related losses over this region. Secondly, the patent indicates that some form of beneficial periodic radial motion can be imparted to the particles as they are spun within a distorted vortex. This effect might be due to the flattening and bunching of the streamlines in the region of interaction, momentarily compressing the orbit of the particles. There are further suggestions that particles can pass from one vortex to another via these collision zones, sharpening classification.

The patent then details further ideas that seem rather doubtful. These include a 4x4 bank of elliptical hydrocyclones, these units being arranged so that the major axes of adjacent ellipses are perpendicular to each other. Here lies a slight paradox. Earlier, the patent clearly stipulated that for the 'wall friction' to be zero, the flow velocity in adjacent hydrocyclones within the collision zone must be equal. The flow velocity will not be uniform around the ellipse as there is not a constant change in direction. This implies that one cyclone's low velocity region will interact with another's high velocity region causing a velocity imbalance in this collision area, the very effect the patent stated it was important to avoid.

In summary, the patent put forward some radical theories, but with no exemplary data backing given to any of them. It gave no design criteria, no performance data and no operational information. However, it is known that supporting work in the general area has taken place in both industry and academia within Finland[4].

Objectives

The main aim of this study was to mount a small scale, experimental evaluation of some of these ideas. The direction of study was twofold. These were:

(i) To design and manufacture a pair of interconnected hydrocyclones that would allow for the staged reduction of the inter-axial spacing in order to progressively increase the margin of interaction.

(ii) To measure the overall separation efficiency and pressure drop of the hydrocyclone pair as the inter-axial spacing was reduced.

SPECULATIONS ON THE FLOW PATTERN

The Finnish patent seemed to be very idealistic when it made its brief references to the flow pattern, especially with regard to turbulence levels within the collision zone. It asserted that 'the vortexes running in to each other at a small angle do **not** create turbulence.....providing that the rotational speeds are equal'. While the patent therefore appeared to visualise a spinning flow free of turbulence, with a smooth distortion of the stream lines, Blackmore[5] and Loader[6] clearly showed during their LDA measurements of fast spinning flows that there was a high degree of turbulence associated with them. Applying this finding to hydraulic interaction, a momentary excess in the spin rate of one vortex over the other would lead to the penetration of one cyclone body by flow from the other. It is therefore not unreasonable to offer that hydrocyclone interaction would increase the level of turbulence, the intensity of which would be dependent on the degree of interaction. This would in turn cause increased particle re-entrainment problems. Having said that however, if the degree of interaction was kept small, these problems would be minimised and perhaps the events described in the patent might be realised

A second consideration was one of the vortex finder position. Since the radial pressure field was distorted within the swirl chamber, it was concluded that the secondary core would be deflected away from the collision zone. From a cyclone design point of view, this became significant in relation to the radial position of the vortex finder.

DESIGN OF EXPERIMENTAL TEST UNIT

The design specification of the twin hydrocyclone had to satisfy two main criteria. These were:

(i) The unit had to allow for the degree of interaction to be varied in order to determine an optimum interaction value.

(ii) Since it was thought that the secondary vortex could be deflected away from the axial centre line as interaction was increased, the unit had to incorporate vortex finders that could move approximately 6% of the diameter of the swirl chamber from the centre line, away from it's opposite cyclone.

The test unit was built up in three main sections. The main bodies of both hydrocyclones which were outside the possible interaction region and the lower section of the cones were machined out of Perspex. To allow for a staged increase in interaction, these sections were divided by three thin sheets each side of one central sheet. The centre sheet served to completely separate the two hydrocyclones; removing this sheet allowed the first slight interaction. The second interaction was then achieved by removing the two most central sheets. Accurate machining was absolutely essential to ensure symmetry of the two units to guarantee the perfect matching of wall edges. The whole unit (shown in Fig.2) was assembled using two tight fitting dowel pins and four tie bolts. As no design data was given as to the optimum degree of interaction, the only guide line that could be used was from the diagram given in the patent (Fig.1), which was assumed to show an optimum geometry. The parameter used to define the degree of interaction is described in the next section but the sheets were designed so that this measured value

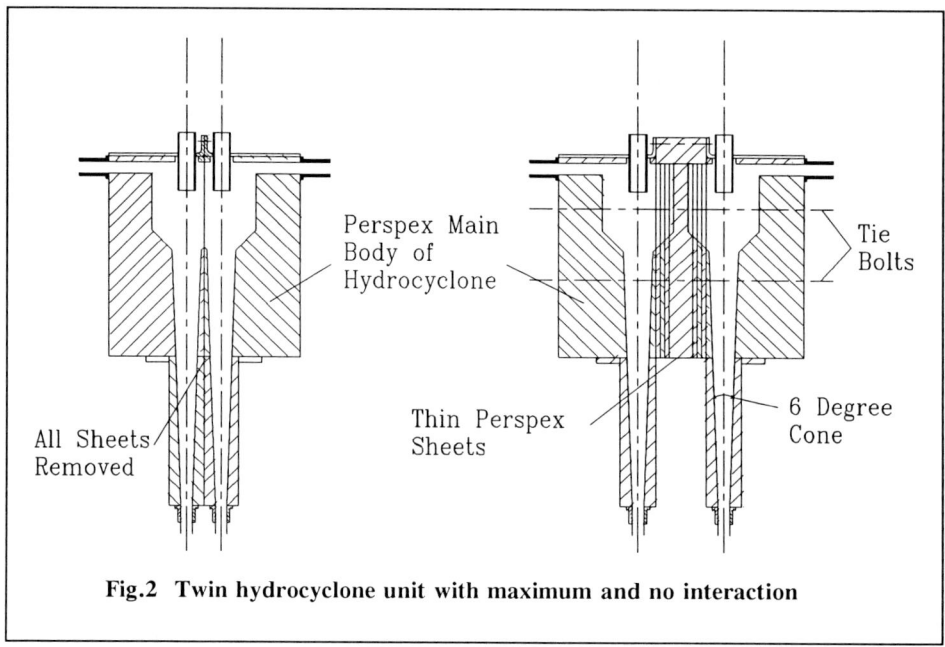

Fig.2 Twin hydrocyclone unit with maximum and no interaction

corresponded to the second stage of interaction, the unit having five distinct degrees of interaction. It was hoped this would allow a clearly defined peak in the performance/interaction data to develop. The particular type of body shape used was based on general Southampton work on liquid-liquid hydrocyclones.

The problem of providing movable vortex finders while maintaining a good seal was solved using two top plates. Each vortex finder was soldered onto a brass top plate which, via a rounded slot, was free to slide by 0.06 D_o over a perspex plate glued to one of the two main hydrocyclone blocks. Four slots were cut into each of the brass plates to allow the passage of four fasteners to ensure a good seal aided by the liberal application of silicone grease.

Considering the high number of components with wetted surfaces, hydraulic sealing of the unit was of prime concern, especially within the region of the thin perspex sheets. During the commissioning of the unit, it could not be described as leak free but it did maintain an acceptable level, largely due to the accurate machining and high polished finish of all sealing surfaces.

EXPERIMENTAL METHOD AND PROCEDURE

The test loop

Smyth and Thew[7] had reported on some very similar light dispersion separation work using a single hydrocyclone of the same dimensions (28.5mm) so much of their experimental procedure was followed here to allow some useful direct comparison and speed of commissioning. They also discuss the prediction of liquid-liquid behaviour from solid particle

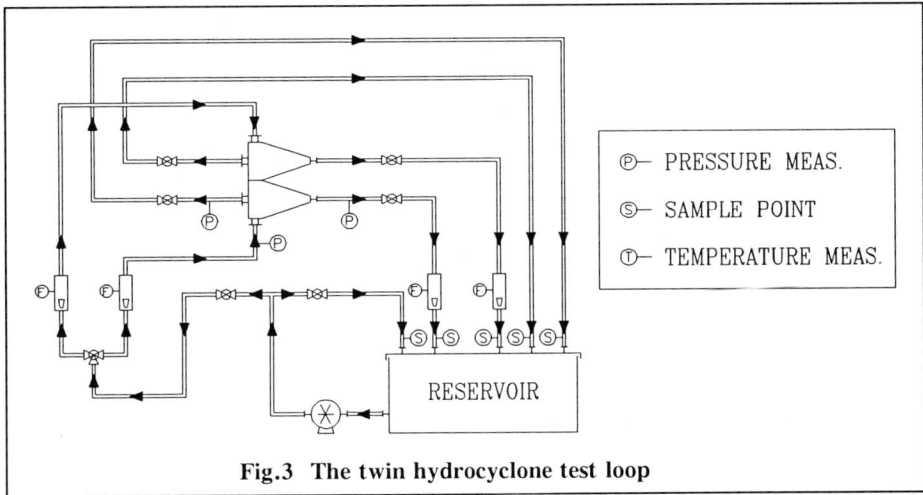

Fig.3   The twin hydrocyclone test loop

analogies[8], using hydrocyclones spanning the unit size employed here. The test loop is shown in Fig.3.

The suspension was maintained in a 30 litre reservoir (R1). The continuous phase used was tap water which was kept at 20°C. The solid component was a nylon powder with a particle size range of 4 - 65 μm, a mean diameter of 31 μm and an approximately normal size distribution by volume. At 20 °C, the nylon had a density $\rho = 1140$ kgm^{-3}. It was coloured red to aid flow visualisation. A dilute dispersion of 0.5 kgm^{-3} of nylon powder was used in these tests. To this was added 0.003 gl^{-1} of dodecyl-benzenesulphonic acid, a surfactant to prevent flocculation and foaming. This suspension was pumped around the hydraulic circuit by a semi-positive displacement pump (P1) which delivered a maximum flowrate of 16 lmin^{-1}. There were two main gate valves controlling the overall flowrate downstream of the pump: one controlling the feed flowrate and the other controlling the flowrate through a by-pass loop back to the reservoir, a necessity due to pump type but also providing a fine tune control of the feed flowrate and jet mixing to preserve suspension homogeneity within the reservoir.

The feed flow was then split via a three way valve which acted as a fine balance control to ensure either equal flowrates at the two hydrocyclone inlets or force cross-flow mixing. Two rotameters were used to monitor each of these inlet flowrates and a further two measured the two underflows. Four ball valves controlled the split ratio. The inlet and outlet ports of one of the two hydrocyclone were fitted with pressure tappings which were connected to a Bourdon gauge (accuracy of ± 7 mb) via a three way rotary switch.

Sampling method

To determine separation performance demanded the measurement of the particle concentrations of all three streams. It was assumed (and experimentally upheld) that if the inlet flowrates were matched, then both hydrocyclones should have identical performance implying that only one unit's flowstreams needed to be sampled. Whole stream sampling (WSS) by definition will always be more accurate than any form of partial stream sampling. WSS was achieved with the two outflows by simply diverting the return lines to a sample bottle. The inlet concentration could only be determined (without affecting the equilibrium of the flow condition) from the discharge of the bypass loop. It was considered that the homogeneity of the feed flow would remain reasonably intact after a T-piece split; but accepted that perfect geometry of a mass produced piping connector was unlikely and thus prone to an increased error.

The act of sampling a flow implies that solids are removed from the system. Consistent sampling from all three streams cannot be guaranteed thereby affecting the overall particle concentration. It was therefore important to determine the number of test runs that could be performed before the amount of solid taken from the system, when compared to the total solid mass within the system, becomes significant. The system was drained and reset after every four runs which implied a total solids removal of 6.9%, with a percentage change of system mass concentration below 1%.

These sampling errors were further minimised by sampling the flow streams in the following order:

    (i) Inlet   - sampling had no effect on system concentration

    (ii) Underflow   - sampled within a period of approx. 10 secs. of each other
        Overflow/   to ensure that the second sample is unaffected by the taking of
                                       the first sample.

Determination of stream concentration

The volume of each sample was measured and the solid filtered off using a Buckner Filter apparatus. This employed a flat disc of filter paper which provided a greater filter area and with aid of a vacuum pump, sped up filtration times considerably. Each disc was then catalogued, dried and weighed.

PERFORMANCE EVALUATION

The classic measure of separation performance by mass of a hydrocyclone relates the overflow concentration as a fraction of the feed concentration.

$$E = 1 - \frac{K_o}{K_i} \qquad \text{Ex.1}$$

It was thought however, that an efficiency expression that only incorporated the outlet flows would be more appropriate as these two streams were whole stream sampled. It was recognised that since the benefits that only one interaction would convey to the separation performance would at best be marginal (the multiple effect becoming more significant within a hydrocyclone grid), an accurate and sensitive performance parameter was vital. To this end, the following efficiency expression was proposed.

$$E = 1 - \frac{K_o}{K_u} \qquad \text{Ex.2}$$

This expression satisfies the majority of the requirements for a suitable efficiency parameter set out by Rietema[9], the one proviso being that neither of the hydrocyclone's two discharge ports are shut off completely. Notably, the effect of split ratio is incorporated, without the need to input flow data. Since both outlet concentrations do not remain constant with changes in operating parameters and, coupled with the wideband size distribution nature of the solid phase, the efficiency, defined in Ex.2, will be very sensitive to small changes in performance. Within the region below the critical split, an increase in the split (which will divert flow from the overflow to the underflow) will allow solid, that was previously forced to leave at the overflow, to exit correctly at the underflow increasing system efficiency. This effect will follow the law of marginal returns; the performance curve will peak (in reality, at a point above that of the ideal critical split, this split being defined as the ratio of the solid to liquid by volume at the inlet) with a subsequent further increase in split merely diluting the underflow, without adversely affecting the value of K_o, provoking the curve to dip.

The proportion of interaction between the two hydrocyclones demanded a defining parameter. The most simple definition would be to express the distance between the two axial centres (d) as a fraction of the diameter of the swirl chambers (D_o).

$$I_D = \frac{d}{D_o} \qquad \text{Ex.3}$$

The patent's concept was one of reducing wall friction to improve efficiency. This implied that an interaction parameter that varied linearly with the reduction in wall area would be more sensitive than I_D, the latter's insensitivity especially apparent when it approaches zero. I_W can thus be defined as the circumferential length of the removed wall expressed as a fraction of the total circumferential length of the swirl chamber, having a value of unity for complete interaction and zero for complete independence.

$$I_W = \frac{2}{\pi} \tan^{-1}\left[\frac{\sqrt{(D_o^2 - d^2)}}{2d}\right] \qquad \text{Ex.4}$$

I_w is the parameter used in figs.8 to 10.

DISCUSSION

Single hydrocyclone test results

It is not the intention here to provide a detailed analysis of the performance results of a single 28.5mm diameter hydrocyclone; further information on this subject can be found in Smyth and

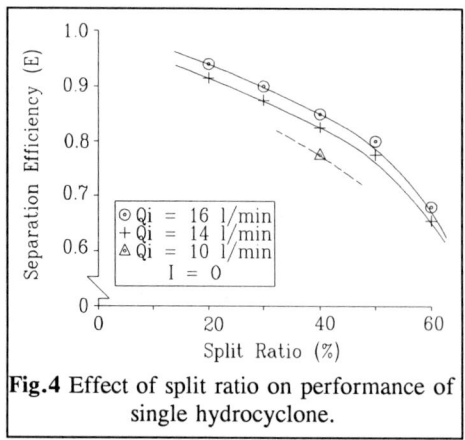

Fig.4 Effect of split ratio on performance of single hydrocyclone.

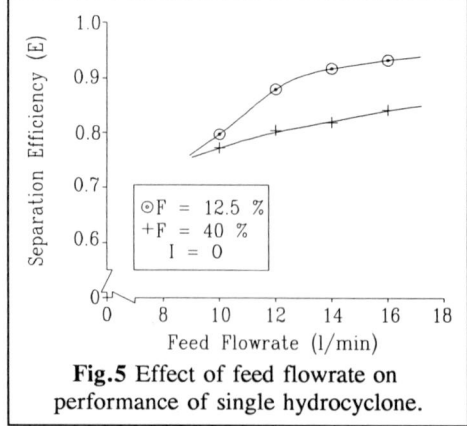

Fig.5 Effect of feed flowrate on performance of single hydrocyclone.

Thew[5]. Figs.4 through 7 show the performance of a one hydrocyclone of the dual unit when

both units were operating in unison, but fully independently. Although this information could be gleaned from the following interaction test results, it is included here in its own right to provide the reader with a 'feel' for the performance of a single hydrocyclone and its behaviour with

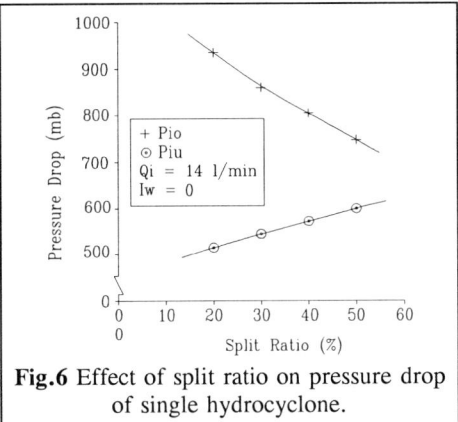

Fig.6 Effect of split ratio on pressure drop of single hydrocyclone.

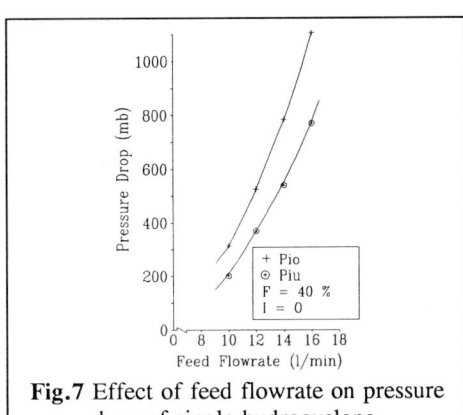

Fig.7 Effect of feed flowrate on pressure drop of single hydrocyclone.

variations in the basic operating parameters. Fig.4 illustrates the effect of split ratio on performance. As the split is increased, the efficiency drops off at an ever accelerating rate. The expected form of the performance curve with split was touched on briefly in the last section. The results upheld the theory that a rise in the split, diverting flow from the overflow to the underflow, causes the dilution of the underflow. K_u will therefore drop, its value tending towards that of K_i, at a higher rate than K_o, whose value increases towards K_i. Both these movements in K_u and K_o are in a direction as to depress E.

The feed flowrate against separation efficiency, shown in fig.5, is more straight forward. An increase in flowrate reaps a greater efficiency. Although the curves begin to plateau out, the optimal flowrate for these units has been shown to be between 22-25 lmin^{-1}. The significance of this is discussed further in the following sections. Typical pressure drop curves for both changes in split and feed flowrate are revealed by figs.6 and 7.

General effect of hydrocyclone interaction

When assessing such parameters such as efficiency or pressure drop, it is important to note that these parameters describe the performance of a single hydrocyclone unit and not the overall performance of an interacted hydrocyclone grid. The plots for efficiency against proportion of interaction are shown in figs. 8 and 9. It is clear that as I_w moved above 0.2, there was a dramatic fall off in the efficiency. This upholds the patent's suggestion that there must be a certain range of angles of incidence the two swirling flows can interact at without causing significant levels of turbulence and intercyclone mixing. This angle, defined as the angle between the tangents to the two swirl chambers at the point where the two flows meet, becomes critical very close to the first point of interaction, having a value of approximately 30°. Below this value of $I_w=0.2$, however, the results were more significant. By extrapolating the performance curve between the experimental points of $I_w=0$ and $I_w=0.17$, it can be shown that the peak in the efficiency lies not at $I_w=0$ but at a point $I_w=0.07$. Unfortunately, these results

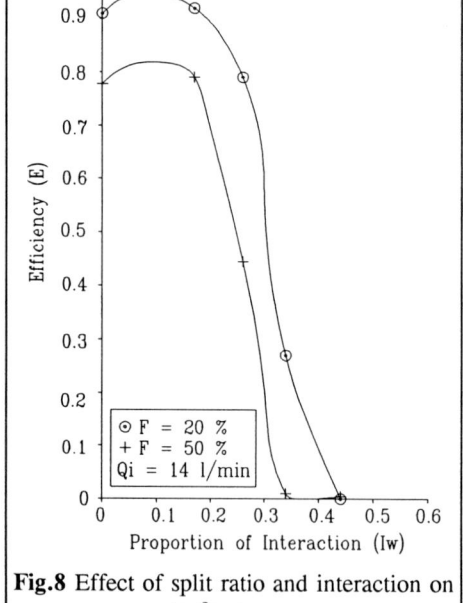

Fig.8 Effect of split ratio and interaction on performance.

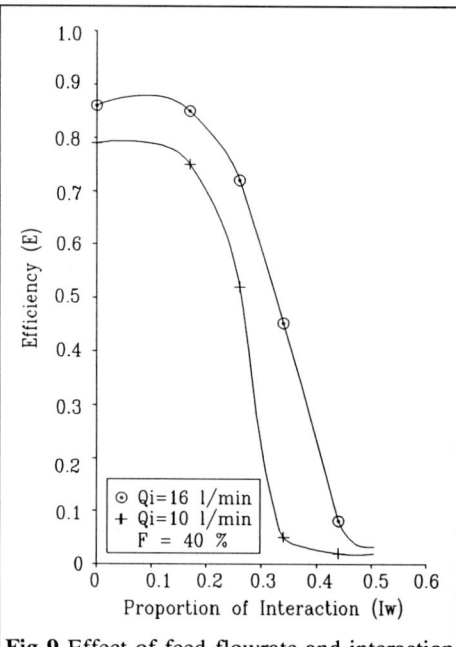

Fig.9 Effect of feed flowrate and interaction on performance.

proved (not surprisingly) that the diagrams given in the patent were drawn 'apparently' arbitrarily and thus did not show an optimised interaction geometry. Since these drawings were used as the basis for the design of the experimental unit, its first interaction of $I_w=0.17$ was too insensitive; a value of 0.05 - 0.1 would have been more appropriate.

However, the case for a proportion of interaction of $I_w=0.07$ providing a positive effect is further strengthened when the pressure drop data, shown in fig. 10, is considered. Taking the worst case, if the region of the efficiency curve between $I_w=0$ to 0.17 is assumed to be a plateau, it is certain from fig.10 that the pressure drop (P_{io}) declines over this region implying that the hydraulic interaction would produce a similar efficiency but at the expense of a lower pressure drop. The data above $I_w=0.3$ becomes widely scattered as the high angle of incidence (in excess of 60°) destroys the swirl. Given the low pressure drops and residence times, the unit actually behaves like a very reasonable mixer.

Effect of split ratio

The results were consistent with those of the single hydrocyclone tests, producing a neat family of performance curves over a range of splits (shown in fig.8 with only the maximum and minimum results illustrated), with a reduction in split generating an increased efficiency. Accepting the error prone process of curve extrapolation, the maximum efficiency obtained was 'measured' at 0.95 at an interaction value of $I_w=0.07$ (cf. E=0.91 at $I_w=0$). A split of lower than 20% might have produced still further improvements in efficiency but due to pump limitations, a 20% split was the lowest achievable when running with an inlet flowrate of 14 lmin^{-1}.

Since an efficiency curve will always follow the law of diminishing returns (ie as the efficiency increases, a further unit increase becomes progressively more difficult to achieve), an improvement, although not experimentally proven, of 0.04 from 0.91 to 0.95 is very significant.

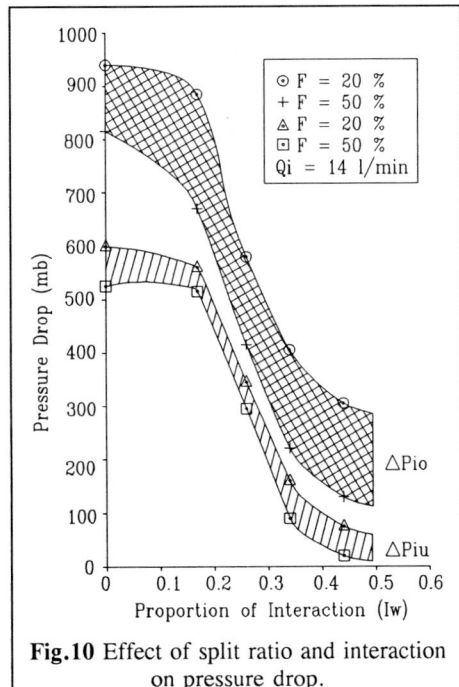

Fig.10 Effect of split ratio and interaction on pressure drop.

The pressure drop (shown in fig.10) falls markedly when $I_w>0.2$. Here the swirling flow begins to break down causing a consequent reduction in the radial pressure gradient. The peak of each of these pressure drop curves (the effect more apparent for P_{io}) occurs quite definitely at $I_w=0$.

Effect of inlet flowrate

The results again produced a family of curves as illustrated in fig.9. However, at the lower flow rates ($Q_i=10$ lmin^{-1}) and at the higher values of I_w, the data became increasingly scattered. The relative spacing between the efficiency data points, for each value of I_w, increases indicating that the mean gradient of a efficiency/inlet flowrate plot no longer flattens off (as seen in fig.5 for the single hydrocyclone tests), but increases as I_w rises. This trend is followed until a critical value of I_w is reached (in this case $I_w = 0.3$) where upon the vortex flow collapses and a new flow regime is created. This implies that the optimum flowrate is higher than the maximum flowrate tested here ($Q_i = 16$ lmin^{-1}) and indeed, as stipulated earlier, Smyth and Thew[5] reported to have obtained the best efficiency for these type of hydrocyclones when running, uninteracted, with an inlet flowrate of 22-25 lmin^{-1}. Since there must be a minimum critical flowrate present to maintain vortex flow, this critical flowrate being directly dependant on I_w, the flow regime must become increasingly unstable at low flows producing the data scatter.

Interacted hydrocyclone grids

The Finnish patent[3] describes the notion of banks of these hydrocyclones deployed in the form of an interacted square grid (see fig.1). If the assumption is made that the improvement in the efficiency of a single hydrocyclone within a grid is directly proportional to the number of interactions that hydrocyclone has with its neighbours, then the overall system efficiency improvement will be directly proportional to the mean number of interactions per hydrocyclone. Therefore, the greater the number of hydrocyclones present within the grid, the greater the overall efficiency. For a 3 x 3 grid, for example, four units have two interactions, four have three interactions and the central unit has four. This produces a mean of 2.67 interactions per hydrocyclone. Fig.11 illustrates the relationship between the mean number of interactions and the grid size. The interaction number increases asymptotically to four as the grid size is enlarged revealing a classic case of the law of marginal returns; the benefits of a further unit increase in grid size will always be less than those of the previous increase. However, an increase in the grid size from a 2 x 2 to a 4 x 4 system allows a 50% rise in the improvement factor so grid sizes of 4 x 4 or 5 x 5 would certainly be theoretically worthwhile. These improvements must always be balanced by the added costs that a larger (and thus more complicated) grid design would entail, although this would depend on the method of manufacture. Due to the speculative nature of the proposed improvement in efficiency that a single interaction will realise, fig.11 is banded to illustrate the possible maximum and minimum improvement for a particular grid configuration.

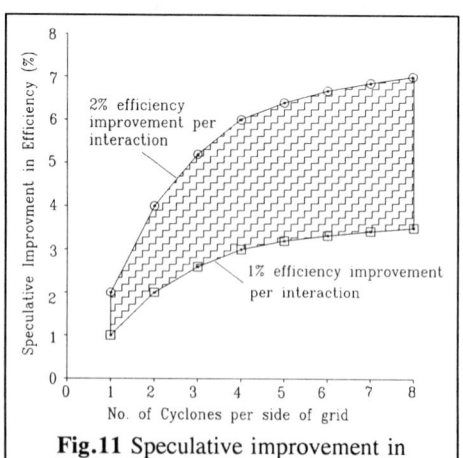

Fig.11 Speculative improvement in efficiency with an increase in grid size.

CONCLUDING COMMENTS

The test results for a single hydrocyclone showed that the efficiency falls off as the split ratio is increased. This is caused primarily by the diversion of carrier fluid from the overflow diluting the underflow. The efficiency curve plateaus as the flowrate rises as the 'pro-separation force', the increasing centrifugal force field, is countered by the strengthening remixing processes induced by the rising turbulence levels.

The interaction results strongly suggest that an interaction of approximately $I_w=0.07$ would be most beneficial to the performance of a hydrocyclone. The improvement in efficiency, caused by this interaction, is certianly dependent on the nature of the graphical interpolation used, but at best, an efficiency peak of 0.96 is plausible and at worst, the results would simply plateau at 0.92. However, even if the interaction provides no efficiency improvement, there will still be a definite reduction in pressure drop.

The time and financial restraints of an undergraduate project imposed a few unfortunate limitations on the work. The pump set aside for the work was inadequate compelling the rig to run sub-optimally both in terms of split and feed flowrate. Time prevented the acquisition of any particle size distribution data which implied that no grade efficiencies could be determined. It would have certainly been advantageous to prove the benefits of swirl chamber interaction on the added basis of cutsize rather than purely by mass.

In summary, the work has shown that a positive effect is certainly created by the interaction of neighbouring hydrocyclones, but the extent and detail of this effect can only be, from purely this work's standpoint, speculated at. Some extra credence, however, can be attached to this principle when it is noted that an application of this technology has been installed in North Wales performing gas cleaning duties.

REFERENCES

1. **Bradley, D.**, The Hydrocyclone. Pergamon Press (1965)
2. **Boadway, J.D.**, An Arrangement for Multiple Hydrocyclones. Paper D2, Proc. 2nd Int. Conf. on Hydrocyclones, Bath 1984. Pub. BHRA (1984).

3. **Ryynanen, S.I.**, A Method of and Apparatus for Separating a Mixture of Several Components of Different Particle Masses. UK patent no. 2 108 409 B. 4th Dec 1985. Prop. Nobar Ky (Finland).

4. **Rantenen, S.**, Multi-stage centrifugal classification process in grinding circuits. Proc. 15th Int. Congress on Mineral Processing, Cannes, France (1985).

5. **Blackmore, C.S.**, An investigation of confined turbulent swirling flow using LDV. Ph.D thesis, Dept. Mech. Eng., University of Southampton. 1975.

6. **Loader, A.J.**, Instability and Turbulence in Confined Swirling Flow Investigated by LDA. Ph.D thesis, Dept. Mech. Eng., University of Southampton. 1982.

7. **Smyth, I.C. and Thew, M.T.**, Comparative Tests on the Solids Removal Performance of Three Hydrocyclones. Report No. ME/82/06, University of Southampton. April 1982.

8. **Smyth, I.C. and Thew, M.T.**, Comparison of the separation of the heavy particles and droplets in a hydrocyclone. Paper G3, Proc 3rd Int. Conf. on Hydrocyclones, Oxford 1987. Pub. BHRA (1987).

9. **Rietema, K. and Verver, C.G.**, Cyclones in Industry. Elsevier Publishing Company, Amsterdam (1961).

AN EVALUATION OF FORMAL EXPERIMENTAL DESIGN PROCEDURES FOR HYDROCYCLONE MODELLING

JJ Cilliers, RC Austin, JP Tucker
Department of Chemical Engineering
University of Cape Town, Rondebosch 7700
South Africa

ABSTRACT

While various robust and general empirical hydrocyclone models have been developed, the modelling of a specific hydrocyclone operation is often required. This paper reviews and compares various experimental design techniques for process modelling on the basis of their rigour, amount of testwork required and ease of implementation. This investigation aims to present a direct, objective comparison of practical experimental design techniques for process modelling.

A number of techniques may be used to find, or the interaction between, and the magnitudes of, the effect of variables. Three experimental design techniques are commonly used for process analysis and modelling, namely, the full factorial, partial factorial and centrally composite rotatable (CCRD) designs. These are described and compared.

It is concluded that the use of the CCRD technique can lead to significant savings in experimental effort. The method is simple to use and yields results in a form that is directly implementable as a quadratic model.

A practical example of the technique is given, as applied to an experimental programme for the evaluation of the effect of three hydrocyclone variables, namely vortex finder diameter, spigot diameter and feed pressure on underflow recovery, % solids and feed rate.

INTRODUCTION

Various robust and general empirical hydrocyclone models have been developed, most notably those of Plitt (1976), Flinthoff et al (1984) and Lynch & Rao (1975). The Plitt model has been found to be general in its applicability (Apling et al, 1980) and is often subjected to parameter estimation exercises when the modelling of a specific hydrocyclone is required (Cilliers & Hinde, 1991). The Plitt model is, however, not bound to practical limits, and can yield unrealistic results (Hinde, 1985). When prediction of the performance of a 50mm Mozley hydrocyclone for the desliming of gold mine tailings was required, this problem was encountered. Rather than embarking on a parameter estimation exercise, it was decided to develop a simple model for the specific exercise that directly predicted the variables of importance, namely the underflow solids recovery and concentration, and the flowrate through the cyclone.

In order to minimise experimental effort, a review of experimental design techniques for process modelling was undertaken. A number of techniques may be used to find optimum operating conditions, or the interaction between, and the magnitudes of, the effect of variables. Three experimental design techniques are commonly used for process analysis and modelling, namely, the full factorial, partial factorial and centrally composite rotatable (CCRD) experimental designs.

This paper describes and compares these techniques on the basis of their suitability, amount of testwork required and ease of implementation. This investigation aims to present a direct, objective comparison of practical experimental design techniques for process modelling.

EXPERIMENTAL DESIGN DEFINITIONS AND PHILOSOPHY

The aim of experimental design is to find the relationship between controlled variables and observer responses. Having these relationships, it is possible to specify a combination of variables that will achieve some practical or financial benefit (Miller, 1986), such as process control, optimisation or fault diagnosis.

The simplest experiment compares two responses that differ in only one attribute. In statistics, this usually takes the form of hypothesis testing. This type of experiment has practical value but offers little insight into the response to some, or all, the attributes likely to affect that response.

In a hydrocyclone size separation process, for instance, the aim is not solely to get a positive or negative response to a test. It is to develop a clearer understanding of how the various attributes tested relate to the response and then, if possible, to develop a mathematical relationship. To gain insight into the relationship between a response and the attributes affecting it requires considering more variables, which, in turn, requires a more sophisticated approach.

Experimental Design is the broad term that defines the process of designing systematic experimentation that will yield clear, useful results with the minimum of effort. Because of the variety of needs that give rise to experimentation, a multitude of techniques have been developed. Although these design methods each relate to achieving a particular goal, generally this is either to understand more about the process variables (modelling) or to find the optimum combination of these variables (optimisation).

Seldom will only one technique be suitable for investigating a particular process, but one is likely to produce the correct results most efficiently in a given situation at the least cost (Diamond, 1989).

All the experimental design procedures available rely to a greater or lesser extent on statistical methods to determine the extent of the error at the end of the test programme. From this the significance of the results may be determined.

When the purpose of the experimentation is to develop an empirical model of the process, either full or partial factorial design will provide a suitable structured approach to experimentation. Depending on the design technique, sufficient levels and combinations of the variables are tested to regress the desired type of model. However, the primary goal of these techniques are not modelling, but significance testing.

The centrally composite rotatable design addresses this by reducing the number of tests required to generate a quadratic model while providing a structured approach to testing the significance of each term in the model and hence the significance of each variable and linear interaction terms.

The following sections consider these design techniques along with their implementation.

FULL FACTORIAL DESIGN

Factorial Design defines a fixed number of levels at which tests are performed on all variables. From this, the mean effect of the variables can be determined as well as individual effects and combined (interaction) effects.

The number of levels at which each of the variables is tested determines the type of factorial design and the order of the interaction terms considered. Using 2 factor-levels, factorial design will allow a first order model in each variable, with interactions, to be determined. If a second order model is to be considered then a third factor level is required; with this third level run between the two extreme versions. Higher order models can be determined, but the number of levels of the variables, and hence the number of experiments required is often prohibitive (Murphy, 1977), since;

$$\text{No. experiments} = (\text{model order} + 1)^{\text{no. of variables}}$$
$$= (\text{factor levels})^{\text{no. of variables}}$$

An experimental design considering linear effects requires a full 2^n factorial design. This requires all combinations of two versions of each of the n variables and would give the significance of the linear variable effects, and of linear interaction between variables.

If a variable is continuous, the two versions are high and low levels of that variable. Alternatively, for qualitative variables, the two versions correspond to two types, which could be the presence and absence of the variable (eg. a reagent) (Box & Hunter, 1961). In minerals processing, higher order interaction terms are seldom significant with respect to the variables and their linear interaction because of the limitations of experimental accuracy (Crozier, 1991).

DESIGN METHOD

In order to resolve interaction effects or other multi-variable relationships, the data must be orthogonal[1] so that there cannot be any hidden correlations among the independent variables being studied (Miller, 1986). To ensure this, the variable levels are first *coded* to convert the high-low levels to the values -1, +1. A coded value of 0 corresponds to the intermediate of the extremes.

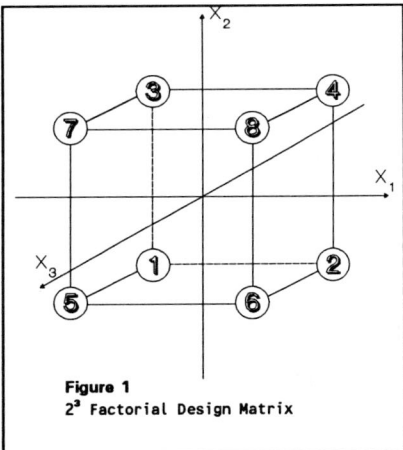

Figure 1
2^3 Factorial Design Matrix

To illustrate, suppose feed % solids high-low values were 10% and 40%. This would then be coded by subtracting the average, 25%, from both and then dividing the result by 15. Each variable would be coded similarly. The coded data is then combined into a table of all combinations of the variable levels. If orthogonal, then both the coded sums, and the coded product (interaction) sums will be zero. Tables for various situations are presented in the literature (for example, Davies, 1963). The 2^3 case (3 variables at 2 levels, giving a first order model with linear interaction effects) is presented below for completeness.

---

[1] The factorial design must be complete. For 3 variables at two levels this implies that all the corners of a cube must be tested.

Trial	Design Matrix 2^3 coded values			Matrix of Independent Variables Main Effects			2nd order interactions			3rd order interaction
	X_1	X_2	X_3	X_1	X_2	X_3	X_1X_2	X_2X_3	X_2X_3	$X_1X_2X_3$
1	-1	-1	-1	-1	-1	-1	+1	+1	+1	-1
2	+1	-1	-1	+1	-1	-1	-1	-1	+1	+1
3	-1	+1	-1	-1	+1	-1	-1	+1	-1	+1
4	+1	+1	-1	+1	+1	-1	+1	-1	-1	-1
5	-1	-1	+1	-1	-1	+1	+1	-1	-1	+1
6	+1	-1	+1	+1	-1	+1	-1	+1	-1	-1
7	-1	+1	+1	-1	+1	+1	-1	-1	+1	-1
8	+1	+1	+1	+1	+1	+1	+1	+1	+1	+1
Σ	0	0	0	0	0	0	0	0	0	0

Once the required number of runs are completed and the response measured, the main effect of each variable, x_i is determined by:

$$x_{i\ main\ effect} = \frac{[\Sigma(responses\ at\ high\ x_i) - \Sigma(responses\ at\ low\ x_i)]}{\frac{1}{2}\ Number\ of\ Runs}$$

The variance (σ^2 or S^2) of individual observations are determined by repeated runs of each test point. To determine if the effect is significant, an interval (standard deviation) is defined as:

$$Interval = \frac{\sigma}{\sqrt{\frac{Number\ of\ Runs}{4}}} = \frac{t_{student\ t} \cdot S}{\sqrt{\frac{Number\ of\ Runs}{4}}}$$

at the confidence level determined by t.

Should the range; $x_{i\ main\ effect} \pm$ Interval; include zero, the effect of the variable (x_i) is considered insignificant. This implies that a number of test sets, each repeating all the trials, must be performed for the factorial design significance tests.

To determine which interactions should be considered, these same tests are applied to the products of those variables that have significant effects.

Further details and statistics of full factorial experimental design can be found in the papers prepared by Box and Hunter (1961), Murphy (1977) and Mular et al (1991), the latter being applied to a flotation system.

Due to the exponential dependence of the number of experiments on the number of variables, factorial design often requires that a large number of tests be performed. In general, this should only be applied to discover more about the variables, their effects on the chosen response, and their interactions. The factorial design technique may also be used to determine the matrix for a grid search. This is suitable for gathering data with which to regress a model of the response function or to calculate plots of 'equal' response contours (Mular & Klimpel, 1991).

A factorial design is not usually applied to generate a response function (process model). Instead, this technique is used to determine which variables are important and to assess the extent of interaction between these variables in a process about which little is known (Miller, 1986).

PARTIAL (Fractional) FACTORIAL DESIGN

In design experiments with continuous variables and a response that varies smoothly, the higher order terms of multi-factor interaction effects are often negligible. In these cases, the matrix of the full factorial design can be simplified to a 2^{n-p} fractional factorial[2], using a smaller number of runs, that exclude the possibility of certain interactions small enough to be ignored. The extreme case exists where the response gives the significance of individual variables (linear effects), while ignoring all interactions between them. These tests would thus detect the main effects (a linear response) within the chosen range. This experimental design is particularly useful if certain variables are known to exhibit no interaction, in which case there is little point in looking at high and low values when a single point will provide the same information (Box & Hunter, 1961). Thus a partial factorial design is a fractional 2^{n-p} design reduced to n + 1 experiments (Murphy, 1977).

---

[2] where n = number of variables, and p = number of variables not varied but instead whose effects are calculated from combinations of interactions.

The technique followed for partial factorial design is the same as for a full factorial design. Design matrices for more than 3 variables are given by Murphy (1977). The three variable case is illustrated below for comparison with the full factorial design.

Trial	Run	x_1	x_2	x_3
1	1	-1	-1	-1
2	4	+1	+1	-1
3	6	+1	-1	+1
4	7	-1	+1	+1
Σ		0	0	0

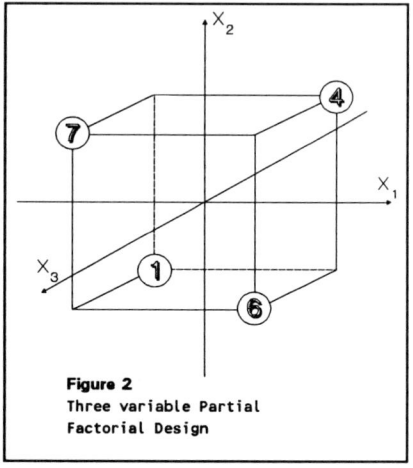

Figure 2
Three variable Partial Factorial Design

The main effect of each x_i and corresponding confidence interval is calculated as before. There is a large reduction in the number of trials required for a partial factorial design relative to a full factorial, as shown in the following table:

	Design Matrix		Number of Trials	Results	
	No. Variables	Factor Levels		Terms Included	No. of Interactions
Full Factorial	3	2	8	7	4
	4	2	16	15	11
	5	2	32	21	26
Partial Factorial	3	2	4	3	0
	4	2	5	4	0
	5	2	6	5	0

This method rapidly determines the significance of variables, but is ambiguous if interactions, assumed negligible, exist. It is important to note that ignoring interactions that are significant will give results where the main effects are confused because of residual (interaction) effects. If this occurs then the significance of individual variables become swamped by the residual term of the statistical analysis. Enough must therefore be known about the process to justify ignoring interactions. In cases where interaction between variables is known to be insignificant, this design technique will quantify the linear significance of individual variables.

CENTRALLY COMPOSITE ROTATABLE DESIGN

When the number of variables or levels to be tested exceeds three, the number of tests required for factorial design at more than two levels is often prohibitive. An effective alternative to factorial design is the centrally composite rotatable design (CCRD) (Crozier, 1992), or five-level fractional factorial. This gives as much information as a three level factorial, but with the minimum of experiments.

The technique uses statistical methods to regress experimental results to a quadratic response model. This is generally sufficient to describe the majority of steady-state process responses.

The general model includes three types of terms in addition to the constant, a_0:

Linear terms in each of the variables
$x_1, x_2, ..., x_n$
Squared terms in each of the variables
$x_1^2, x_2^2, ..., x_n^2$
First order interaction terms for each paired combination
$x_1x_2, x_1x_3, ..., x_{n-1}x_n$

Using the three-variable example illustrated in Figure 3, the results of experiments are reduced to a regressed function of the form:

$$y = a_0 + a_1x_1 + a_2x_2 + a_3x_3 \\ + a_{11}x_1^2 + a_{22}x_2^2 + a_{33}x_3^2 \\ + a_{12}x_1x_2 + a_{13}x_1x_3 + a_{23}x_2x_3$$

This approach limits interactions to first order. This is justified in most industrial applications because of experimental accuracy.

DESIGN METHOD

The design is illustrated below for 3 variables. This discussion applies to 2, 3, and 4 variables. For higher order cases, refer to the detailed discussion in Diamond (1989) or Crozier (1992).

The region of interest is first defined and the values coded as illustrated in Figure 3 for 3 variables. The factorial points (± 1) are the same as in a full factorial design; however, it is the axial points ($\pm \beta$) that allow easy regression if they are correctly placed. The centre point, and not all points as is the case with factorial design, is replicated to estimate variance due to experimental error.

The properties of the CCRD depend on β being $2^{n/4}$ (n = 2,3 or 4), and there being 5 (n = 2), 6 (n = 3) or 7 (n = 4) centre replications.

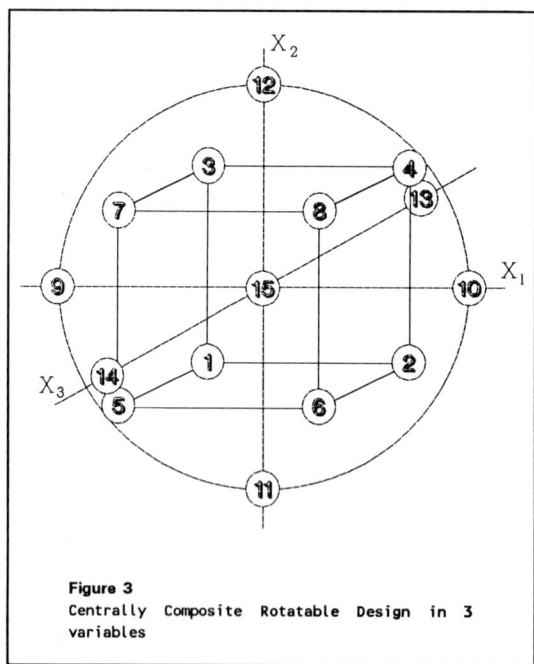

Figure 3
Centrally Composite Rotatable Design in 3 variables

The design matrix of the trials in Figure 3 expressed in coded terms:

Trial	x_1	x_2	x_3
1	-1	-1	-1
2	+1	-1	-1
3	-1	+1	-1
4	+1	+1	-1
5	-1	-1	+1
6	+1	-1	+1
7	-1	+1	+1
8	+1	+1	+1
9	$-\beta$	0	0
10	$+\beta$	0	0
11	0	$-\beta$	0
12	0	$+\beta$	0
13	0	0	$-\beta$
14	0	0	$+\beta$
15-20	0	0	0

In three dimensions, the co-ordinates of each experiment are located at a centre point surrounded by two cubes rotated 90° in each axis. The arrangement is illustrated in Figure 3.

When testing is complete, a regression analysis is carried out to determine the coefficients of the response function (refer to the example below). An analysis of variance then determines the statistical significance of terms in the model. If this model is significantly representative of the system, an analysis can be made of the shape of the response surface. This could include contour or response surface plotting.

The following table compares the experimental effort required, and the results obtained, for CCRD and factorial designs. Note that the factorial trials include no statistical analysis, and that this analysis will increase the number of runs required.

	Design Matrix			Results
	No. Variables	No. of Trials	Terms Included	Type of Terms
Centrally Composite Rotatable Design	2	13	5	2 x_i, 2 x_i^2, 1 x_ix_j
	3	20	9	3 x_i, 3 x_i^2, 3 x_ix_j
	4	32	14	4 x_i, 4 x_i^2, 6 x_ix_j
	5	32	20	5 x_i, 5 x_i^2, 10 x_ix_j
2 Level Factorial	3	8	7	3 x_i, 3 x_ix_j, 1 $x_ix_jx_k$
3 Level Factorial	2	8	5	2 x_i, 2 x_i^2, 1 x_ix_j
	3	27	16	3 x_i, 3 x_i^2, 3 x_ix_j, 6 $x_i^2x_j$, 1 $x_ix_jx_k$
	4	64	31	4 x_i, 4 x_i^2, 6 x_ix_j, 12 $x_i^2x_j$, ...
	5	125	55	5 x_i, 5 x_i^2, 10 x_ix_j, 20 $x_i^2x_j$, ...

It is important to note that the experimental design for CCRD requires exact positioning of the axial test points; not simply a high-low variation. This is, therefore, often difficult to apply in processes where variables are not necessarily continuous. For example, in the hydrocyclone testwork shown below, vortex finders and spigots with very specific dimensions had to be used. This inconvenience was, however, offset by the considerable reduction in experimental effort, providing definite advantages over a pure factorial design.

APPLICATION OF THE CENTRALLY COMPOSITE ROTATABLE DESIGN TO HYDROCYCLONE TESTWORK AND MODELLING

The model development was aimed at establishing the relationship between three dependant variables; the underflow solids recovery and concentration and the flowrate, on three independent variables; the vortex finder and spigot diameters, and the operating pressure. This model is for a specific ore and feed size distribution, and a constant feed solids concentration, as treated by a 50mm hydrocyclone.

In the design of the trials, the variable ranges are coded to lie at ±1 (the factorial points) and the β points are determined as being at $\pm 1 \bullet (2^{3/4})$ or ±1.68.

The variable values, in both their coded and true form, are shown in the following table for each of the 20 experiments in the design. Note the very specific values required for the test points. This coded design matrix forms the basis of the analysis procedure.

Factorial Trials				Axial Trials				Centre Trials			
Trial	Vortex	Spigot	Pres.	Trial	Vortex	Spigot	Pres.	Trial	Vortex	Spigot	Pres.
1	-1(17)	-1(5.4)	-1(68)	9	-β(15.4)	0(7.5)	0(80)	15	0(19.7)	0(7.5)	0(80)
2	+1(22)	-1(5.4)	-1(68)	10	+β(24)	0(7.5)	0(80)	16	0(19.7)	0(7.5)	0(80)
3	-1(17)	+1(9.6)	-1(68)	11	0(19.7)	-β(4)	0(80)	17	0(19.7)	0(7.5)	0(80)
4	+1(22)	+1(9.6)	-1(68)	12	0(19.7)	+β(11)	0(80)	18	0(19.7)	0(7.5)	0(80)
5	-1(17)	-1(5.4)	+1(92)	13	0(19.7)	0(7.5)	-β(60)	19	0(19.7)	0(7.5)	0(80)
6	+1(22)	-1(5.4)	+1(92)	14	0(19.7)	0(7.5)	+β(100)	20	0(19.7)	0(7.5)	0(80)
7	-1(17)	+1(9.6)	+1(92)								
8	+1(22)	+1(9.6)	+1(92)								

Experiments were performed at each point in random order. Timed samples of 5 seconds were taken simultaneously from the underflow and overflow streams, weighed, dried and re-weighed. From this the recovery (R), underflow solids concentration (U) and flowrate (F) could be calculated. A check was made of the feed solids concentration; this was found to be 19% ± 1.5%. For illustration, the analysis procedure is given in detail for the recovery (R) response. The results for underflow solids concentration (U) and flowrate (F) were similarly found and are presented without calculation.

Experimental Results											
Factorial Trials				Axial Trials				Centre Trials			
Trial	R	U	F	Trial	R	U	F	Trial	R	U	F
1	66.2	66.5	44.6	9	77.3	53.4	43.8	15	69.8	64.6	52.3
2	54.4	66.2	50.8	10	67.3	66.2	59.3	16	70.8	64.3	52.3
3	76.7	50.8	42.1	11	27.3	66.8	56.5	17	69.7	64.1	50.8
4	69.6	65.0	48.6	12	73.4	61.8	52.8	18	70.0	65.0	51.5
5	63.8	66.4	55.9	13	69.8	62.7	43.9	19	70.6	65.8	49.9
6	57.1	65.9	62.4	14	71.2	66.0	62.8	20	70.3	64.7	52.6
7	75.9	53.5	51.1								
8	69.5	66.8	58.5								

The coded design matrix is expanded as illustrated in the following table (columns x_i, x_i^2 & $x_i x_j$), and the results tabulated (columns R). Each entry in the coded columns is then multiplied by the response, and the columns summed. These two steps are illustrated in the following table:

CCRD Design Matrix with Results (Recovery expanded)										
	Design Matrix			Quadratic Terms			Linear Interaction Terms			
T	x_1	x_2	x_3	x_1^2	x_2^2	x_3^2	$x_1 \cdot x_2$	$x_1 \cdot x_3$	$x_2 \cdot x_3$	R
1	-(66.2)	-(66.2)	-(66.2)	+(66.2)	+(66.2)	+(66.2)	+(66.2)	+(66.2)	+(66.2)	66.2
2	+(54.4)	-(54.4)	-(54.4)	+(54.4)	+(54.4)	+(54.4)	-(54.4)	-(54.4)	+(54.4)	54.4
3	-(76.7)	+(76.7)	-(76.7)	+(76.7)	+(76.7)	+(76.7)	-(76.7)	+(76.7)	-(76.7)	76.7
4	+(69.6)	+(69.6)	-(69.6)	+(69.6)	+(69.6)	+(69.6)	+(69.6)	-(69.6)	-(69.6)	69.6
5	-(63.8)	-(63.8)	+(63.8)	+(63.8)	+(63.8)	+(63.8)	+(63.8)	-(63.8)	-(63.8)	63.8
6	+(57.1)	-(57.1)	+(57.1)	+(57.1)	+(57.1)	+(57.1)	-(57.1)	+(57.1)	-(57.1)	57.1
7	-(75.9)	+(75.9)	+(75.9)	+(75.9)	+(75.9)	+(75.9)	-(75.9)	-(75.9)	+(75.9)	75.9
8	+(69.5)	+(69.5)	+(69.5)	+(69.5)	+(69.5)	+(69.5)	+(69.5)	+(69.5)	+(69.5)	69.5

	CCRD Design Matrix with Results (Recovery expanded)									
	Design Matrix			Quadratic Terms			Linear Interaction Terms			
T	x_1	x_2	x_3	x_1^2	x_2^2	x_3^2	$x_1 \cdot x_2$	$x_1 \cdot x_3$	$x_2 \cdot x_3$	R
9	$-\beta(77.3)$	0(77.3)	0(77.3)	$\beta^2(77.3)$	0(77.3)	0(77.3)	0(77.3)	0(77.3)	0(77.3)	77.3
10	$+\beta(67.3)$	0(67.3)	0(67.3)	$\beta^2(67.3)$	0(67.3)	0(67.3)	0(67.3)	0(67.3)	0(67.3)	67.3
11	0(27.3)	$-\beta(27.3)$	0(27.3)	0(27.3)	$\beta^2(27.3)$	0(27.3)	0(27.3)	0(27.3)	0(27.3)	27.3
12	0(73.4)	$+\beta(73.4)$	0(73.4)	0(73.4)	$\beta^2(73.4)$	0(73.4)	0(73.4)	0(73.4)	0(73.4)	73.4
13	0(69.8)	0(69.8)	$-\beta(69.8)$	0(69.8)	0(69.8)	$\beta^2(69.8)$	0(69.8)	0(69.8)	0(69.8)	69.8
14	0(71.2)	0(71.2)	$+\beta(71.2)$	0(71.2)	0(71.2)	$\beta^2(71.2)$	0(71.2)	0(71.2)	0(71.2)	71.2
15	0(69.8)	0(69.8)	0(69.8)	0(69.8)	0(69.8)	0(69.8)	0(69.8)	0(69.8)	0(69.8)	69.8
16	0(70.8)	0(70.8)	0(70.8)	0(70.8)	0(70.8)	0(70.8)	0(70.8)	0(70.8)	0(70.8)	70.8
17	0(69.7)	0(69.7)	0(69.7)	0(69.7)	0(69.7)	0(69.7)	0(69.7)	0(69.7)	0(69.7)	69.7
18	0(70.0)	0(70.0)	0(70.0)	0(70.0)	0(70.0)	0(70.0)	0(70.0)	0(70.0)	0(70.0)	70.0
19	0(70.6)	0(70.6)	0(70.6)	0(70.6)	0(70.6)	0(70.6)	0(70.6)	0(70.6)	0(70.6)	70.6
20	0(70.3)	0(70.3)	0(70.3)	0(70.3)	0(70.3)	0(70.3)	0(70.3)	0(70.3)	0(70.3)	70.3
Σ	-48.8	128	1.75	941	817	931	5.00	5.80	-1.20	1341

<u>The coefficients of the regression equation</u> are computed from the total of each column. For the equation:

$$y = a_0 + a_1 x_1 + a_2 x_2 + a_3 x_3 + a_{11} x_1^2 + a_{22} x_2^2 + a_{33} x_3^2 + a_{12} x_1 x_2 + a_{13} x_1 x_3 + a_{23} x_2 x_3,$$

the procedure is:

a_0 = 0.1663402267•(R) - 0.05679210581•$(x_1^2 + x_2^2 + x_3^2)$
a_0 = 0.1663•(1341) - 0.05679•(941 + 817 + 931)
   = 70.30

a_i = 0.0732233047(x_i)
a_1 = 0.07322•(-48.8)
   = -3.57
a_2 = 9.37
a_3 = 0.128

a_{ii} = 0.062500•(x_i^2) + 0.00689003779•$(x_1^2 + x_2^2 + x_3^2)$ - 0.05679210581•(R)

a_{11} = 0.06250•(941) + 0.006890•(941 + 817 + 931) - 0.05679•(1341)

= 1.18

a_{22} = -6.57

a_{33} = 0.559

a_{ij} = 0.125000•$(x_i x_j)$

a_{12} = 0.1250•(5.00)

= 0.625

a_{13} = 0.725

a_{23} = -0.150

<u>An estimate of the variance</u> is obtained from the centre points using the formula

$$S^2 = \frac{\sum (x_i - \bar{x})^2}{N_c - 1}$$

where N_c is the number of centre points, x_i the response at each of the centre points, and $\bar{x}$ the average value of the centre point responses.

$$S^2 = \frac{(69.8 - 70.2)^2 + (0.36 + 0.25 + 0.04 + 0.16 + 0.01)}{6 - 1}$$

S^2 = 0.196, S = 0.443

<u>A test of significance</u> is made on each term in the regression equation using the following formulae for the standard errors (SE) and t-test.

SE(a_i) = 0.271•S

SE(a_1) = 0.271•(0.443) = 0.120

SE(a_2) = 0.120

SE(a_3) = 0.120

SE(a_{ii}) = 0.263•S

SE(a_{11}) = 0.263•(0.443) = 0.117

SE(a_{22}) = 0.117

SE(a_{33}) = 0.117

SE(a_{ij}) = 0.354•S

SE(a_{12}) = 0.354•(0.443) = 0.157

$SE(a_{13}) = 0.157$

$SE(a_{23}) = 0.157$

Significance testing and comparison with tabulated 1-sided t-test values with $N_c - 1$ degrees of freedom at the required level of significance:

$$t(a_k) = \frac{|a_k|}{SE(a_k)}$$

a_k = equation coefficient in coded form

single-sided t-test values for 3 Variables with 5 degrees of freedom, at:		
99%	95%	90%
3.37	2.02	1.48

$t(a_1)$ =	|-3.57|/0.120	
=	29.8 > 2.02	∴ Significant @ 95%
$t(a_2)$ =	78.1	Significant @ 95%
$t(a_3)$ =	1.07	Insignificant
$t(a_{11})$ =	10.9	Significant @ 95%
$t(a_{22})$ =	56.2	Significant @ 95%
$t(a_{33})$ =	4.78	Significant @ 95%
$t(a_{12})$ =	3.98	Significant @ 95%
$t(a_{13})$ =	4.62	Significant @ 95%
$t(a_{23})$ =	0.955	Insignificant

If any term is not significant at t = 2.02 (95% confidence level), it is eliminated from the response equation.

The governing equation of the hydrocyclone % solids recovery to underflow (R), therefore is:

$$R = 70.30 - 3.57 x_1 + 9.37 x_2 + 1.18 x_1^2 - 6.57 x_2^2 + 0.559 x_3^2 + 0.625 x_1 x_2 + 0.725 x_1 x_3$$

where x_1 = vortex finder diameter (mm),

x_2 = spigot diameter (mm),
x_3 = feed pressure (kPa)

Similarly, the equations for the underflow solids concentration (U) and the flowrate through the cyclone (F) are as follows (at 95% confidence):

$$U = 64.74 - 3.54\, x_1 - 2.73\, x_2 + 0.693\, x_3 - 1.77\, x_1^2 + 3.52\, x_1 x_2 + 0.628\, x_2 x_3$$

$$F = 51.63 + 3.86\, x_1 - 1.42\, x_2 + 5.39\, x_3 + 0.767\, x_2^2$$

Figure 4 gives a comparison of the equation predictions versus the experimental results for the data.

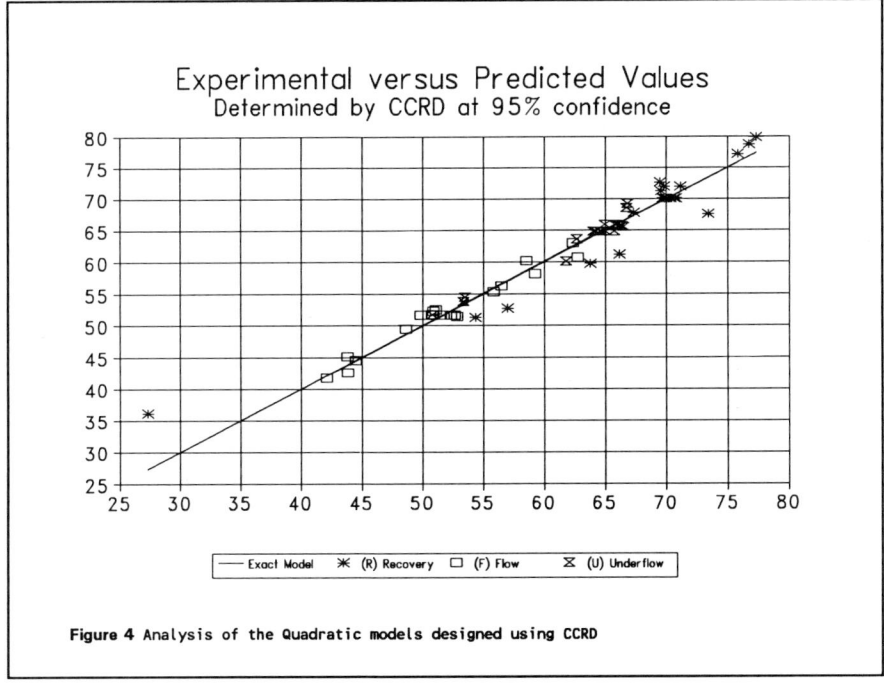

Figure 4 Analysis of the Quadratic models designed using CCRD

It can be noted that the recovery predicted at the very low end of the scale (36%) is considerably higher than that measured (27%). This point, however, was generated under roping conditions and would correspond to a discontinuity in the operating regime.

The response equation can be used to compute the estimated response at any point in the system. It can also be used to generate a response surface.

CONCLUSIONS

The use of the Centrally Composite Rotatable Design (CCRD) can have tremendous benefits in reducing the experimental effort required for process modelling. The CCRD directly yields a quadratic equation with linear interactions, usually adequate for the modelling of steady-state processes.

The full factorial experimental design may be more suitable when no prior understanding of the process is available, while the partial factorial design should only be applied when considerable *a priori* knowledge about variable interactions and response forms is available.

The CCRD is suitable for hydrocyclone experiments, however, very specific test conditions are required. This may prove impossible under industrial conditions. Under laboratory conditions, the technique is readily applied.

The form of the equations produced by the CCRD is suitable to model, in general, the response functions of hydrocyclones. In cases where discontinuity exists in the response, e.g. when roping occurs, the model form may not adequately fit the response.

The CCRD procedure is well defined, and is readily implementable in various formats, especially spreadsheets.

REFERENCES

Apling, A.C., Montaldo, D., Young, P.A., "Hydrocyclone models in an ore-grinding context", International Conference on Hydrocyclones, Cambridge, UK, pp 113 - 125, 1980.

Box, G.E.P., Hunter, W.G., "The 2^{k-p} Fractional Factorial Designs, Part I & II", Technometrics, Vol 3, pp 311 - 351 & pp 449 - 458, 1961.

Crozier, R.D., Flotation Theory, Reagents and Ore Testing, Pergamon Press, New York, 1992.

Davies, O.L. ed, The Design and Analysis of Industrial Experiments, Imperial Chemical Industries Ltd, Oliver & Boyd, London, 1963.

Diamond, W.J., Practical Experimental Design for Engineers and Scientists, Edition 2, Van Nostrand Reinhold, New York, 1989.

Flinthoff, B.C., Plitt, L.R., Turak, A.A., "Cyclone modelling: a review of present technology", CIM Bull., Vol. 80, No. 905, pp 39 - 50, 1987.

Hinde, A.L., "Classification and concentration of heavy minerals in grinding circuits", 114th AIME Annual Meeting, New York, pp 85 - 126, 1985.

Lynch, A.J., Rao, T.C., "Modelling and scale-up of hydrocyclone classifiers", Proc. Int. Miner. Process. Congr., 11th, Cagliari, pp 245 - 269, 1975.

Miller, R.E., "Experimental Design", Chemical Engineering, pp 113 -116, June 23, 1986.

Mular, A.L., Burkert, A., Cifuentes, E.I., "Experimental Design for Decision Making", in Evaluation and Optimization of Metallurgical Performance, Malhotra, Klimpel, Mular, Eds, AIME, Littleton, Colorado, 1991.

Mular, A.L., Klimpel, R.R., "Optimization Procedures in Minerals Processing Plants", in Evaluation and Optimization of Metallurgical Performance, Malhotra, Klimpel, Mular, Eds, AIME, Littleton, Colorado, 1991.

Murphy, T.J. Jnr, "Design and Analysis of Industrial Experiments", Chemical Engineering, pp 165 - 182, June 6, 1977.

Plitt, L.R., "A mathematical model of the hydrocyclone classifier", CIM Bull, Vol. 69, No. 776, pp 114 - 123, 1976.

THE EFFECT OF THE AIR CORE ON THE FLOW FIELD WITHIN HYDROCYCLONES

LUO QIAN and XU JI RUN

Northeast University of Technology,
Department of Mineral Engineering, Shenyang
110006, P. R. China

ABSTRACT

It is essential for a hydrocyclone to keep a stable folw field during its operation. In the paper, the unstable air core and its effect on the flow field in conventional hydrocyclones have been analysed and some new conclusions made. The research shows that the constant change of the air core size and the frequent oscillation of the air core position, yielded mainly by the turbulent feed pressure and the variable effective cone angle and much reinforced by the compressibility and fluidness of the air core, strengthen the turbulence of the three—dimensional velocity, disturb the regular distribution of the classified particles and cut down the separation efficiency in hydrocyclones. Then a new kind of hydrocyclone with no air core is reported.

INTRODUCTION

It is no doubt that a stable flow field in a hydrocyclone is the prerequisite for the highly effective performance. The stability of the flow field is, on great extent, dependent uponthe feed parameters, such as the feed pressure, the feed concentration and so on, but the internal structure of the flow field itself has also important effect on the stability. The former has been enough recognized, but little attention so far has been paid to the latter which is just the focus of our research in this paper.

As we know, there is always a compound movement of the forced vortex and semi—free vortex within a conventional mineral processing hydrocyclone. It is necessary here to point out first that the forced vortex is the inevitable outcome of the vortex movement(1), not the prerequisite for the hydrocyclone operation. Secondly, the forced vortex in a conventional hydrocyclone, in fact, is the air core. Although former researcher thought that the forced vortex can extend to a rather large area outside the air core (2), authors' theoretical analysis and experimental measurements (3) show that it is more reasonable to think the area between the air core and the semi—free vortex as the transition of the two parts than as the forced vortex only. Then the distribution of the tangential velocity along the radius in a conventional hydrocyclone can be illustrated in Fig. 1. Thirdly, some researches (4, 5) verify that no separation process takes place within the forced vortex i. e. the air core, because the treated particles can not enter into the forced vortex where the radial velocity is zero in theory. Therefore, the energy contributed to the air core area is useless to the performance of the hydrocyclone.

In addition, the instability is another important feature of the air core in conventional hydrocyclones. As mentioned above, the forced vortex consists of the air core inhaled from the apex. So we only need to take the air core into account in order to investigate the unstable forced vortex. In following paragraphs, the unstable size, shape and position of the air core will be studied and the effection of the instability on the flow field discussed then a new mineral processing hydrocyloe with no forced vortex introduced. All experiments are undertaken in hydrocyclones filled with clean water.

INSTABILITY OF THE AIR CORE

Certainly, there have been many researches on the air core, but few people have studied its instability. Boadway (6) observed the unstable air core with twisting shape, but no further information was reported. Authors (4) pointed out that the unstable air core in hydrocyclones is harmful to classification, but did not give explanation in detail. Here, the unstable size, shape and position of the air core in conventional hydrocyclones will be discussed. The diameter of the air core is obtained by photogrammetry.

Unstable air core size

The size, i. e. the diameter of the air core, determined mainly by feed pressure and constructure parameters of the hydrocyclone, is difficult to be kept stable.

Feed pressure: The diameter of the air core is the function of the feed pressure in a given hydrocyclone. Fig. 2 shows the general relation of the air core diameter and feed pressure. It can be seen that the air core does not appear until the feed pressure reaches a certain value which varies with the geometrical parameters of the hydrocyclone, then the diameter increases rapidly with the pressure and lastly becomes relatively stable when the feed pressure is greater than a critical point which is about 0. 98 bar in the experiments. The feed pressure in practice is usually 0. 49—0. 98 bar for the hydrocyclones treating coarse materials and 0. 98—2. 94 bar for that dealing with fine ones(7). At least for the first group of hydrocyclones, the feed pressure just lies in the range where a small turbulence of the pressure will bring out a large change of the air core diameter. Unfortunately, the pressure turbulence is unavoidable in practice because of the pump performance and the variable feed concentration, so that the size of the forced vortex, i. e. the diameter of the air core is always in the constant extension and contraction.

<u>Cone angle</u>: Cone angle is an important factor which determines the size of the air core and effects the stability. The variation of the air core diameter with the cone angle is shown in Fig. 3 from which an interesting fact can be seen, i. e. the air core has its maximum diameter at a certain cone angle (This fact may be explained in view of the energy transition in hydrocyclones but is not important to the subject discussed here). In theory, the cone angle should have no effection on the stability of the air core within a hydrocyclone with given geometric parameters but in fact, the effective cone angle, determined by the pile sutuation of the solid particles near the apex (See Fig. 4), is changable at all times and then the size of the air core becomes unstable.

As for other geometrical parameters, such as diameters of apex and vortex finder, will not be discussed here because of their little influence on the stability of the air core although they determine to a great extent the air core size.

Unstable air core shape and position

In addition to the unstable size, an air core has also unfixed shape and space position. Fig. 5 illustrates how the shape of the air core changes from a cone to a column with the increase of feed pressure. Similar to the size, the shape of the air core becomes a relatively stable column only when the feed pressure gets greater than the critical value (about 0.98 bar). Meanwhile it is easily observed with a transparent hydrocyclone in laboratory that the air core shakes laterally and the air—liquid interface changes its position frequently.

Air core and its compressibility and fluidness

The main external factors which result in the instability of the air core, such as the turbulent feed pressure and the unpredictable effective cone angle have been analysed above, but the internal factors, i. e. the compressibility and fluidness of the air core have even more important influence on the stability of the forced vortex and should be paid enough attention. In a conventional hydrocyclone, the forced vortex made up of the air core is compressible. According to Pericleous et al. , the air concentration is inversely proportional to the radius within the air core (5). The pressure distribution in radius is drawn qualitatively in Fig. 6 where the curve in liquid area is obtained by referring to experimental results and that in forced vortex depicted according to theoretical analyses. As predicted, the gauge pressure is zero and the pressure gradient is maximum on the air—liquid interface where a discontinuity in pressure gradient also exists due to the differing fluid densities. Because of the compressibility of the air core, the high pressure gradient and its discontinuity, a small disturbance may bring forth a rather large turbulence of the interface position. In the water—sealed hydrocyclone, the apex is sealed with water and the air core replaced by a water core then the stability improved (4). However, the shaking of the forced vortex composed of incompressible water is not eliminated though limited to some extent, which indicates that the fluidness is also an important factor which makes the forced vortex unstable. It is because of the compressibility and fluidness of the air core itself in conventional hydrocyclones that a stable forced vortex is difficult to be formed.

EFFECT OF UNSTABLE AIR CORE ON FLOW FIELD

On one hand, the instability of the air core is the reflection of the turbulent flow field in the separation space; on the other hand, the unstable air core strengthens the turbulence and asymmetry of the flow field and disturbs the regular distribution of solid particles within hydrocyclones.

Strengthening the turbulence of the flow field

The effection of the unstable air core on the axial velocity is shown qualitatively in Fig. 7 where r_a is radius of the air core in conventional hydrocyclones. In the figure, the fact that the point at which the axial velocity is zero does not change its position with the turbulent feed pressure has been verified by Authors' measurements.

If r_a kept unchangeable when a turbulence is given to the feed pressure, the axial velocity would be in the shade area shown qualitatively in Fig. 7. But in fact, a correspondent change of r_a is produced by the turbulent pressure (the change may be from the size or the interface position of the air core), then the actual axial flow will lie in a larger turbulent range between curves 1 and 2. In other words, the instability of the axial velocity is reinforced by the changeable air core.

Fig. 8 is an illustration of the unstable air core and the turbulent tangential velocity. Similar to that in Fig. 7, the shaded area in Fig. 8 stands for the change range of the tangential velocity at the presumption that the air core radius does not vary with the feed pressure, and the area between curves 1 and 2 is the actual turbulence range of the tangential flow. As to the radial velocity, an analysis similar to Fig. 8 can be made.

Strengthening the asymmetry of the flow field

A totally symmetrical flow field within single inlet hydrocyclone is impossible to be abtained. Generally speaking, the axis of the air core does not coincide with that of the hydrocyclone, in other words, the physical centre of the hydrocyclone usually deviates from the geometrical centre. In conventional hydrocyclones, the symmetry of the flow field may not be disturbed seriously because the diameter of the hydrocyclone is much larger than the air core size. But within the vortex finder, the effection of the shaking air core on the symmetry of the flow field (See Fig. 9) has to be taken into account because the air core size is usually greater than 60% of the vortex finder diameter (9). when the air core lies in the center of the vortex finder, the flow area within the finder is symmetrical and particles with same size distribution report to overflow in every directions; but when the air core moves to one side, the symmetry is destroyed and particles entering into the vortex finder in different directions have different size distribution, leading to a low classification efficiency.

It should be pointed out that the analysis above is also suitable for the water-sealed and, probably, the lock-bottom (8) hydrocyclones where the air core is replaced by water core but the position of the water core is still unstable.

INVESTIGATION OF ELIMINATING FORCED VORTEX

It is easy to be seen that the unfavourable reaction of the turbulent size and position of the air core to the stability and symetry of the flow field is achieved by changing the main separation area which is below the vortex finder and outside the air core Therefore, if the forced vortex boundary is fixed, the unfavourable reaction may be effectively limited or eliminated. But, as long as the air core or water core exists in a hydrocyclone, the fixed boundary can never be obtained because at least the turbulence within the forced vortex unavoidably makes the vortex change its position frequently. In order to get a flow field not disturbed by forced vortex, a possible method is to occupy the air core or water core with an appropriate solid core. Considering the characteristics of the forced vortex mentioned at the beginning of the paper, the method at least has no adverse effects on the hydrocyclone performance(The actual particle experiments will be reported elsewhere).

The tangential and radial turbulence intensity distributions, measured by LDA on a same cross-section in the cone part of the hydrocyclone with a solid core occupying the forced vortex area and the conventional hydrocyclone with air core, are illustrated in Fig. 10 and Fig. 11 respectively. It can be seen from the figures that the turbulence, especially the radial turbulence has certainly been decreased in the new hydrocyclone. At the same time, the flow area within the hydrocyclone is symmetric, which is beneficial to the regular distribution of the classfied particles. As for other advantages of the new hydrocyclone with no forced vortex, such as the higher tangential and lower radial velocities as well as the lower internal loss have been reported in other papers(10,11)

CONCLUSIONS

There is always a forced vortex composed of the air core in conventional hydrocyclones. The air core is not the prerequisite of the hydrocyclone operation. It not only consumes energy unnecessarily because of no separation process in it but also strengthens the instability and asymmetry of the flow field due to its turbulence in size and postion.

The size, shape and position of the air core in conventional hydrocyclones are unfixed because of the turbulent feed conditions and the changeable effective cone angle. Moreover, the compressibility and fluidness of the air core itself make the forced vortex more unstable.

The forced vortex in commercial hydrocyclones has unfavourable effection on the main flow field. It intensifies the turbulence of the main flow field, disturbing the regular distribution of solid paricles and leading to poor performance.

In the hydrocyclone with no forced vortex, the stability and symmetry of the flow field get improved and some other advantages obtained.

NOMENCLATURE

D	hydrocyclone diameter, mm
d_o	vertex finder diameter, mm
d_s	apex diameter, mm
h	distance below the upper cover of the cyclone, mm
α	cone angle, degree
α_e	effective cone angle, degree
r_a	radius of air core, mm
P_f	feed pressure, bar
P_{cr}	critical feed pressure, bar

REFERENCES

1. Douglas, J. F., Gasiorek, J. M. and Swaffield, J. A., Fluid Mechanics, Pitman, Bath, 1979, P. 192.
2. Kelsall, D. F., A study of the motion of solid particles in a hydraulic cyclone. Trans. Chem. Eng., 1952, 30. 87—104
3. Xu, J. R., Research on the forced vortex and internal loss in hydrocyclones (Ph. D. Thesis). NEUT, 1989 (in Chinese with English abstract).
4. Luo, Q., Deng, C. L., Xu, J. R., Yu, L. X. and Xiong, G. A., Comparison of the performance of water— sealed and commercial hydrocylones. Int. J. of Miner. Process., 1989, 25, 297—310.
5. Pericleous, K. A., Rhodes, N. and Cutting, G. W., A mathematical model for predicting the flow field in a hydrocyclone classifier. In 2nd Int. Conf. on Hydrocyclones. paper B1, Bath, England, 1984.
6. Boadway, J. D., A hydrocyclone with recovery of velocity energy, In 2nd Int. Conf. on Hydrocyclones, paper D1, Bath, England, 1984.
7. Sun, Y. B., Gravitational Separation, PHMI, Beijing, 1982, p. 115 (in Chinese).
8. Lin, I. J., Technological innovations in the separation of particulates in solid—solid and solid—solid—liquid systems. Sep. Purif. Methods, 1984, 13, 1, pp. 1—42.
9. Bowaloph, A. N., Hydrocyclones, PHMI, Beijing, 1982, p. 9.
10. Xu J. R., Luo, Q. and Qiu. J. C., Studying the flow field in a hydrocyclone with no forced vortex, part 1 : average velocity. Filt. and Sep., 1990, 27, 9, 226—228.
11. Xu, J, R., Luo, Q. and Qiu, J. C., The investigation of the internal pressure loss in hydrocyclones. Sep. Sci. and Tech., 1989, 24, 14, 1167—1178.

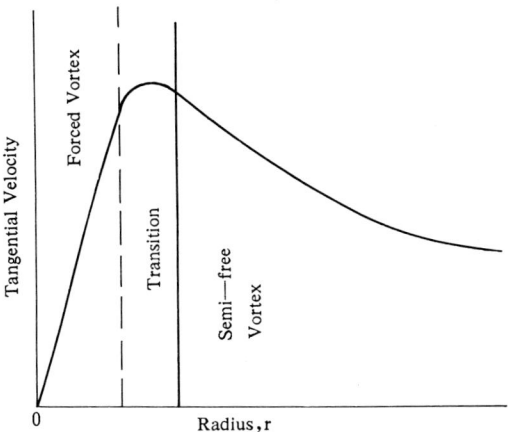

Fig. 1 Distribution of Tangential Velocity in Conventional Hydrocyclones

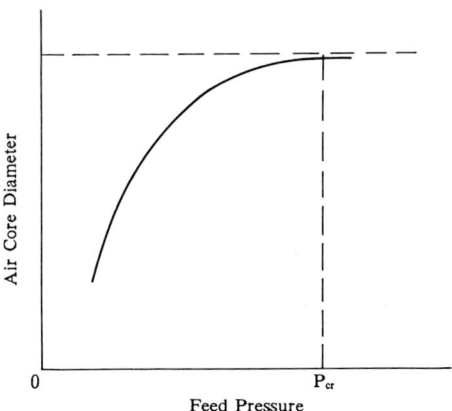

Fig. 2 Air Core Diameter vs Feed Pressure

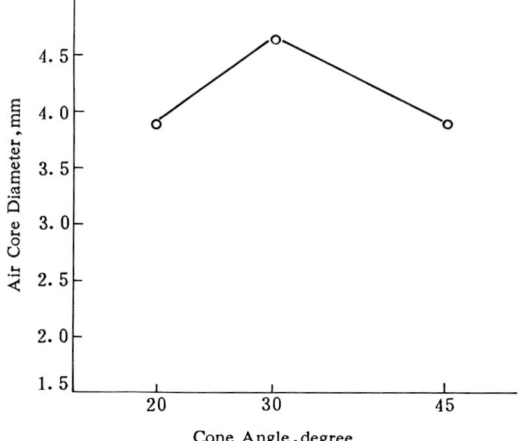

Fig. 3 Air Core Diameter vs Cone Angle
($D=33$mm, $d_o=7$mm, $d_s=4$mm, $d_t=6$mm, $P_t=1.08$ bar)

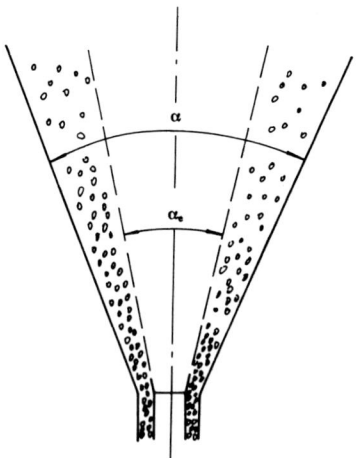

Fig. 4 Changing Effective Cone Angle
  α ——— Geometrical Cone Angle
  α_e ——— Effective Cone Angle

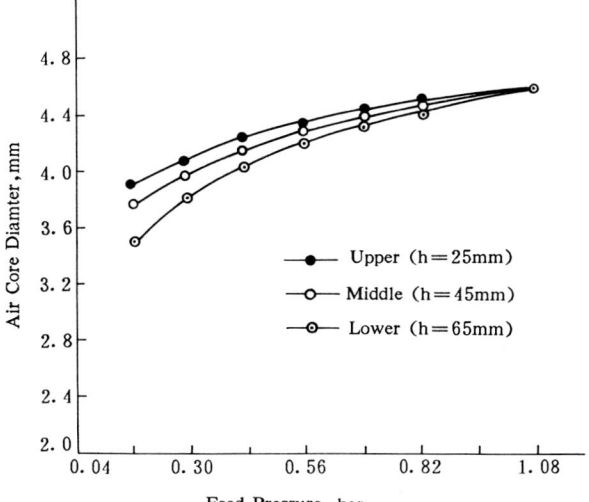

Fig. 5 Air Core Shape vs Feed Pressure (Illustrated by Air Core Diametres at Different Heights)
($D = 33mm, d_o = 7mm, d_s = 4mm, d_t = 6mm$)
$\alpha = 30°$

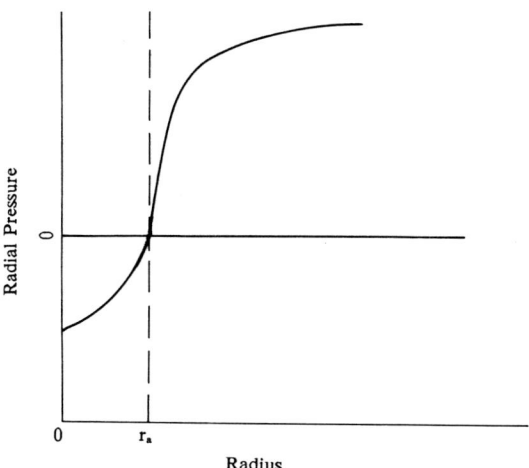

Fig. 6 Radial Pressure Distribution

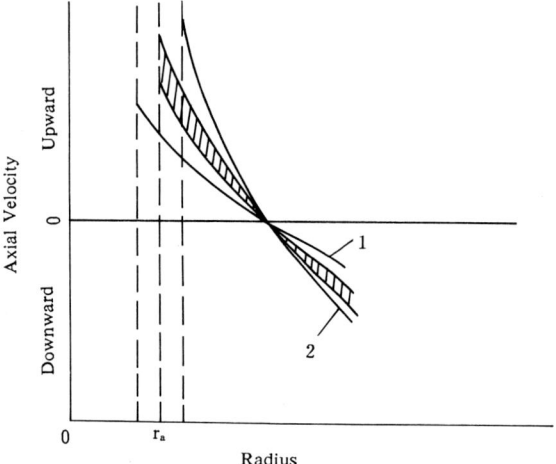

Fig. 7 Axial Velocity vs Unstable Air Core

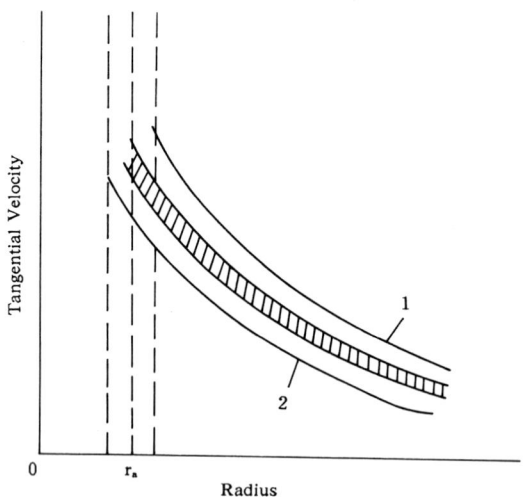

Fig. 8 Tangential Velocity vs Unstable Air Core

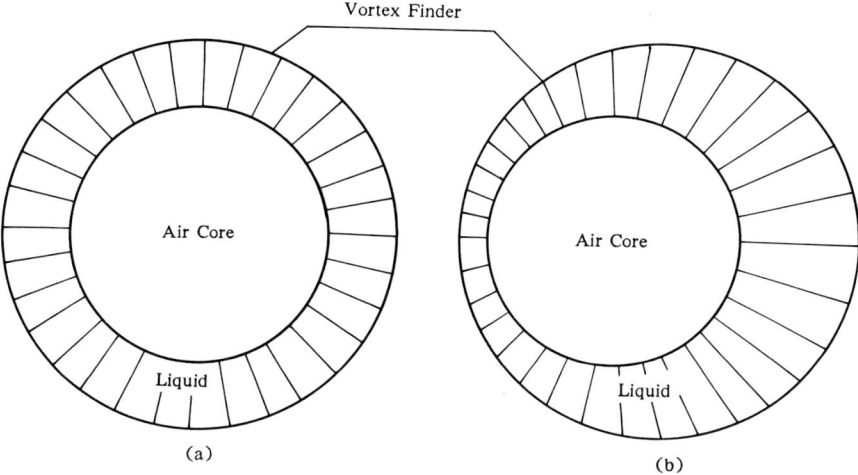

Fig. 9 Illustrations of (a) Symmetrical and (b) Asymmetrical Flow Field within Vortex Finder.

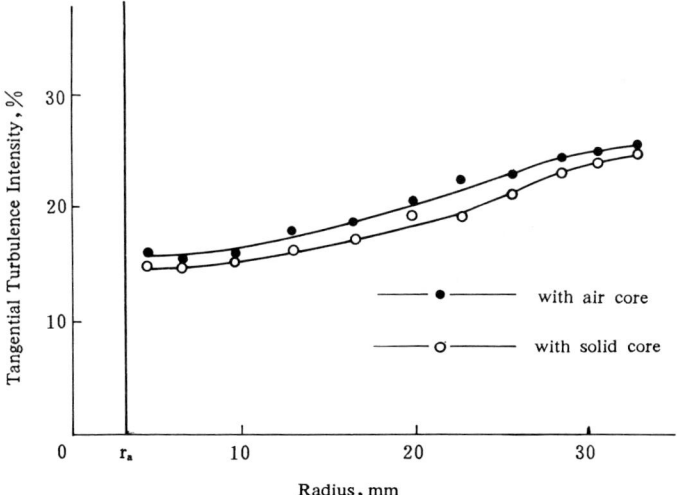

Fig. 10 Tangential Turbulence Intensity Distribution
(D=80mm, d_o=12mm, d_s=6mm, d_f=10mm, α=20°, h=155mm, P_f=0.98 bar)

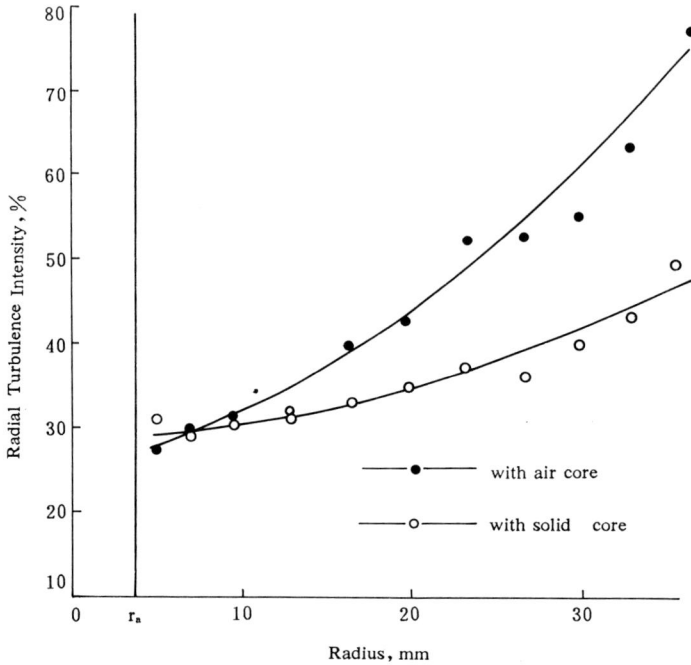

Fig. 11 Radial Turbulence Intensity Distribution
($D=80mm, d_o=12mm, d_s=6mm$ $d_f=10mm$,
$\alpha=20°, h=155mm, P_f=0.98$ bar)

AN EVALUATION OF THE PLITT AND LYNCH & RAO MODELS FOR THE HYDROCYCLONES

M.A.Z. COELHO and R.A. MEDRONHO*
* Escola de Química/UFRJ, CT, Bloco E
Federal University of Rio de Janeiro
21941 - Rio de Janeiro - RJ, Brazil

ABSTRACT

The Plitt and Lynch & Rao empirical models are widely used for prediction of hydrocyclones performance. In this work, these models were used to predict the performance of five different hydrocyclones: Rietema, Bradley, DEMCO, Mozley, and AKW. The cut size, the flowrate, and the feed liquid reporting to underflow predicted by Lynch & Rao model showed large errors in comparison with the experimental values. The same has occurred with the Plitt model, apart from the flow rate which showed a reasonable average error (± 13.5%).

NOMENCLATURE

C_v	Feed solid concentration by volume (per cent)
C_w	Feed solids concentration by weight (per cent)
d'_{50}	Reduced cut size (μm)
D_c	Hydrocyclone body diameter (cm)
D_i	Inlet equivalent diameter, by area (cm)
D_o	Overflow diameter (cm)
D_u	Underflow diameter (cm)
H	Pressure drop expressed in head of feed slurry (m)
L	Total length of hydrocyclone (cm)
ℓ	Vortex finder length (cm)
ΔP	Pressure Drop (kPa)
Q	Volumetric flow rate of feed slurry (ℓ/min)
R_w	Recovery of feed liquid to the underflow
S	Underflow-to-overflow ratio (flow split)
W	Mass flow rate of the feed slurry (t/h)
μ	Liquid viscosity (cP)
ρ	Liquid density (g/cm³)
ρ_s	Solids density (g/cm³)

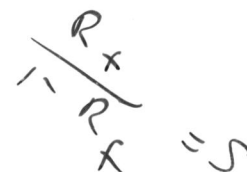

INTRODUCTION

Hydrocyclones are very simple in design. Despite of this, it is quite difficult to predict their performance using theory only. That is why empirical models as the ones due to Plitt [1] and Lynch and Rao [2] are widely used [3,4,5,6,7].

Plitt [1] has used three custom-made cyclones with interchangeable parts, with diameters of 3.2 (1¼ "), 6.4 (2½ ") and 15.2 cm (6"). He was able to vary D_i, D_o, D_u and $h = L - \ell$ for each cyclone. Apart from these geometric variables, he has also varied the pressure drop and the feed concentration. When developing his model, Plitt has added to his 174 experimental points another 132 tests from Lynch and Rao [8]. His words about his model are: "The work in this project was undertaken to develop a more universally applicable model which would give reasonable predictions over a wide range of operating conditions" and more: "The model which has been formulated enables the performance of a hydrocyclone to be calculated with reasonable accuracy when no actual experimental data are avaiable". This model is formed by the following equations:

$$d'_{50} = \frac{50.5 \, D_c^{0.46} \, D_i^{0.6} \, D_o^{1.21} \, \mu^{0.5} \exp(0.063 \, C_v)}{D_u^{0.71} \, (L - \ell)^{0.38} \, Q^{0.45} \, (\rho_s - \rho)^{0.5}} \qquad (1)$$

$$\Delta P = \frac{1.88 \, Q^{1.78} \exp(0.0055 \, C_v)}{D_c^{0.37} \, D_i^{0.94} \, (L - \ell)^{0.28} \, (D_u^2 + D_o^2)^{0.87}} \qquad (2)$$

$$S = \frac{1.9 \, (D_u/D_o)^{3.31} (L - \ell)^{0.54} \, (D_u^2 + D_o^2)^{0.36} \exp(0.0054 \, C_v)}{H^{0.24} \, D_c^{1.11}} \qquad (3)$$

The expression $\mu^{0.5}$ was added by Plitt to equation (1) in a later work [9].

Lynch and Rao [2] have used three Krebs hydrocyclones with diameters of 10.2 (4"), 15.2 (6"), 25.4 (10") and 38,1 cm (15"). Changing D_o and D_u in some of their cyclones, they were able to obtain a data set of 220 experimental points.

Based on these data, Lynch and Rao [2] have developed a model which, according to them, requires the determination of constants which must be experimentally evaluated for every specific application. Their model is formed by the equations below:

$$\log d'_{50} = 0.0400\, D_o - 0.0576\, D_u + 0.0366\, D_i + 0.0299\, C_w - 0.0001\, Q \tag{4}$$

$$Q = 6.00\, D_o^{0.73}\, D_i^{0.86}\, \Delta P^{0.42} \tag{5}$$

$$R_w(\%) = 193.0\, D_u/W - 271.6/W - 1.61 \tag{6}$$

The objective of the present work is to test the above models, with their original constants, for different geometries of hydrocyclones.

MATERIALS AND METHODS

Experimental points arising from five different geometries of hydrocyclones have been used in this work. The tests with Rietema, Bradley and DEMCO 4H came from previous work of some of the authors [10,11,12,13]. As these experiments did not include any data with feed concentrations higher than 10% by volume, the results from an earlier work of Marasingle and Svarovsky [14] were added to the data set. The total number of hydrocyclone tests resulted in 174 experiments, including the 15 points obtained by Marasinghe and Svarovsky with AKW RW 2515 and Mosley 2" hydrocyclones.

Table 1 gives the hydrocyclones diameters, the operating conditions and materials used in this work. Figure 1 shows the particle size distributions of the materials tested. The geometric proportions of the hydrocyclones listed in Table 1 can be found elsewhere [15], and details of test rigs and methods can be found in the respective references.

TABLE 1

Hydrocyclones diameters, range of operating conditions, and materials used in this work

Hydrocyclone	D_c (cm)	ΔP (kPa)	C_v (%)	Material	ρ_s (g/cm^3)	Nº of Tests
Bradley [10]	1.5	69-276	1-10	CaCO$_3$ I	2.45	20
Bradley [10]	3.0	69-276	1-10	CaCO$_3$ I	2.45	20
Bradley [10]	6.0	69-276	1-10	CaCO$_3$ I	2.45	20
Bradley	3.0	69-276	1	Barite	3.75	12
Rietema [11,12]	2.2	69-276	1-10	CaCO$_3$ II	2.78	15
Rietema [11,12]	4.4	69-207	1-10	CaCO$_3$ II	2.78	15
Rietema [11,12]	8.8	69-138	1-10	CaCO$_3$ II	2.78	12
DEMCO 4H [13]	12.2	69-172	1-10	Barite	3.75	18
Mozley 2" [14]	4.5	207	3-28	CaCO$_3$ III	2.71	8
AKW RW 2515 [14]	12.5	103	3-28	CaCO$_3$ III	2.71	7

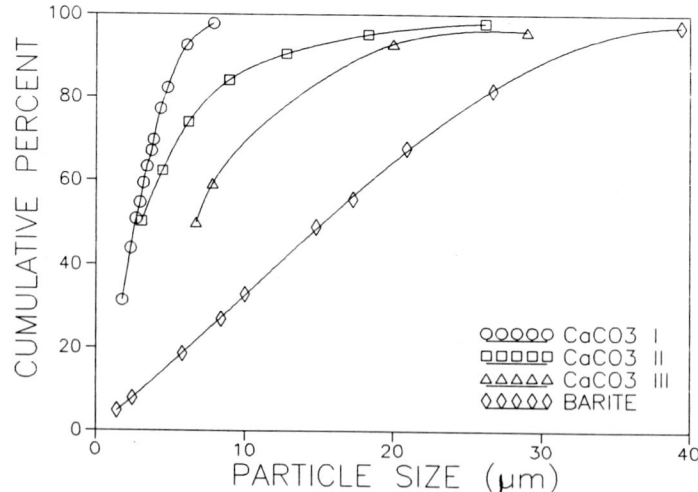

Figure 1. Particle size distribution of the materials used in this work.

RESULTS AND DISCUSSION

Table 2 gives the ranges of diameters, geometric proportions, and operating conditions used by Plitt [1], Lynch and Rao [2], and this work.

TABLE 2

Range of the variables used by Plitt [1], Lynch and Rao [2], and this work

Variable	Plitt	Lynch & Rao	This work
D_c (cm)	3.2-15.2	10.2-38.1	1.5-12.5
D_i/D_c	0.15-0.30	0.20-0.25	0.14-0.28
D_o/D_c	0.12-0.40	0.27-0.52	0.20-0.34
D_u/D_c	0.06-0.40	0.13-0.25	0.04-0.28
$(L - \ell)/D_c$	0.95-5.52	—	2.75-7.14
ΔP (kPa)	17-207	17-207	69-276
C_v (%)	0-50	6-50	0-28

Figures 2 and 3 show a comparison of the observed values for the reduced cut size and the ones calculated through Plitt and Lynch and Rao models. As can be seen in Figure 2, equation (1) tends to overestimate d'_{50}. The average error given by this equation is ±49%. The Lynch and Rao equation (4) is practically insensible to changes in operating conditions, except for the feed concentration.

A comparison of observed values for the feed flow rate and the ones calculated by Plitt and Lynch and Rao models can be seen in Figures 4 and 5. Equation (2) predicts the flow rate reasonably well (see Figure 4). The average error was equal to ±13.5%, and 76% of the experimental points were predicted with less than ±20% error.

The maximum error found was +45%.

Equation (5) always overestimates the flow rate.

Equations for R_w or S predictions are in general very unreliable. This can be seen in Figures 6 and 7. The average error found when using equation (3) was ±58%, with a maximum error as high as +500%. Equation (6) showed to be useless since most of its results were negative.

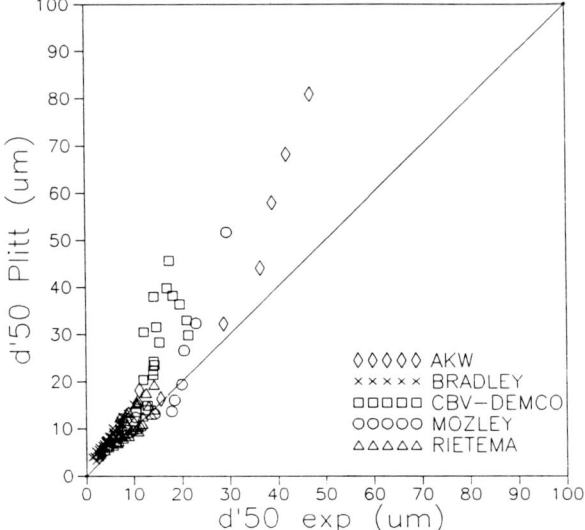

Figure 2. Comparison of calculated (equation (1)) and observed values for reduced cut size.

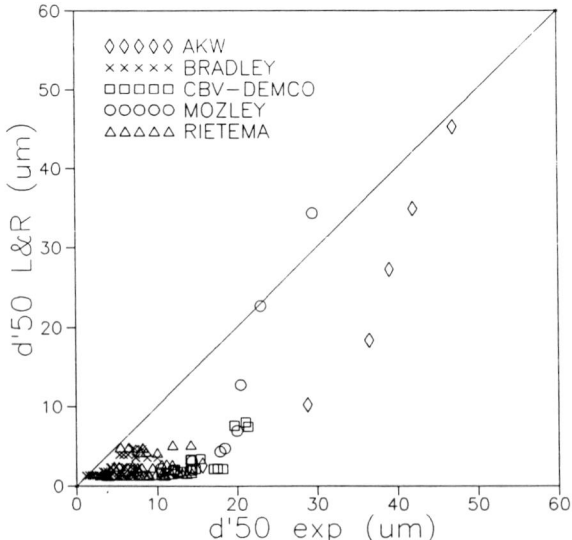

Figure 3. Comparison of calculated (equation (4)) and observed values for reduced cut size.

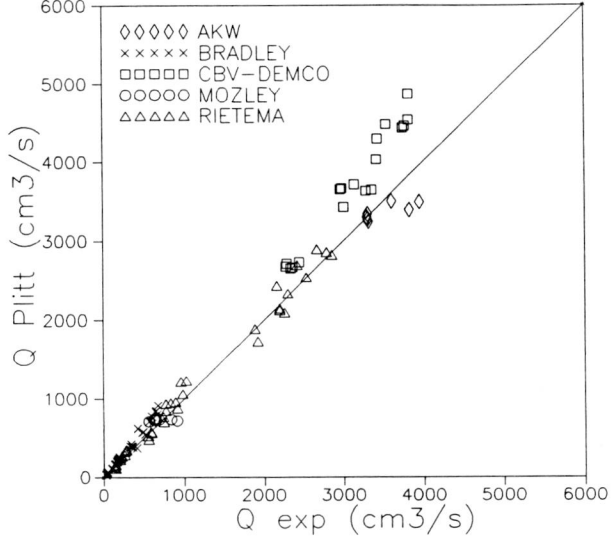

Figure 4. Comparison of calculated (equation (2)) and observed values for hydrocyclone capacity.

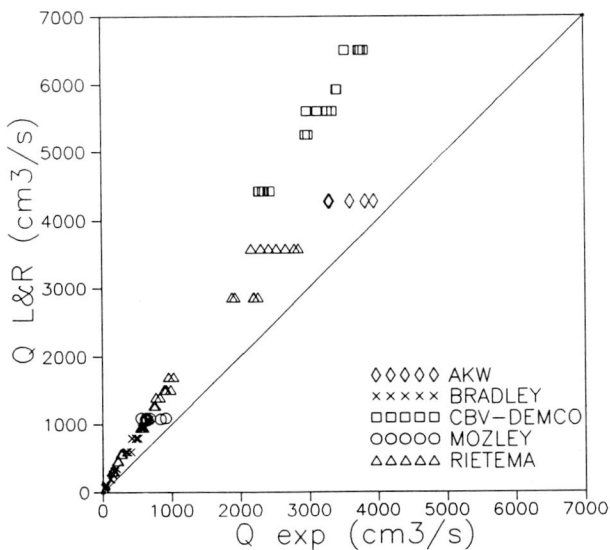

Figure 5. Comparison of calculated (equation (5)) and observed values for hydrocyclone capacity.

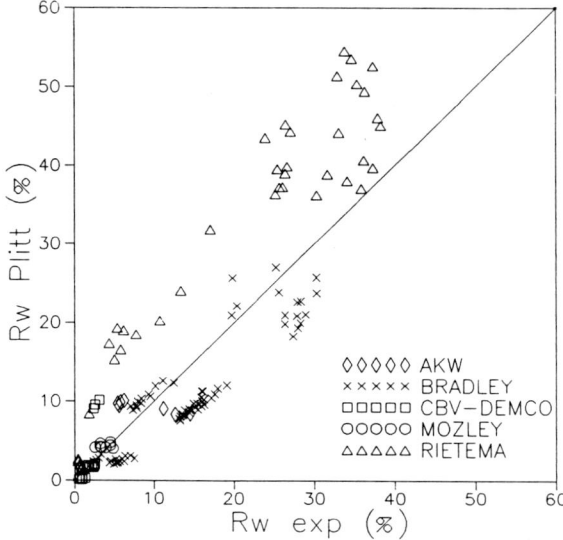

Figure 6. Comparison of calculated (equation (3)) and observed values for fraction of water entering coarse product.

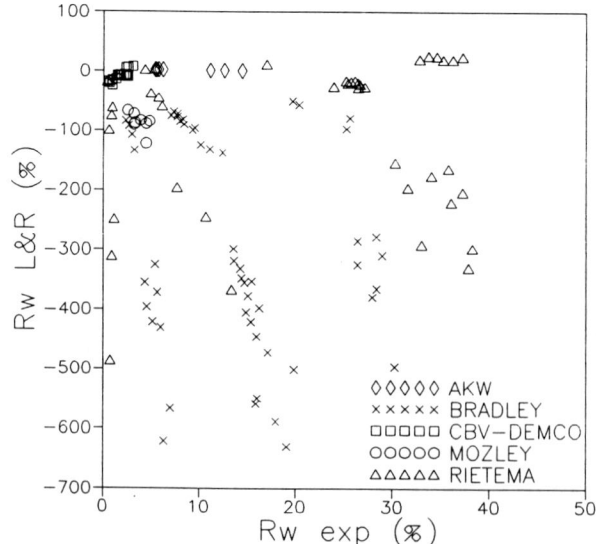

Figure 7. Comparison of calculated (equation (6)) and observed values for fraction of water entering coarse product.

CONCLUSIONS

This work has shown that Lynch and Rao model [2] can not be used with its original constants, either to predict hydrocyclone performance or to design it.

This conclusion is in accordance with Lynch and Rao recommendation that the constants of their model should be recalculated for each specific data set.

The Plitt model [1] gives a reasonable flow rate prediction (average error of ±13.5%), but it tends to overestimate the reduced cut size (this has happened to 90% of the experimental points), and gives unreliable predictions of the flow split. Therefore, if it is desired to predict hydrocyclone performance, equation (1) would give a reasonable good value for the flow rate. However, the reduced total efficiency would have a great chance to be underestimate since d'_{50} was probably overestimated (see chapter 3 of reference 16 for reduced total efficiency calculation). Therefore, the Plitt model [1] would probably give a conservative value of the reduced total efficiency.

REFERENCES

1. Plitt, L.R., A mathematical model of the hydrocyclone. CIM Bulletin, Dez. 1976, 114-23.

2. Lynch, A.J. and Rao, T.C., Modelling and scale-up of hydrocyclone classifiers. Proc. 11th Int. Mineral Processing Congress, Gagliari, 1975, 1-25.

3. Colussi, I., High-purity silica from silicate rocks: a study on leucite-bearing rocks. Advanced Separation Science (Process Symp.), Trieste, 1978, 280-6.

4. Adams, R.W., Sand and Mica separation from kaolinite in the United Kingdom China Clay Industry using 350 mm hydrocyclones. Proc. 1st Int. Conference on Hydrocyclones, BHRA, Cambridge, 1980, 99-112.

5. Apling, A.C., Montaldo, D. and Young, P.A., Hydrocyclone models in an ore-griding context. Proc. 1st Int. Conference on Hydrocyclones, BHRA, Cambridge, 1980, 113-25.

6. Pericleous, K.A., Rhodes, N. and Cutting, G.W., A mathematical model for predicting the flow field in a hydrocyclone classifier. Proc. 2nd Int. Conference on Hydrocyclones, BHRA, Bath, 1984, 27-40.

7. Rao, T.C., Vanangamudi, M. and Sufiyan, S.A., Modelling of dense-medium cyclones treating coal. Int. J. Miner. Process., 1986, **17**, 287-301.

8. Lynch, A.J. and Rao, T.C., The operating characteristics of hydrocyclone classifiers. Ind. J. Tech., 1968, **6**, 106-14

9. Plitt, L.R., Finch, J.A. and Flintoff, B.C., Modelling the hydrocyclone classifier. European Symp. Particle Technology, Amsterdam, 1980, 790-804.

10. Antunes, M. and Medronho, R.A., Bradley hydrocyclones: design and performance analysis. 4th Int. Conference on Hydrocyclones, Southampton, Elsevier for BHRA, 1992.

11. Medronho, R.A., Scale-up of hydrocyclones at low feed concentrations, Ph.D. Thesis, University of Bradford, 1984.

12. Medronho, R.A. and Svarovsky, L., Tests to verify hydrocyclone scale-up procedure. Proc. 2nd Int. Conference on Hydrocyclones, BHRA, Bath, 1984, 1-14.

13. Azevedo, R., Veloso, C.O. and Medronho, R.A., Performance of CBV-DEMCO 4H hydrocyclones. Anais do XVIII Encontro sobre Escoamento em Meios Porosos - Vol. II, Nova Friburgo/Brazil, 1990, 481-488 (in Portuguese).

14. Svarovsky, L. and Marasinghe, B.S., Performance of hydrocyclones at high feed solids concentration. Proc. 1st Int. Conference on Hidrocyclones, BHRA, Cambridge, 1980, 127-42.

15. Svarovsky, L., Hydrocyclones, Holt, Rinehart & Winston, Eastbourne, 1984, pp. 60.

16. Svarovsky, L., Efficiency of separation of particles from fluids. In Solid-Liquid Separation, ed. L. Svarovsky, Butterworths, London, 1990, pp. 42-73.

DETERMINATION OF PHASE BEHAVIOUR IN DE-WATERING HYDROCYCLONES USING RESISTIVITY MEASUREMENTS: FEASIBILITY STUDY

M T THEW
M F BECKER
Mechanical Engineering Department
University of Southampton
Southampton SO9 5NH, UK

ABSTRACT

A small scale experimental study of phase inversion in a hydrocyclone using brine in kerosine dispersions at flow rates up to 30 ℓ/min is described. An array of stainless steel electrodes was used with diameter 1.5mm fitted through the wall to end flush with the wall, also there were three pairs of diametrically opposed electrodes. Resistivity was measured with wave recorders energising the electrodes with a 3kH$_z$ alternating current square wave supply at very low current and with a 1% brine solution. Precautions to obtain valid results using this technique are described; variations in flow rate, split ratio and water drop size were explored. Whilst the technique requires further developments, it provided clear evidence of near wall localised changes in composition. It emerged that radial concentration distributions can be measured and thus there is a good prospect of obtaining three-dimensional isoconcentration contours. These contours would be of considerable help in understanding the behaviour of cyclones where phase inversion occurs.

NOTATION LIST

$\overline{di}$	Mass mean brine drop diameter in feed	
D	Hydrocyclone body diameter	
F	Split ratio (Qd/Qi) . (1-F) = Split ratio Qu/Qi)	
Fcrit	Value of F at the 'knee' point	
Ki	Inlet brine concentration	%
Ku	Overflow (upstream) brine concentration	%
ℓ	Distance of electrode from end of taper at underflow	mm
Qd	Flow rate from underflow (downstream)	ℓ/min
Qi	Feed flow rate	ℓ/min
Qu	Flow rate from overflow (upstream)	ℓ/min
R	Hydrocyclone radius at any plane normal to the axis of the hydrocyclone	
R_c	Notional radius of oil core in this plane	
V_r	Voltage ratio (dispersion voltage/voltage with 100% brine)	

INTRODUCTION

Within the realm of liquid-liquid hydrocyclones the de-oiling design, that is with a light dispersion, has been well developed and is now extensively employed in oil fields, particularly offshore. Designs have also been developed at Southampton, with other similar designs elsewhere, to explore the possibility of removing a dispersion of a relatively heavy liquid. Several field trials of so-called de-watering cyclones have been held. With a de-watering hydrocyclone the dispersion is a heavy one so there are some differences in geometry from the de-oiling hydrocyclone.

The range of concentrations within de-watering cyclones is very different from that encountered in de-oiling. In the latter the concentration of dispersion is seldom likely to range up to 1%, except for specialised uses. Within the de-watering hydrocyclone the concentration may range up to 20% brine, 30% or even to substantially large figures. At such concentrations phase inversion may occur and has indeed been seen in laboratory tests. When the phase inversion occurs the interfacial behaviour exercises a much larger influence than in the de-oiling hydrocyclone. Experimental work has shown very significant limits to the concentration of brine in oil at which a reasonable separation efficiency may be obtained as can be seen in figure 1 from reference 1.

To improve the performance of de-watering cyclones it is believed that we need to know about the relative concentrations of the two phases and ideally it would be good to obtain three dimensional concentration contours within the hydrocyclone. The present work described therefore has three objectives.

Objectives

Within low budget experimentation the first objective was to develop a practical technique and this was to be developed with an electrode array which was non-invasive and of simple geometry. The second objective was to conduct some experiments to examine the near wall behaviour and to obtain preliminary quantitative information. The third objective was to examine the feasibility of determining by means of radial measurements of the two phases within the hydrocyclone, their relative concentration.

BACKGROUND INFORMATION

Extensive work has been carried out at Southampton on the brine/kerosine system and because of this considerable information we decided to continue with this particular combination for the present work. It also has the advantage that the oil, with care, may be re-used whereas mixtures of brine and crude oil are notorious for the large amount of waste than can be generated.

A simple model of a two-component arrangement or two-phase arrangement with a dispersed heavy liquid is very much determined by whether the concentration Ki is greater or less than the split ratio, that is,F. If Ki > F then in the model where the performance of the hydrocyclone is optimised, all the brine will pass out in the underflow up to the value of the split ratio, but the excess value of the brine ie (Ki - F) will move into the overflow, which is clearly undesirable.

The second (favourable) regime is where Ki < F. In the idealised situation there will be no brine in the overflow but the underflow will contain oil of fractional amount $(1 - (Ki/F))$.

We found that experimental results with brine/kerosine systems are fairly near the idealised model, see figure 2 [1] where the separation efficiency is measured by the dimensionless ratio, Ku/Ki. Using this parameter there is a fairly marked change in behaviour between the regimes but now F/Ki in the favourable regime has to be at least several percent greater than 1, because very fine drops of brine will not pass into the underflow, but are convected and emerge in the overflow. The change occurs at a knee point as shown in figure 2 and this point may be denoted as a critical point and as shown in figure 3 the knee point is not very dependant on the input concentration Ki.

With this background information one can look and see if there are

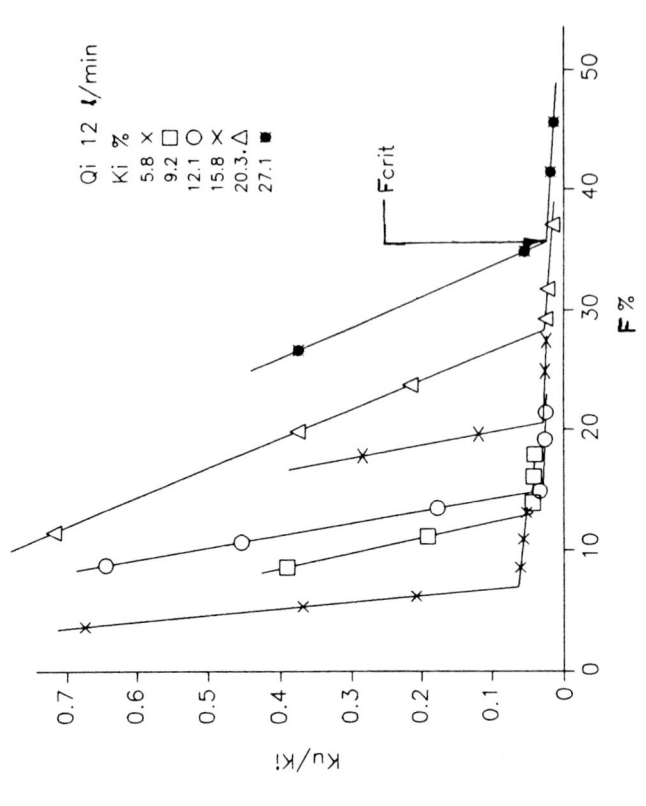

VARIATION OF SEPARATION WITH SPLIT RATIO FOR DIFFERENT INLET CONCENTRATIONS. [1]   FIG. 2

EFFECT OF FEED CONCENTRATION ON SEPARATION [1] : FIG.1

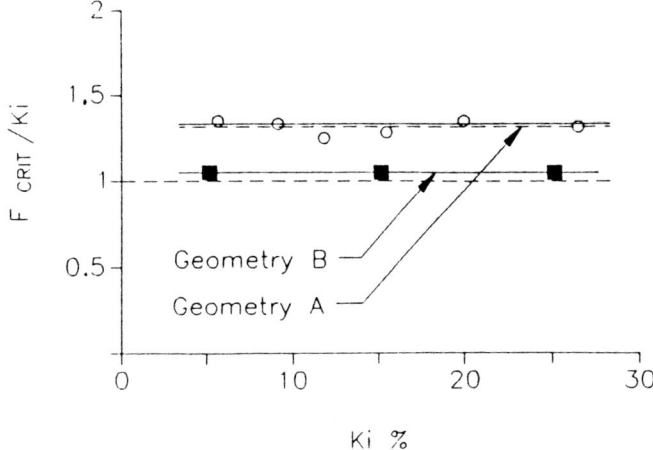
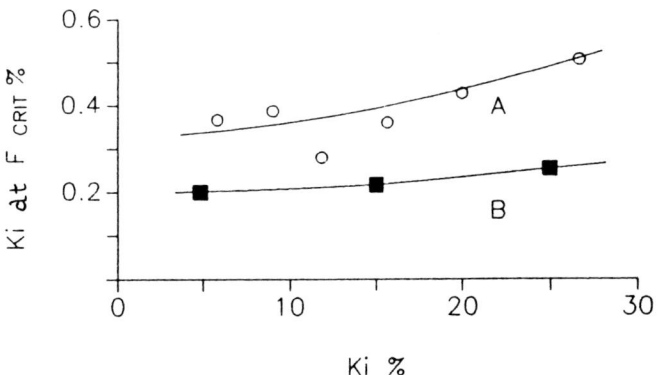

CONDITIONS AT CRITICAL SPLIT RATIOS [1] : FIG.3

measurements of local concentration that can be obtained and are related to this simple model.

DEVELOPMENT OF THE EXPERIMENTAL TECHNIQUE

The ratio of resistivity between kerosine and brine is typically several orders of magnitude. It is therefore a suitable parameter to try and differentiate between the two components or two phases. Previous hydrocyclone work using the measurement of resistivity at Southampton, was to determine residence time as shown in references 2 and 3. Wave monitors were employed since they operate on alternating current supply and have a very small effect near to the electrode. In figure 4 we show the set up of the instrumentation that was employed.

In the previous work described in references 2 and 3 the brine concentration normally used was small, about 0.04% by weight sodium chloride. In the present work the electrode area was markedly less, therefore it seemed likely that we would have to use a higher concentration of brine to obtain an acceptable signal to noise ratio. We carried out a series of tests at different brine concentrations and as figure 5 shows, at concentrations previously used, the voltage output from the wave recorders was small and to obtain a reasonable signal to noise ratio we decided to use a 1% brine concentration, since higher concentrations did not improve the signal to noise ratio.

The electrode configuration and hydrocyclone geometry is shown in figure 6 and the electrodes are non-invasive as previously noted, of low cost and made from 1.5mm diameter stainless steel rod. We did however find that the end of the electrode where it protrudes through the hydrocyclone wall has to be extremely carefully shaped and minor protuberances or a small dip below the surface can significantly alter the output voltage.

Arbitrarily we chose to use a 9 volt output at high conductivity, that is with 100% brine, and set up the wave recorders such that the output with 100% kerosine was approximately 0 volts. Within this voltage (or more properly, impedance) range, a log-log plot of recorder output (α resistance) against distance along the hydrocyclone is shown in figure 7. Whilst

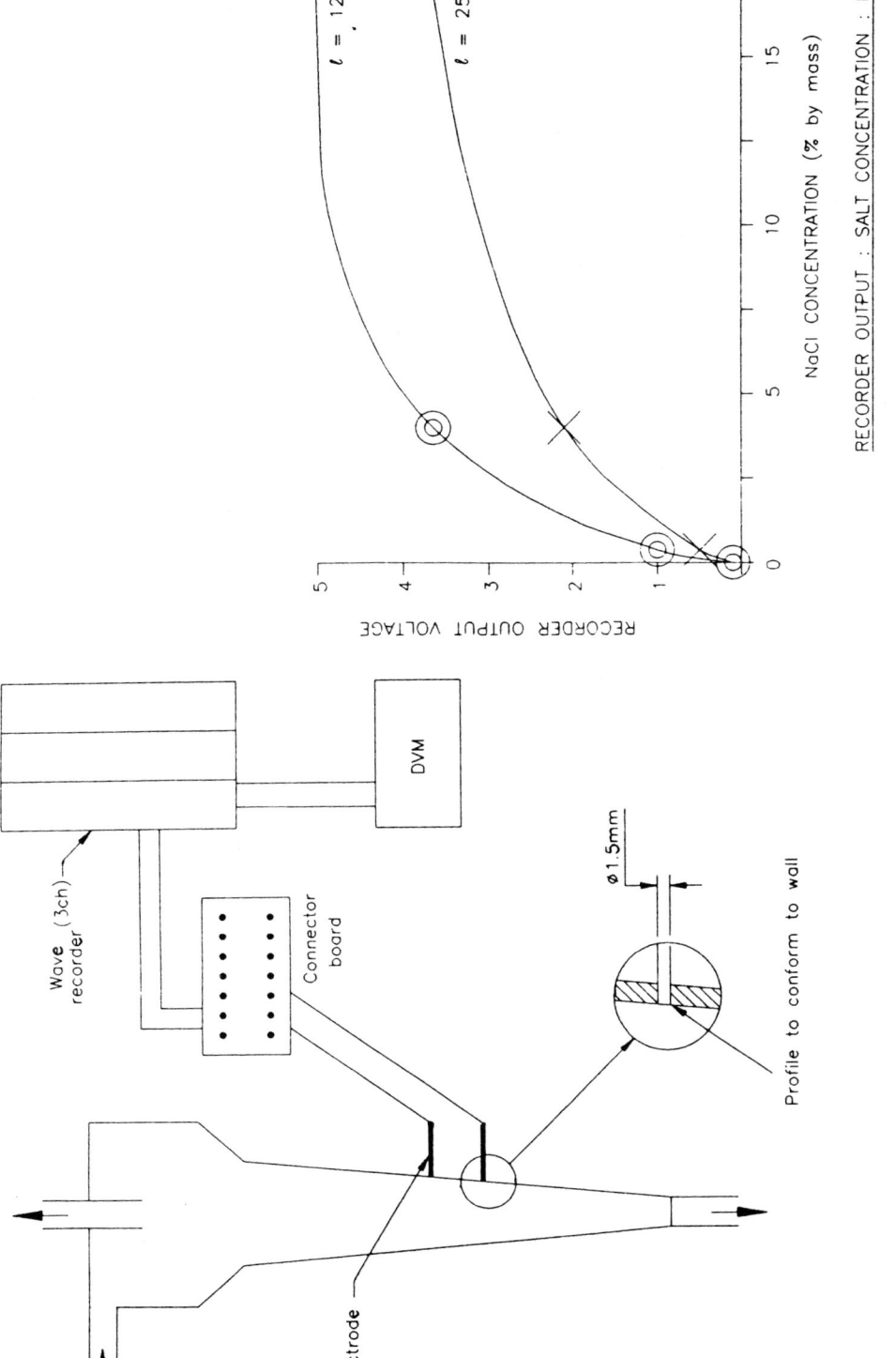

RECORDER OUTPUT : SALT CONCENTRATION : FIG. 5

INSTRUMENTATION   FIG. 4

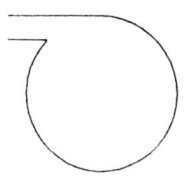

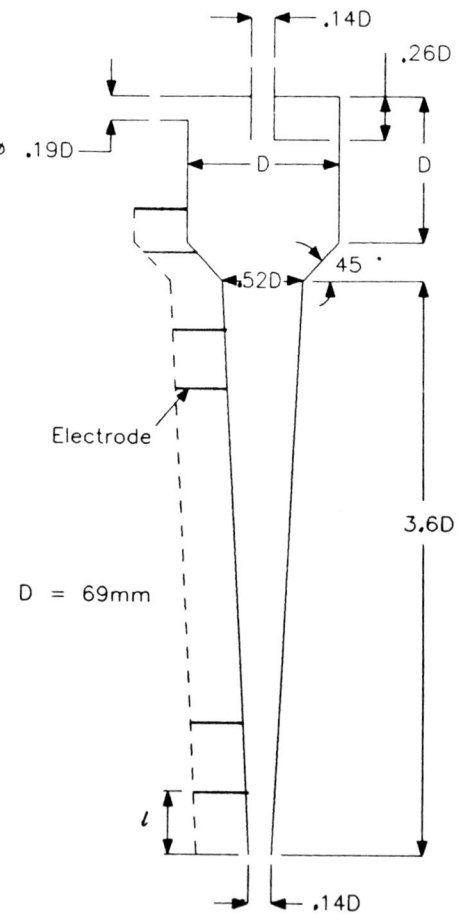

Electrode Spacing
l = 16 mm
l = 36 mm
l = 52 mm
l = 70 mm
l = 93 mm
l = 116 mm
l = 141 mm
l = 161* mm
l = 181 mm
l = 201* mm
l = 221 mm
l = 233* mm
l = 261 mm
l = 281 mm
* Diametral pairs

HYDROCYCLONE PROPORTIONS : FIG 6

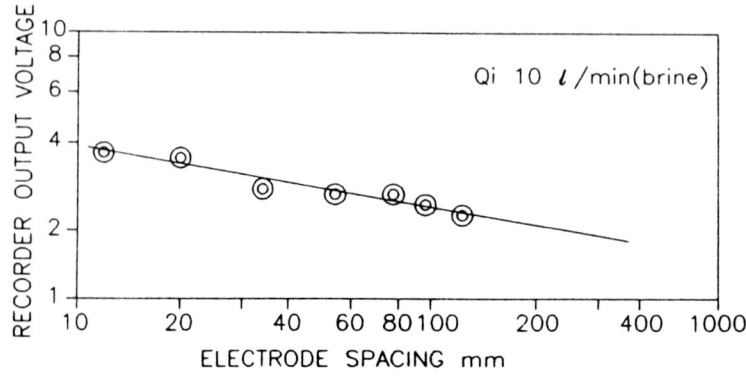

RECORDER OUTPUT : ELECTRODE SPACING : FIG 7

approximately linear, the slope is surprisingly small, showing that conductivity decreased approximately with the distance along the electrode array to the negative power of 0.2. This seems extremely small, with almost the same characteristic for static brine and flowing brine. If the resistance to the current flow was linearly a function of path length then a gradient of -1 would be anticipated for figure 7. This suggests that the resistance is made up from two components, the major resistance occurring near to the electrode and a minor part (α path length) due to the intervening distance. The lack of change in resistance produced by brine motion suggests negligible change in the ion transport mechanism.

Preliminary tests with dispersions of brine in kerosine indicated that we did obtain a range of voltages spanning the whole values between zero voltage and 9 volts. An important detail of experimental procedure was to conclude each run with a period without kerosine in the flow, to avoid oil contamination of the electrode surface. Since this work was exploratory at this stage the programme moved to an exploration of various parameters, without clarifying further the physical meaning of V_r. It seems clear that high values mean that the brine is the continuous component and that to some extent the highest values indicate relatively little kerosine is dispersed in the brine. It is not yet clear how linear the relationship is between brine content and V_r, nor what the zone size is from which the value of V_r is obtained.

TEST RIG

The test rig had to provide control of inlet feed rate up to a maximum flow rate of about 30ℓ/sec, a brine concentration that could be varied between zero and approximately 60% and a variation in split ratio approximately between 0.3 up to 0.7. The rig which is shown in figure 8 was primarily designed and is described in a report by Warburton [4]. Preliminary work by Becker and Thew on the resistivity measurements, with detailed evaluation of the experimental technique is described in reference 5.

This rig did not give major control over the brine feed drop size. However, at the flow rate of 20ℓ/min the use of a mixing valve did permit some

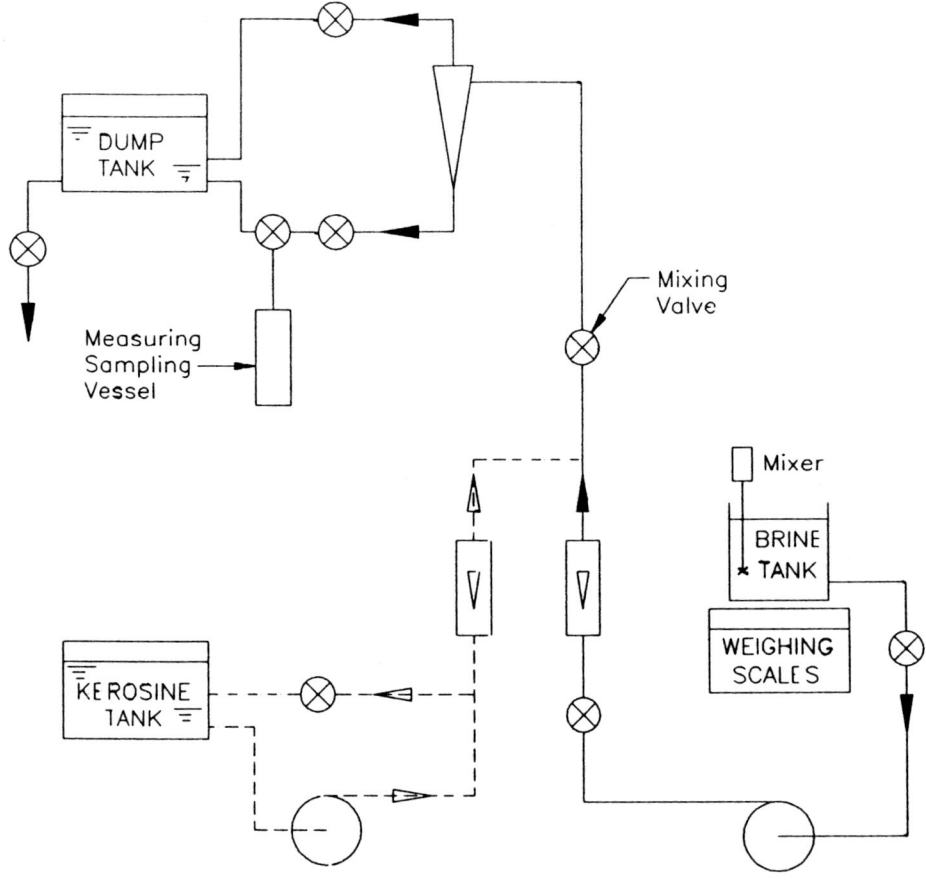

(Arrangement for kerosine recirculation not shown)

TEST RIG LAYOUT   FIG. 8

variation in drop sizes. The drop sizes were measured by obtaining high magnification of photographs and looking at them with a semi-manual counter. The drop sizes used were generally significantly larger than in previous separation work, for example in reference 1.

The split ratio F was checked using whole stream sampling and the ratio Ki/F was taken up to 1.25 when substantial water break-through in the overflow would be anticipated.

DISCUSSION OF RESULTS

Axially placed electrodes

On figure 9 is shown a series of results for the axial variation of V_r for various conditions at a flow rate of 20ℓ/min. Two values of split ratio were used and variation in drop size was obtained with the mixing valve.

It can be seen that there is a marked variation between results for concentrations up to typically 40% and those for concentrations of 45 or 50%. The effect of position along the hydrocyclone, examining the ratio V_r between adjacent pairs of electrodes moving away from the vertex, is that the presence of brine is markedly reduced. As can be seen by referring back to figure 6, the electrode pairs were placed all the way up the slow taper with one pair in the cylindrical entry portion of the hydrocyclone.

At low feed concentrations the proportion of brine near the wall though apparently small, seems to have reached a maximum at a point part way along the slow taper, but once one gets to high concentrations the proportion of brine as judged by the voltage ratio increases largely monotonically from positions near the inlet down towards the underflow. It can be seen that the change in split ratios had quite a significant effect, a rise in F reflecting the dominant downstream flow, but rather surprisingly producing lower values of V_r eg at F = 0.4, V_r ~ 0.9 for ℓ 100-200mm but at F = 0.6, V_r ~ 0.8.

However, the larger water droplets produced when no mixing valve was used, do seem to have to have reached the wall more readily giving the same V_r at Ki = 45% as for Ki = 50% with small droplets.

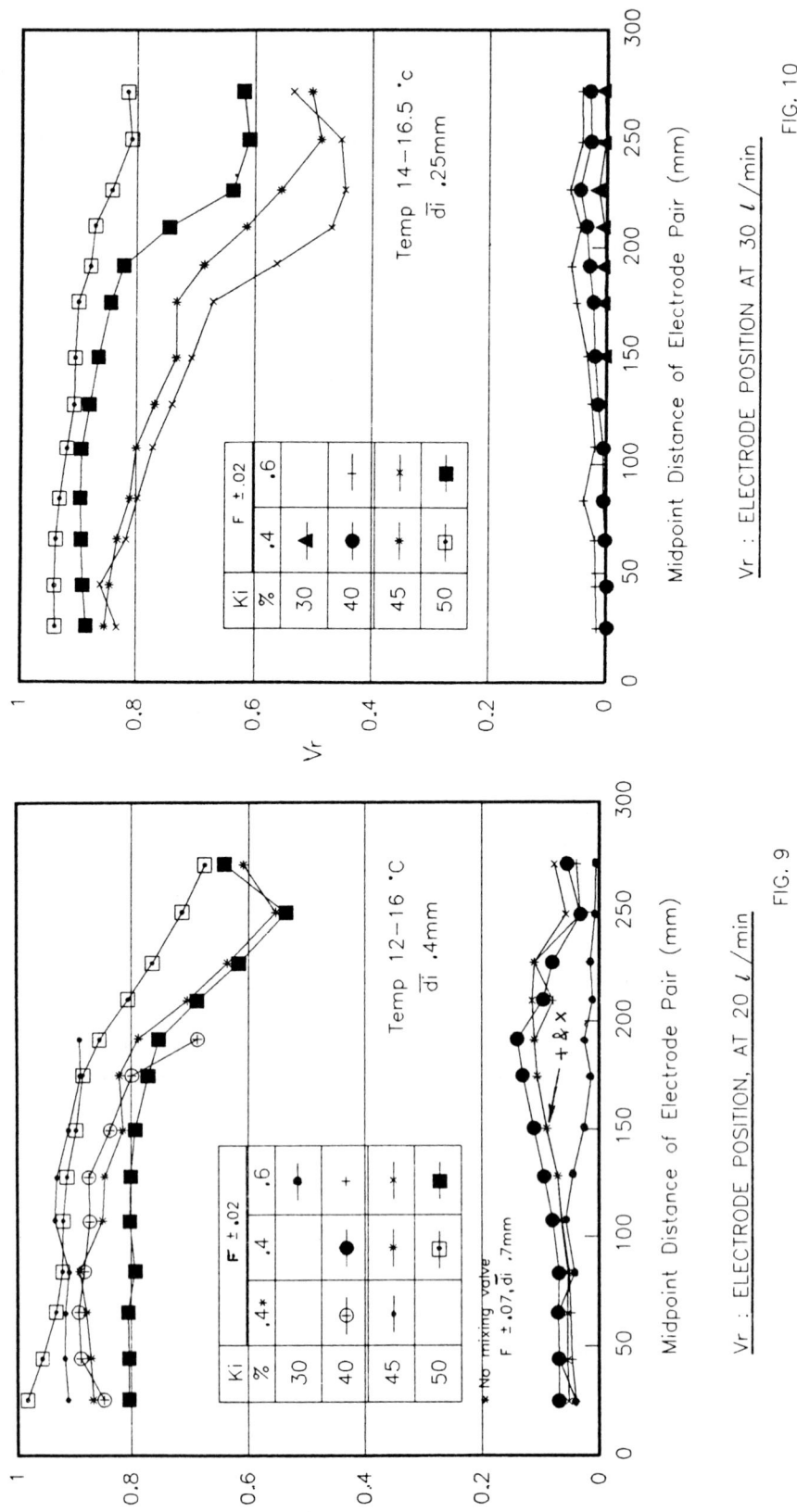

FIG. 10 Vr : ELECTRODE POSITION AT 30 l/min

FIG. 9 Vr : ELECTRODE POSITION, AT 20 l/min

On figure 10 are the sets of results at the higher flow rate of 30ℓ/min and again there is a fairly sudden jump in V_r from low values at feed concentrations of the order 40% or so up to much higher levels at 45 or 50%. There is again a discernible difference between the split ratio of 0.4 and the split ratio of 0.6, with the higher values of V_r for the lower proportion of the feed leaving at the downstream (underflow) part. The larger acceleration field produces a more structured flowfield as judged by values of V_r, since they are smaller than those on figure 9 for Ki ≤ 40% but higher than those on figure 9 for Ki ≥ 45%.

It is more informative to cross-plot the results to examine how the percentage of brine in the feed alters the voltage ratio, this is done on figures 11 and 12. Figure 11 shows some early results at a small flow rate of 10ℓ/min with only two axial positions of electrode pairs. It must be remembered that this is the voltage ratio between electrode pairs that were approximately 25mm apart on their centre lines. It can be seen that there is a sudden jump with one electrode pair position at about 40% brine up to 45%, but at the other position the increase in voltage ratio is much slower. This suggests that at the position nearer the underflow a slight increase in feed brine concentration causes clear cut inversion, but further away inversion is only completed when there is somewhat more brine in the feed. The drop size for these preliminary experiments at this small flow rate was relatively large: see Table 1 below.

On figure 11 are shown further results at F = 0.4 for flow rates of 20ℓ/min and 30ℓ/min. It can be seen that the mixing valve has a significant effect in terms of where the jump in voltage ratio occurs due to the change in drop size. There are some minor variations depending where the electrode pairs are considered along the cyclone, but perhaps rather surprisingly the changes occur along most of the length of the slow taper part of the hydrocyclone over the same range of Ki at both flow rates. Results at the higher split ratio (F = 0.6) are on figure 12; at this split ratio Ki/F is 0.83, yet there is a difference of about 6% in Ki for the jump. It looks as if inversion is hastened by the greater acceleration field produced by the large flow rate.

As the figures show, there was some variation in temperature during the

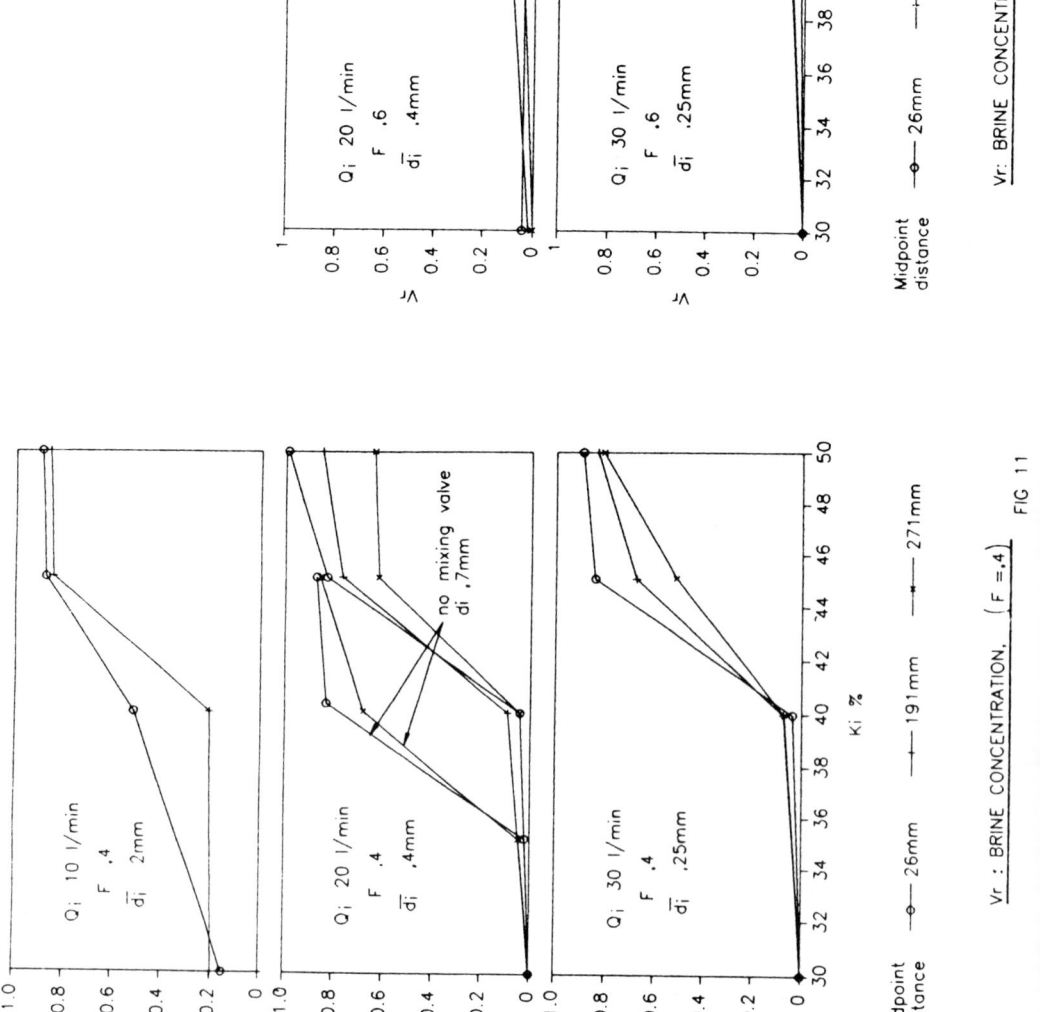

Vr: BRINE CONCENTRATION, (F =.6): FIG.12

Vr : BRINE CONCENTRATION, (F =.4) FIG 11

runs, but it is not believed that this has a major effect on the results achieved. Arbitrarily if one takes a voltage ratio of 0.5 as indicating the phase change, then in Table 1 it is shown how the values of Ki vary at the voltage ratio of 0.5 for two split ratios and the values of Ki minimum and maximum are taken to correspond with one of the three electrode pairs that have been used.

TABLE 1: Values of Ki at a V_r ratio of 0.5 (rounded)

Flow rate ℓ/min	F	d_i µm	Ki_{min}%	Ki_{max}%	Comments
10	.4	2000	39*	42*	no mixing valve
20	.4	700	38*	39*	no mixing valve
20	.4	400	43	44	mixing valve
30	.4	250	43	45	no mixing valve
20	.6	400	48	49	mixing valve
30	.6	250	43	45	no mixing valve

* electrode N/A at 271mm

It can be seen that at a voltage ratio of 0.5 the brine concentrations vary between 38% as a minimum and 49% at maximum and that most fall in a rather narrower range. At the highest flow rate, and therefore highest acceleration field the other effects (drop size and split ratio) appear to be less important. It does appear that the technique is capable of giving us a quantitative picture of what is happening, although the limitation must be observed that the results discussed so far relate to near wall conditions and to obtain the information we really desire it is necessary to try and obtain some idea of the phase variation across a diameter. So the last part of the work that is to be described discusses a very preliminary attempt to have a look at the concentration variation across a diameter.

Diametrally placed electrodes

In Appendix 1 a very approximate theory is developed which relates the voltage ratio, V_r and a calculation of the oil core diameter. It is a very crude

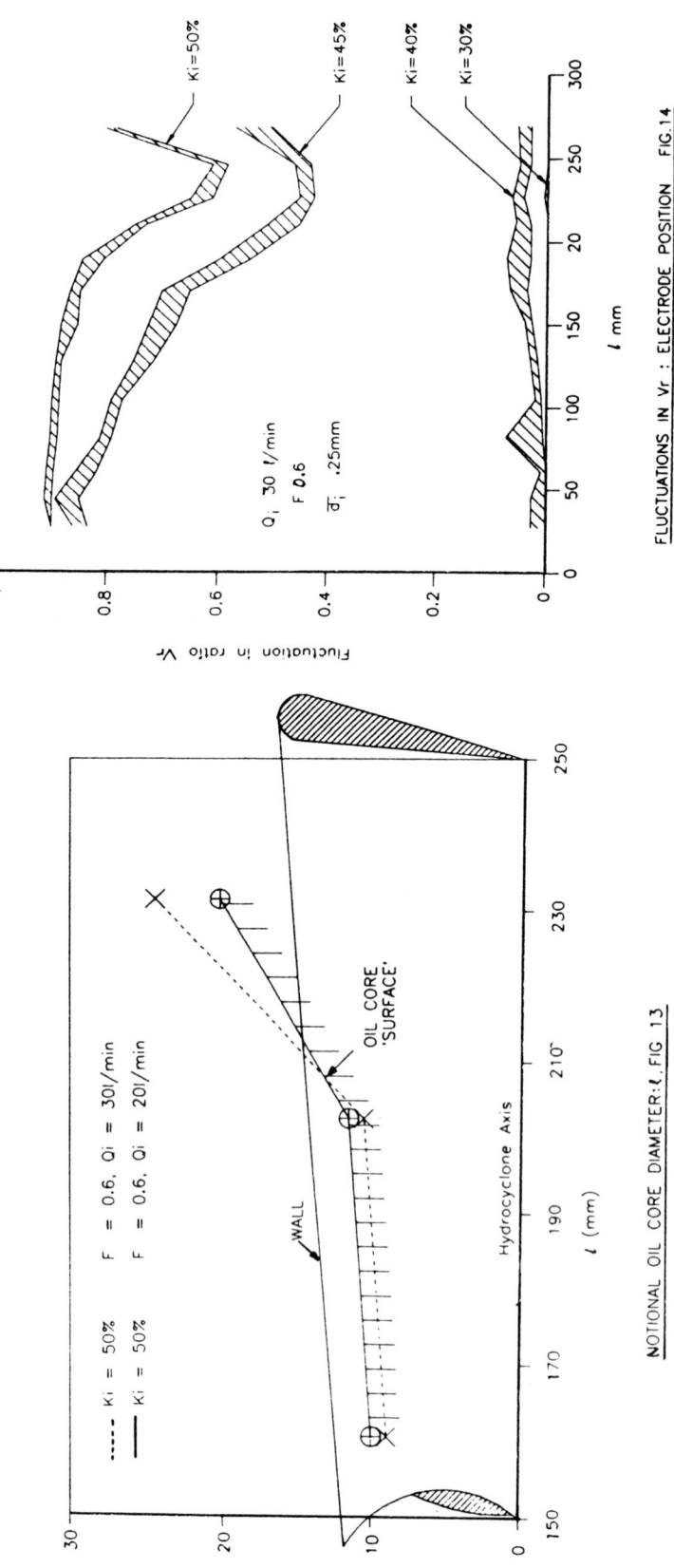

FLUCTUATIONS IN Vr : ELECTRODE POSITION FIG.14

NOTIONAL OIL CORE DIAMETER:ℓ FIG.13

model since it assumes that the change between the oil core and the water is indeed sharp, rather than gradual.

The results of the calculation are shown on figure 13 and it can be seen that at two of the axial electrode spacings, notional core diameter does come within the hydrocyclone wall, but for the third axial spacing of the electrode pairs the apparent core diameter is outside the hydrocyclone wall which indeed may induce some scepticism! The interpretation of this is clearly going to be somewhat imperfect because of the crude model we have.

Nevertheless the apparent increase in core size as the inlet region of the hydrocyclone is approached, is given some support by visual observation and apart from the fact that this is a very crude model, the main defect of the apparent core diameter is that it is too large. Clearly this technique needs refinement to allow for the fact that the core is not going to have a discontinuity in oil concentration and it probably needs some form of actual calibration. It would also be helpful to have a multiple electrode array to allow for non-diametral current paths. This sort of technique has been pioneered in work at UMIST [6].

Voltage fluctuation

The wave recorder output fed to the DVM was only capable of showing fluctuations in signal at up to about 10Hz and a very definite fluctuation could be seen for all the readings.

In figure 14 the maximum and minimum V_r values recorded at different conditions have been shown, figure 14 being solely an exemplary model for one particular set of conditions. It can be seen that for Ki 50% the fluctuations are rather narrower than at Ki 45% and at Ki = 40% then again one has fairly small fluctuations. It is believed that these fluctuations are real but if an approximate velocity is put in, then the typical eddy size that results is considerably larger than the local hydrocyclone diameter. It therefore seems that if the wave recorder output is indeed a true reflection of the frequency, what is visible is a gross disturbance produced by a fluctuation in the mixing conditions upstream of the hydrocyclone. At 50% inversion is assumed to be complete in much of the hydrocyclone and at 40%, little inversion has occurred.

It is believed that with better instrumentation capable of a higher frequency response these variations in signal are capable of giving some indication of the local structure of concentration variation and the wave recorder does provide an auxiliary output which should be capable of showing fluctuations up to about 100Hz.

FURTHER DEVELOPMENTS

The most pressing need to develop the technique further is to understand the calibration, that is to understand how the voltage ratio V_r varies with the local concentration of water. It is unlikely that there is going to be a linear variation; we may perhaps be able to use pipeflow which is well mixed to enable better calibration, and also intend to search for analytical models. The centrifugal field does impose an order on the flow and therefore in relation to the hydrocyclone wall it is believed there is a substantially time stationary variation in concentration. The largest spatial gradients in V_r occur with a feed brine concentration in the inversion criticality region (40-45%) at the start of the slow taper.

Apart from the need for understanding calibration, the other need to enable better use of the technique is to have a more sophisticated electrode array. This particularly applies to looking at the radial distribution of concentrations for the water. Using a number of electrodes round the periphery in any given plane along the hydrocyclone and with better signal processing, it is believed the technique pioneered for a different sort of measurement [6] may be applied to measure a gradual radial variation in concentration. An understanding of radial variation for several planes would enable attainment of the eventual goal of mapping out contours of time-averaged concentration for water.

Another development is to employ higher response rate instrumentation to give a picture of the local domain size for fluctuations in concentrations.

It is also clear that if the method can be understood rather better, that there is a need to move to systems that are more representative of field conditions; water/kerosine is, after all, a very easy system to separate.

CONCLUSIONS

i) The technique clearly justifies further work. This will need a refinement of instrumentation and electrode configuration and above all a better understanding of the physical significance of the voltage ratio.

ii) Even with the small scale piece of work carried out, it seems fairly clear that dynamic phase inversion is occurring for an inlet brine concentration of 40-45% and this indeed fits the substantial fall off in separation efficiency.

Variation in feed drop size and flow rate produced changes in V_r that were anticipated, but changes in split ratio produced a puzzling effect with less water near the wall for the larger value of F.

iii) The technique also needs better faster signal processing to enable an insight into the microstructure of concentration fluctuations.

ACKNOWLEDGEMENTS

The authors would like to acknowledge the early work done by Warburton on the design and construction of the rig and this was supported jointly by the SERC and by the BP Research Centre. We also like to acknowledge the contribution given by the Central Design Service, and in particular David Price, for preparation of the figures and for Sylvia Allison for producing order out of chaos.

REFERENCES

1. Thew, M.T. Hydrocyclones for liquid-liquid separation. I.Chem.E. Course on Two-Phase Separation with Cyclones, Amsterdam. Course manual supplement, pp. 67-102, April 1988.

2. Thew, M.T., Silk, S.A. and Colman, D.A. Determination and use of residence time distributions for two hydrocyclones. Proc. 1st Int. Conf. on Hydrocyclones, pp. 225-248, Cambridge UK, 1980. Pub BHRA, Cranfield, 1980.

3. Thew, M.T., Wright, C.R. and Colman, D.A. R.T.D. Characteristics of hydrocyclones for the separation of light dispersions. Proc. 2nd Int. Conf. on Hydrocyclones, pp.163-176, Bath Uk, 1984. Pub BHRA, Cranfield, 1984.

4. Warburton, S.R. Phase inversion in hydrocyclones. Unpublished report ME/89/18, Dept. Mech. Eng., University of Southampton, 1989.

5. Becker, M.F. and Thew, M.T. Preliminary experiments on conductivity measurements to determine the behaviour of an oil-water mixture in a hydrocyclone. Unpublished report ME/91/16, Dept. Mech. Eng., University of Southampton, Sept. 1991.

6. Barber, D.C and Brown, B.H. Applied potential tomography. J. Phys. E: Sci. Instrum. 1984, v. 17, pp. 723-733.

Appendix 1: calculation of nominal oil core diameter

At values of ℓ = 161, 201 and 233mm ie in the third of the slow taper nearest the inlet, further electrodes were inserted diametrically opposed to the linear array. It was believed that if a reasonably coherent oil core was present in this region, then the path length for current to flow in the brine-rich liquid between the electrodes would be longer. In turn this would raise the resistivity of the path as compared with the situation when the hydrocyclone was filled with (flowing) brine. Time did not permit a calibration, for example by inserting a cylindrical insulator on the hydrocyclone axis, so for simplicity it was assumed that resistivity α path length.

The above assumption does not agree with the resistance α (path length)$^{0.2}$ demonstrated by figure 8, but in this latter case the shortest conduction path was wholly in the boundary layer, as opposed to the diametral path here.

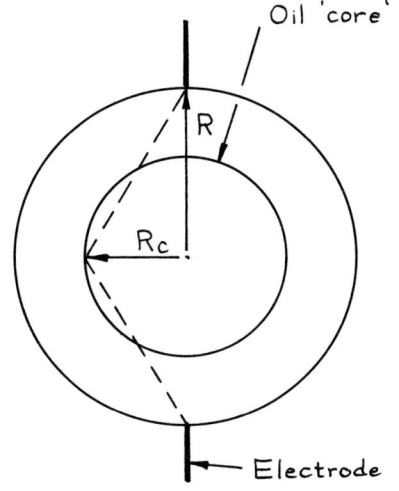

A very simple geometric path was assumed for this crude model: the current path is the dashed line. Whilst this cuts the core, of radius R_c, the discrepancy is not large.

Let V_r be voltage ratio for:

$$\frac{\text{voltage with oil core present}}{\text{voltage without core}}$$

Then $V_r \, \alpha \, \dfrac{1}{\text{path length ratio}}$

or $V_r = R/(R^2 + R^2_c)^{\frac{1}{2}}$, giving $R_c = R(\dfrac{1}{V_r^2} - 1)^{\frac{1}{2}}$

Measured values of V_r were used to produce the results shown in figure 13. It will be seen that the assumptions have led to an apparent value of $R_c > R$ at the electrode pair nearest to the inlet; the assumptions need refining.

FLUID-FLOW MODEL OF THE HYDROCYCLONE FOR CONCENTRATED SLURRY CLASSIFICATION

RAJ K. RAJAMANI AND LUDOVIC MILIN
Comminution Center
115 EMRO
University of Utah
Salt Lake City, Utah 84112
U.S.A.

ABSTRACT

The mathematical model of the hydrocyclone based on the Navier-Stokes momentum balance, though very difficult, has advanced over the last two decades. The availability of the laser-Doppler velocimeter has further helped validate the model. However, the extension of the model to the size classification of concentrated slurries is relatively new.

The solution of the momentum-balance equation set requires viscosity at the nodal points, which depends on the concentration of particles in the volume surrounding the node. The force balance on a particle yields its internal trajectory which determines the spatial concentration profile as well as the size-classification curve. However, the force balance requires fluid velocity at the node. Thus, the model solution is an iterative calculation. Experimental velocity profiles and size-classification curves determined up to 40% solids in a 75-mm hydrocyclone confirm the model predictions.

NOTATION

C_D	=	drag coefficient
$C'_{i,j,k}$	=	volumetric concentration of kth particle (from the inlet) of a specified size in cell (i,j)
$C'_{i,j}$	=	volumetric concentration of all particles of a specified size in cell (i,j)
$C_{i,j}$	=	normalized volumetric concentration of all particles of a specified size in cell (i,j)
D_p	=	particle diameter, cm
i	=	index number of cell (i,j) in the vertical direction
j	=	index number of cell (i,j) in the horizontal direction
R	=	radial distance from the axis of symmetry in cylindrical coordinates, cm
R_c	=	radius of the hydrocyclone, cm
Re	=	Reynolds number of the hydrocyclone defined as $R_c U_0/\nu$
r	=	dimensionless radial distance from the axis of symmetry in cylindrical

	coordinates
t	= dimensionless time
U_0	= mean inlet velocity, cm/s
U_s	= particle radial-slip velocity, cm/s
u	= dimensionless radial velocity of the fluid
V	= tangential velocity of the fluid, cm/s
v	= dimensionless tangential velocity of the fluid
v_{ij}	= volume of cell (i,j), cm^3
v_p	= volume of particle of size p, cm^3
W	= axial velocity of the fluid, cm/s
W_s	= particle axial-slip velocity, cm/s
w	= dimensionless axial velocity of the fluid
z	= dimensionless axial distance from the roof of the hydrocyclone in cylindrical coordinates
Z	= axial distance from the roof of the cyclone, cm
η	= dimensionless vorticity
λ	= Prandtl mixing length, cm
μ_m	= slurry viscosity (poises)
μ_0	= suspending medium viscosity (poises)
μ_t	= turbulent viscosity, g/cm.s
ρ_ℓ	= density of the suspending medium, g/cm^3
ρ_m	= density of the slurry, g/cm^3
ρ_p	= density of the particle, g/cm^3
υ	= kinematic viscosity, cm^2/s
$\tau(i,j,k)$	= residence time of kth particle of size p in cell (i,j), s
$\tau(ij)$	= residence time of all particles of size p in cell (i,j), s
ψ	= dimensionless stream function
ϕ_v	= volume fraction of solids in the slurry
$\phi_{i,j}$	= total volume fraction of solids in cell (i,j)
$\phi_{i,j,p}$	= volume fraction of particles of size p in cell (i,j)
$\phi_{p,feed}$	= volume fraction of particles of size p in the feed
Ω	= dimensionless angular spin velocity

INTRODUCTION

The extractive metallurgical industry, which is a significant part of the world economy, processes iron, copper, zinc and lead ores to extract the respective metals. This industry uses crushing and grinding circuits to produce a minus 200 mesh product, before the ore is processed in one of the concentration operations. Grinding circuits use a bank of 12 to 16 hydrocyclones of diameters 65 cm to 75 cm. The slurry from the grinding mill discharge is diluted to approximately 40% by wt. before it is pumped to the hydrocyclone bank. Consequently the hydrocyclone underflow stream is over 60% by wt. and the overflow stream is approximately 20%. Often the feed-percent solid is increased to produce a higher concentration of solids in the overflow stream as this practice would eliminate the need for a dewatering unit ahead of the ore concentration process. Thus, modeling of hydrocyclones processing a high concentration of solids is very useful both for design and for improving existing operations. This paper addresses the problem of fluid-flow modeling of such hydrocyclones.

Most of the currently available models that can handle a high concentration of solids in the feed are empirical, and a typical one in this category would be that of Plitt [1]. Fluid mechanics based models, that is the models that solve the Navier-Stokes equation set, have not fully addressed the concentrated slurry flows. This is entirely due to the fact that the non-Newtonian nature of mineral suspensions has not been studied in detail. Furthermore, fluid-flow models are currently considering the problems involved in turbulent closure and the mechanics of the numerical solution itself. Nevertheless, there have been two isolated attempts in the literature [2,3] regarding non-Newtonian fluid flow within the hydrocyclone. Both used Ostwald-de Waele power-law to compute the spin velocity. Here we present a simplified approach to model the size-classification characteristics of the hydrocyclone. While doing so, the concentration of particles at all locations within the hydrocyclone emerges. However, a rigorous modeling approach would be more appropriate.

A Brief Description of the Fluid-Flow Model
A detailed exposition of the model may be found in Hsieh [4] or Hsieh and Rajamani [5], so we present a very brief description here. In view of the fact that particle-slip velocities are likely to be of small magnitude in the hydrocyclone, we assume that particle-fluid momentum coupling is absent and thus break the modeling work into two parts. First, the liquid-phase momentum-balance equation set is solved for axial, radial and tangential velocities; then the particle motion, with respect to the fluid, is computed by balancing the forces acting on the particle itself. Next, the trajectories of particles of each size, from their entry into the unit until their exit via one of the outlets, are calculated, which in turn yields the particle-separation curve of the hydrocyclone. Under the axisymmetric assumption, the Navier-Stokes equation set in the vorticity-stream function form is:

$$\frac{\partial \eta}{\partial t} = \frac{1}{r^3}\frac{\partial \Omega^2}{\partial z} - \frac{\partial u\eta}{\partial r} - \frac{\partial w\eta}{\partial z} + \frac{1}{Re}\left(\frac{\partial^2 \eta}{\partial r^2} + \frac{1}{r}\frac{\partial \eta}{\partial r} - \frac{\eta}{r^2} + \frac{\partial^2 \eta}{\partial z^2}\right) \quad (1)$$

$$\frac{\partial^2 \psi}{\partial r^2} - \frac{1}{r}\frac{\partial \psi}{\partial r} + \frac{\partial^2 \psi}{\partial z^2} = -r\eta \quad (2)$$

$$\frac{\partial \Omega}{\partial t} = -\frac{\partial u\Omega}{\partial r} - \frac{u\Omega}{r} - \frac{\partial w\Omega}{\partial z} + \frac{1}{Re}\left(\frac{\partial^2 \Omega}{\partial r^2} - \frac{1}{r}\frac{\partial \Omega}{\partial r} + \frac{\partial^2 \Omega}{\partial z^2}\right) \quad (3)$$

and

$$\frac{1}{r}\frac{\partial \psi}{\partial r} = w, \quad -\frac{1}{r}\frac{\partial \psi}{\partial z} = u, \quad \frac{\Omega}{r} = v. \quad (4)$$

The normalization constant used here is based on the advective time scale R_c/U_0. The turbulence closure is affected by a modified Prandtl mixing-length model of the form:

$$\mu_t = \rho_m \lambda^2 \left(\left|\frac{\partial V}{\partial R} - \frac{V}{R}\right| + \left|\frac{\partial W}{\partial R}\right|\right) \quad (5)$$

Finally, for the mixing length λ, which is a characteristic length scale of turbulent motion, a correlation of the form

$$\lambda_\eta = 0.010 R_c (\mu_m/\mu_0)^{1/5} (R_z/R_c)^{1/4} (R/R_c)^{1/2} \qquad (6)$$

$$\lambda_\Omega = 0.015 R_c (\mu_m/\mu_0)^{1/5} (R_z/R_c)^{1/4} (R/R_c)^{1/2} \qquad (7)$$

was used. The numerical problem is the solution of the set of coupled parabolic (Eqs. 1 and 3) and elliptic (Eq. 2) differential equations. Equations (1) and (3) are solved using the Hopscotch method, while Eq. (2) is solved by the successive over-relaxation method [6].

The prediction of tangential and axial velocity of the 75-mm hydrocyclone under consideration in this study is shown in Figures 1 and 2. Experimentally, the velocities were measured with a laser-Doppler velocimetry system. In general, the free vortex is very well predicted for the tangential velocity but a slight discrepancy is seen in the forced vortex region. However, the magnitude and the shape of the tangential-velocity profiles are predicted accurately. For the axial component, the flow reversal and the magnitudes are predicted very precisely in the cylindrical section. In the conical part, some deviations are observed. Extensive discussions of the velocity prediction may be found in Hsieh and Rajamani [5] or in Monredon et al. [7] where velocity predictions for five different hydrocyclones up to a diameter of 150 mm are shown. The accuracy of velocity predictions everywhere assures that the computed particle trajectories would be accurate also.

Particle Trajectories
A solid particle of diameter D_p and density ρ_p experiences a centrifugal force directed outward from the axis of symmetry and a radial drag force that opposes the centrifugal force. Similarly in the axial direction, the particle experiences the gravitational force and the opposing drag force. A force balance in the radial direction yields the radial slip velocity U_s as,

$$U_s = \left| \frac{4}{3} \left(\frac{\rho_p - \rho_m}{\rho_l} \right) \frac{V^2}{R} \frac{D_p}{C_D} \right|^{1/2} \qquad (8)$$

Likewise, a force balance in the axial direction yields the axial slip velocity W_s as,

$$W_s = \left| \frac{4}{3} \left(\frac{\rho_p - \rho_m}{\rho_l} \right) g \frac{D_p}{C_D} \right|^{1/2} \qquad (9)$$

There are no significant forces on the particle in the azimuthal direction, hence, it is considered to move with the fluid in that direction. Equations (8) and (9) include a particle-particle interaction effect in as much as this effect is accounted for by the local viscosity. For moderate inlet solid concentrations, the concentration within the central region of the hydrocyclone is expected to be "dilute" except near the conical wall where the concentration is very high. Due to the high centrifugal field, coarse particles reach the outside wall quickly in the cylindrical region and so most of the central region may be classified as "dilute." Hence, Equations (8) and (9) would be expected to work satisfactorily under these

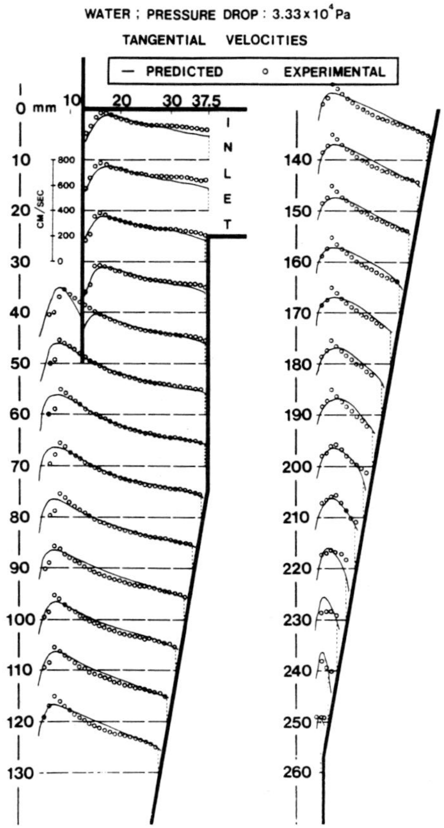

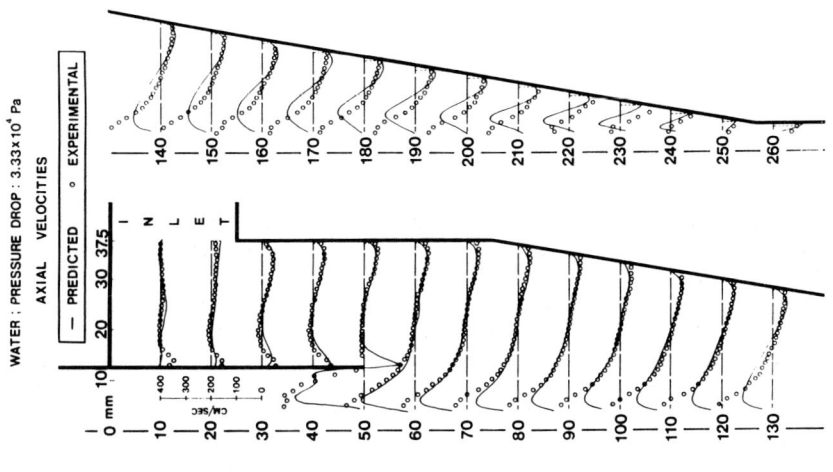

Figure 2. Measured and predicted axial velocities in a 75-mm hyrdocyclone.

Figure 1. Measured and predicted tangential velocities in a 75-mm hydrocyclone.

circumstances.

A solid particle of a given size can enter the hydrocyclone at any location across the inlet tube. It is assumed that the particle has a uniform probability of entering at any location across the inlet. Furthermore, it is assumed that particles follow the computed path across the entire hydrocyclone volume. With these assumptions and knowing the magnitudes of radial and axial velocities at every node point, it is possible to calculate the trajectories of any particle of a given size. Typically, 100 particle entry points, equally spaced on the inlet diameter, were simulated.

The particle trajectories for very dilute inlet feed conditions are shown in Figure 3. The fluid streamlines show that most of the streamlines report to the overflow, which is consistent with the fact that about 95% by wt. of the water reports to the overflow in experiments with the 75-mm hydrocyclone for a very dilute feed. The two boundary-layer flows are readily seen: first, a stream hugs the roof and moves downward over the outer wall of the vortex finder; secondly, another stream moves along the cylindrical wall and then over the conical wall. The first stream joins the overflow and the second stream joins the underflow; these are streams that leave the hydrocyclone without being classified. Figure 3 shows the trajectories of five different diameter particles of specific gravity 2.7. Although all the particles enter at the same location, some report to the underflow and others to the overflow, depending on their size. A strong inward flow in the conical section brings the fine particles toward the axis of symmetry. These fine particles reverse their direction after passing their respective equilibrium orbits and move toward the vortex finder along with the central upward flow. As for the coarse particles, due to their inertia, they are pressed to the wall by the centrifugal force despite the strong inward flow. Then they join the boundary-layer flow along the conical wall and exit through the spigot. If the size of the spigot is not large enough to handle all of the flow moving toward it, a part of the fluid stream would reverse direction at the vicinity of the spigot. In this case some of the classified coarse particles would again join the central upward flow. This crowding effect is included in our trajectory calculations via a simple scheme, as discussed in the next section.

Model Prediction for Dilute Feed

First, the size-classification predictions made with the particle trajectory approach without including the iterative calculations for concentration are presented. In other words, after computing the fluid-velocity profiles for water, flow the particle trajectories are computed. The fraction of the 100 particles of a size joining the spigot discharge stream represents the size classification brought about by the hydrocyclone. A number of predictions made for feed slurries in the range of 5-10 percent by wt. is found in Hsieh and Rajamani [5] and Monredon et al. [7]. Here a particularly high percentage of solids, 20 percent by wt., is shown in Figure 4. The predictions are far more than satisfactory except near the coarse end of 20 to 30 μm. This is due to the axisymmetric assumption, which simply does not hold good for the tangential-inlet hydrocyclone. The viscosity of the 20 percent by wt. limestone slurry is 1.3 centipoises and since this is close to that of water, it is expected that the model would predict satisfactorily. Even more, as soon as the slurry enters the hydrocyclone it redistributes itself, due to the centrifugal field, into a region of high concentration along the conical walls, and a region of low concentration in the central part. The concentration in the central region is approximately equal to the overflow concentration. For the 20 percent by wt. feed, the overflow solids are 10 percent by wt., which is very dilute. Hence, the excellent quality of model predictions. How this very simplified picture breaks down is demonstrated in Figure 5. Here, at a feed concentration of 40 percent by wt. the model is unable to predict the experimental trend even after setting the viscosity at 1.9 and 3.7

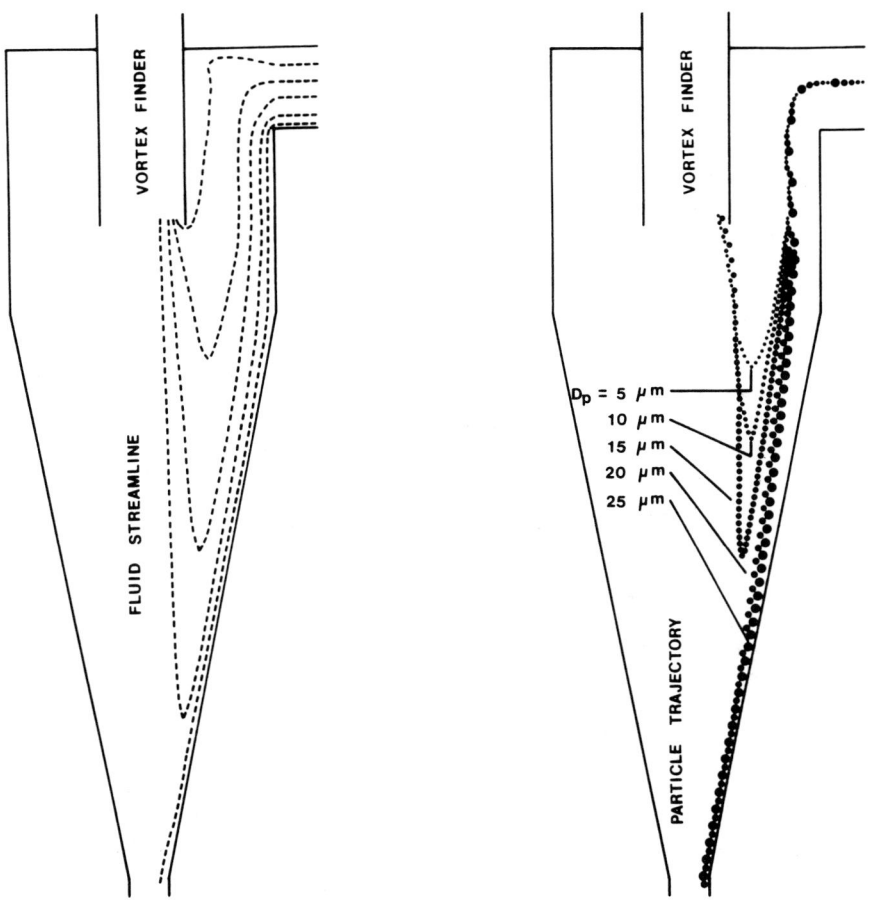

Figure 3. Predicted fluid streamlines and particle trajectories in a 75-mm hydrocyclone.

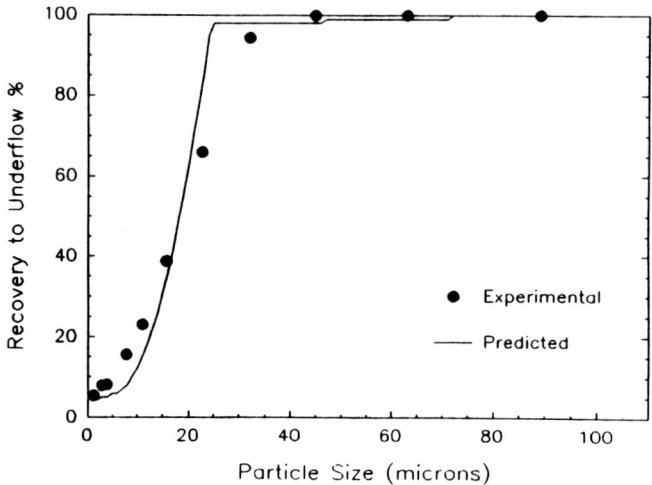

Figure 4. Experimental and predicted classification curves at 20% limestone by wt. in the feed.

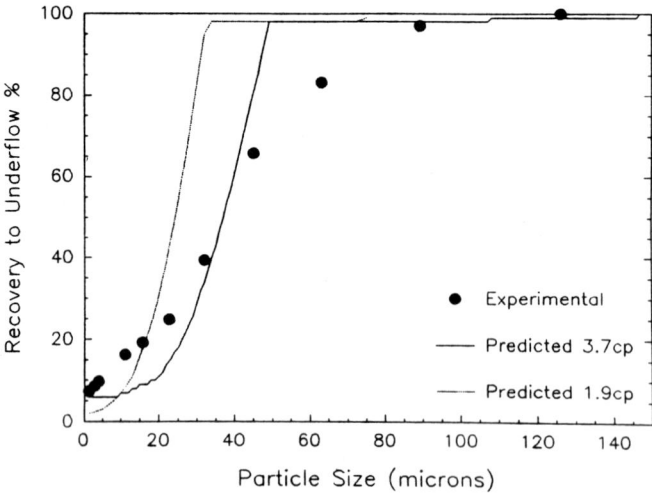

Figure 5. Experimental and predicted classification curves at 40% limestone by wt. in the feed.

centipoises in the fluid-flow calculations. The experimental data are pointing out that the gradual variation in concentration from the central region toward the walls must be accounted for in the calculations.

Extension to Concentrated Suspension

The net velocity of the slurry at any node point depends on the solids concentration and the density of the fluid at that location. The concentration, in turn, influences the viscosity of the slurry. A simplified approach to dealing with the variable solid concentration is presented in the following.

It is assumed that concentrated suspensions influence fluid velocities via fluid viscosity alone; the non-Newtonian nature of the suspension is ignored. The Thomas expression [8] for viscosity is:

$$\mu_m/\mu_0 = 1 + 2.5\phi_v + 10.05\phi_v^2 + 0.00273\exp(16.6\phi_v) \quad (10)$$

Next the concentration variable is uncoupled from the fluid dynamic-equation set: first the fluid dynamic set alone is solved with the values of water viscosity and density at each of the node points, followed by a separate calculation for concentration at the node point. Then, knowing the concentration, the viscosity and fluid density at each node are calculated. The calculations are iteratively repeated until the velocity values and concentration values converge to a single set, at which point the predicted fluid velocities are consistent with the concentrations at all node points.

As stated earlier, for dilute suspensions, the trajectory of 100 particles entering at equally spaced intervals across the inlet is computed. Then, the time spent by a particle, of size p, entering at the *k-th* location at the inlet, in a rectangular cell *(i,j)*, termed as the residence time, is noted. Depending on the size of the particle, some cells are traversed by numerous particles and some by none at all. Let $\tau(i,j,k)$ represent the retention time of the *k-th* particle in cell *(i,j)*. The primary assumption is that the concentration of a particle in a cell is proportional to its residence time in that cell and the particle volume. This is logical since, if the residence time is zero, the particle did not pass through that particular cell and so that particle's contribution to concentration in that cell would be zero; on the other hand, if the residence time is very long, the probability of finding particles of that type in the cell is finite, and hence, the concentration contribution would be proportional to the residence time. The volume of the cell and particle volume are given by:

$$\text{cell (i,j) volume:} \quad v_{ij} = \pi[Z(i) - Z(i-1)][R^2(j) - R^2(j-1)]$$

$$\text{particle volume:} \quad v_p = (1/6)\pi D_p^3 \quad (11)$$

Now the contribution of concentration to cell *(i,j)* due to the *k-th* particle is proportional to:

$$C'_{i,j,k} \propto \tau(i,j,k) v_p/v_{ij} \quad (12)$$

There are 100 locations at the inlet, hence, the overall contribution due to a particle of specified size in the feed is:

$$C'_{ij} \alpha \frac{v_p}{v_{ij}} \sum_{k=1}^{100} \tau(i,j,k) \qquad (13)$$

The particle of a given size entering all across the inlet cross section distributes its feed volumetric concentration as per the fraction:

$$C_{ij} = C'_{ij} / \sum_i \sum_j C'_{ij} \qquad (14)$$

Since the net concentration of all particle sizes within the hydrocyclone, excluding the air-core, is nearly equal to the feed concentration, the contribution to the cell due to a given particle of size p is:

$$\phi_{i,j,p} = \phi_{p,feed} \times C_{ij} \qquad (15)$$

where $\phi_{p,feed}$ is the volumetric concentration of particle size p in the feed. Therefore, the overall volumetric concentration in cell (i,j) is the sum over all particle sizes in the feed:

$$\phi_{ij} = \sum_p \phi_{i,j,p} \qquad (16)$$

Thus, the volumetric concentration at any location is determined by all particles sizes entering at all locations across the inlet.

In the iterative size-classification calculation, we solve fluid-dynamic equations with the value of water viscosity at all node points, and then compute the expected volumetric concentration at each node, as outlined above. Next, the viscosity at each node is calculated with the Thomas formula of Eq. [10] and the fluid-dynamic calculations are repeated. The tangential velocities calculated in the first three iterations are shown in Figures 6 and 7. As seen in these figures, both near the vortex-finder region and the mid-region of the cone, the velocities converge in a few iterations. This is to be expected because the volumetric concentration at each node must increase gradually, and hence, the velocities will vary accordingly. The size-classification curve predicted with the iterative procedure for an inlet concentration of 35 percent by wt. is shown in Figure 8. In this figure, the predicted curve for the first iteration corresponds to 1.0-centipoise viscosity at all node points. Calculations made with slightly higher viscosity would be nearly parallel to the first iteration, and so it would not follow the experimental trend at all. Whereas, the iterative model predictions converge after the third iteration and the conformity with the experimental data, also shown in Figure 8, are excellent. The corresponding concentration profiles are shown in Figure 9. Here, the gradual variation of concentration from the central line toward the wall, the high concentration along the conical wall leading to the spigot, the dilute region in the vicinity of the vortex finder and the two isolated islands of high concentration at the tip of the vortex finder are seen. Furthermore, a high concentration region emanating from the spigot and extending almost half way into the vortex finder is also seen. The reason being, in the particle-trajectory computations, a particle was made to circumvent a cell that was already filled with 50 percent by vol. of solids. This numerical device was implemented to simulate the crowding, and hence, blocking of the spigot discharge, which causes coarse particles to

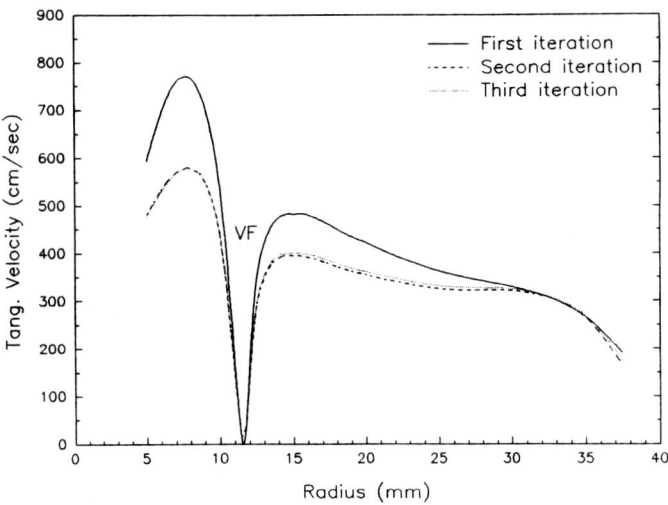

Figure 6. Predicted tangential velocities in the first three iterations near the vortex finder.

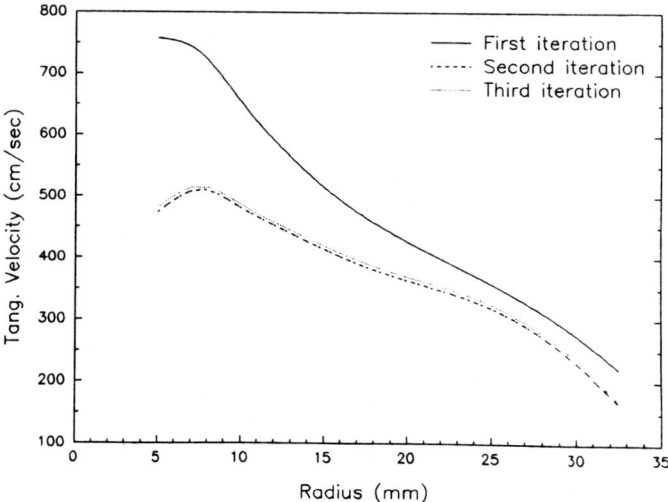

Figure 7. Predicted tangential velocities in the first three iterations in the mid-cone region.

be entrained in a secondary flow toward the vortex finder.

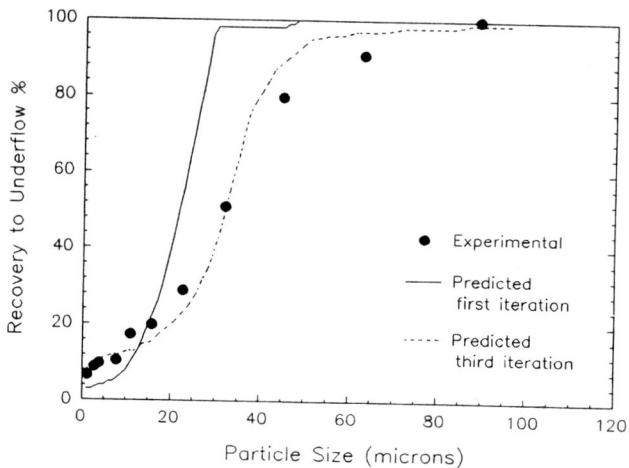

Figure 8. Experimental and predicted classification curves for 35% limestone by wt. in the feed.

CONCLUSION

An iterative calculation that uncouples the fluid-momentum balance from the concentration or particle mass balance is shown. It is seen that up to 20 wt. percent solids in the feed, the model prediction of size-classification is adequate even without the iterative calculations because the classification occurs in a dilute concentration regime in the central part of the hydrocyclone. However, the influence of particle concentration begins to exert itself at higher concentrations. The iterative calculation also produces the unknown concentration profiles within the hydrocyclone. Furthermore, the existence of a secondary flow emanating from the spigot and carrying the already classified coarse particles to the vortex finder is seen in these calculations.

ACKNOWLEDGMENTS

This research has been supported by the Department of the Interior's Mineral Institute program administered by the Bureau of Mines through the Generic Mineral Technology Center for Comminution under Grant number G1105149.

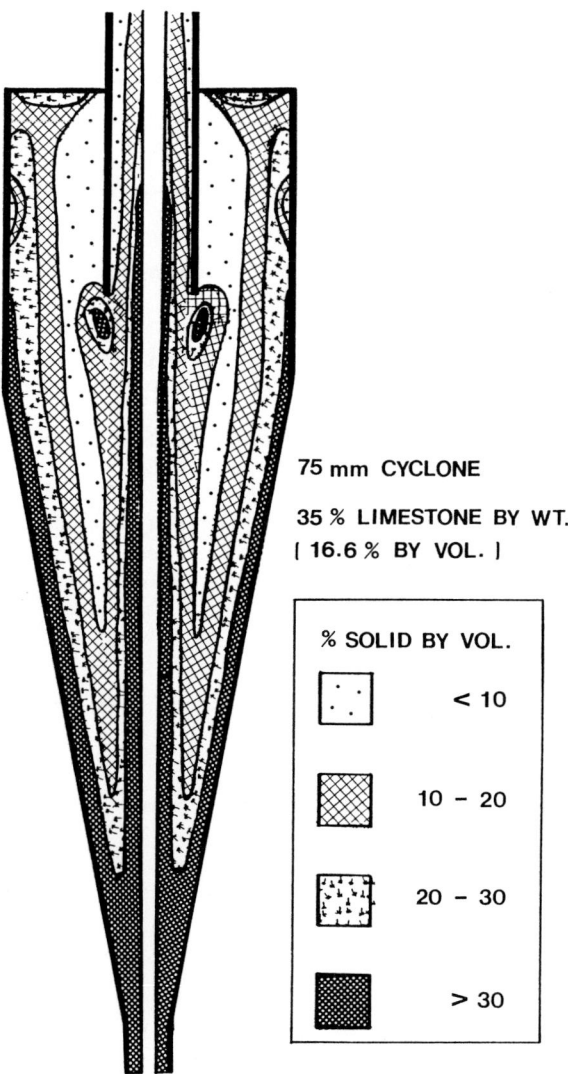

Figure 9. Predicted volumetric concentration map for 35% limestone by wt. in the feed.

REFERENCES

1. Plitt, L.R., Flintoff, B.C. and Suffco, T.J., Roping in Hydrocyclones, Third International Conference on Hydrocyclones, Bath, 1987, paper A3.

2. Upadrashta, K.R., Ketcham, V.J. and Miller, J.D., Tangential Velocity Profile for Pseudoplastic Power-Law Fluids in the Hydrocyclone–a Theoretical Derivation, Int. J. Miner. Process., 1987, **20,** pp. 309.

3. Ferguson, J.W.J., Theoretical Aspects of a Pulp Suspension Flowing in a Conventional Hydrocyclone, Tappi, J., 1988, **71,** pp. 125.

4. Hsieh, K.T., Phenomenological Model of the Hydrocyclone Based on the Physics of Fluid Flow, Ph.D. Thesis, University of Utah, Salt Lake City, Utah, USA, 1988.

5. Hsieh, K.T. and Rajamani, R.K., Mathematical Model of the Hydrocyclone Based on the Physics of Fluid Flow, AIChE J., 1991, **37,** pp. 735-46.

6. Roache, P.J., Computational Fluid Dynamics, Hermosa Publishers, Albuquerque, NM, 1972.

7. Monredon, T.C., Hsieh, K.T. and Rajamani, R.K., Fluid Flow Model of the Hydrocyclone: An Investigation of Device Dimensions, Int. J. Miner. Process., 1992, in print.

8. Thomas, D.G., Transport Characteristics of Suspensions: VIII, J. Colloid Sci., 1965, **20,** pp. 267-77.

PROSPECTS FOR THE USE OF HYDROCYCLONES FOR BIOLOGICAL SEPARATIONS

D. Rickwood[1], J. Onions[1], B. Bendixen[1] and Ian Smyth[2]

[1] Department of Biology, Essex University, Colchester

[2] Mechanical Engineering Department, Southampton University, Southampton.

INTRODUCTION

The development of the hydrocyclone at Southampton University for separating oil/water emulsions for the oil industry during the last 15 years would also appear to offer a possible new dimension for biological separations since both types of separation are similar in terms of their shear sensitivity and low differential density. The peak centrifugal forces generated in small hydrocyclones can be several thousand times the force of gravity which is similar to the centrifugal force used to sediment cells in centrifuges. Therefore, there would seem to be a potential for separating different types of cells in hydrocyclones. In addition, hydrocyclones are low cost, low inventory separators that are easy to isolate and clean; this last property is important for biotechnological applications such as production of genetically-engineered proteins in microorganisms. This paper reports on the preliminary assessment of the degree of compatibility of biological samples with separation in hydrocyclone systems and describes the initial results of dense separations in a small hydrocyclone test rig originally designed for low shear liquid-liquid separations.

SHEAR IN HYDROCYCLONE SYSTEMS

In addressing the problem of using hydrocyclones to separate shear-sensitive material including suspensions of biological material such as cells, the characteristics of both the separator and pressurising pump need to be considered.

Oil industry experience with low pressure oily water treatment applications of the hydrocyclone has shown the efficacy of using low shear progressive cavity pumps to provide an effective operating head without breaking up the oil droplets (Flannigan et al., 1989). Hence this type of pump was selected for use for this hydrocyclone test rig.

The rapidly spinning, turbulent flows in hydrocyclones themselves have substantial potential for shearing action. Possible sources of damage to suspended particulate material include:
a) viscous shear due to time-averaged velocity gradients;
b) transient shears and local pressure fluctuations caused by
   turbulence.

Three areas can be identified in the hydrocyclone where shear forces are likely to be high, these are:
a) in the boundary layer next to the hydrocyclone wall;
b) at the inner edge of the free vortex where steady state shears peak and away from the inlet at least, turbulence intensities are also at their highest;
c) on entry into the hydrocyclone due to the turbulence resulting from the interaction of the feed with the main liquid body.

Of these three, probably the most significant is the entry turbulence as all the suspension has to pass through the feed and inertial forces will be dominant at the scale of the suspension (tens of microns) (Smyth, 1989). Generating swirl with minimal energy loss has represented a key element in the development of liquid-liquid hydrocyclones and the geometry used for the experiments reported here, which came out of that programme, is characterised by a low kinetic energy feed through dual tangential inlets into a comparatively large diameter swirl chamber (see Figure 1). Hydrocyclone size also has a bearing on the turbulence levels with the average power dissipated per unit mass through the separator increasing as the diameter decreases (Johnson et al. 1976).

MATERIALS AND METHODS
Model particle systems

For this work a number of carefully selected model systems, including different types of cells were chosen. The properties of the different types of particles are summarised in Table 1. The densities of particles were determined using isotonic Nycodenz gradients (Ford and Rickwood, 1982). The viscosities of sample suspensions were not significantly different from that of the suspending liquid, essentially the same as water. All separations were carried out at room temperature (18-22°C).

Sephadex G-25 superfine, a synthetic polysaccharide, was swollen overnight in isotonic saline (0.9% w/v NaCl) to give spherical particles 10 - 130μm in diameter. The density of the particles was determined to be 1.19g/ml.

Yeast (*Saccharomyces cerevisiae*) was grown in 1% yeast extract, 1% bactopeptone at 30°C in a rotary incubator. Yeast divides by budding and the ovoid cells, approximately 3μm in diameter, form clumps 6-33μm in diameter. The density of yeast cells growing singly or in clumps was determined to be 1.13g/ml.

Blood was obtained from freshly slaughtered calves or lambs at the abattoir and mixed with sodium citrate to give a final concentration of 0.32% to prevent the blood from clotting. For most experiments the blood was diluted by the addition of four volumes of isotonic (0.9%) NaCl. While dilution of the blood did reduce its viscosity markedly it had little effect on the results of the experiments designed to examine the effect of shear on red blood cells. The density of the red blood cells determined to be 1.09g/ml.

Hydrocyclone rig

The test hydrocyclone used was a comparatively small (diam.=15mm) version of an early oil dewatering prototype developed at Southampton University using water in kerosine dispersions (Smyth et al. 1984) and its geometry is shown in Figure 1. The design of the test rig is shown in Figure 2. The pump used was a Mono, low shear J series pump with a stainless steel chamber running at 325rpm.

Using the carrier liquid supply the rig operating pressure was adjusted to 4 bar using the bypass valve which was a 1/2" ball valve (Figure 2). The split was adjusted to either 20% or 10% underflow using the a 1/2" gate valve on the underflow line from the hydrocyclone. Using these conditions a flowrate through the hydrocyclone of 5 litres/min was obtained and the input velocity was calculated as 3.7m/s. The flowrates were checked each time the hydrocyclone was used. The working liquids used throughout these experiments were homogeneous suspensions of samples in either distilled water or, for osmotically sensitive samples, isotonic saline (0.9% NaCl).

For experimental work the hydrocyclone was run and a constant flow rate was taken as a sign of equilibrium conditions being established. At this point the sample suspension was fed to the pump by means of a three way tap and pumped through the hydrocyclone.

Samples were taken from the bypass stream, the underflow and overflow of the hydrocyclone and compared to the original sample.

Analysis of particles
The concentrations of Sephadex particles in the bypass, underflow and overflows were determined volumetrically. The concentrations of yeast cells were determined by light scattering at 600nm. The concentration of red blood cells was determined by absorption at 410nm after complete lysis of the cells. The sizes of particles were analysed by light microscopy using a graduated graticule which gave measurements accurate to within $0.2 \mu m$. Microscopic analysis was used to determine the morphology of cells before and after passage through the hydrocyclone rig; yeast cells were viewed under bright field and red blood cells were visualised with Wright's stain.

The degree of lysis of red cells was determined by measuring the amount of absorbance at 410nm not pelleted by centrifugation at $2000g$, that is to say the amount of haemoglobin released into the suspending medium after passage through either the pump alone or the pump and the hydrocyclone.

RESULTS AND DISCUSSIONS
Studies of shear effects on cells
Initial experiments were aimed at determining the sensitivity of biological samples to the effects of shear generated by the pump and hydrocyclone. For these experiments samples (2-4 litres) were pumped through the rig under the standard operating conditions as described in the Materials and Methods and samples of the bypass, overflow and underflow were taken. Yeast cells have very robust cell walls and so might be expected to be resistant to shear but they also grow as clumps which are sensitive to shear forces. There was no evidence for disruption either of yeast cell clumps as a result of passing through the pump or hydrocyclone. Red blood cells which have no rigid cell wall might be expected to be sensitive to shear forces but when they were passed through the rig again there was no evidence for cell lysis after passage through the pump or hydrocyclone as judged by the release of haemoglobin into the isotonic medium in which the cells were suspended (Table 2), very similar results were found with four different batches of blood. The conclusion from these results is that the passage of samples through the pump and hydrocyclone, both of which have been designed to generate minimal amounts of shear, had negligible effects on the morphology or integrity of cells even when they were not protected by a rigid cell wall.

Fractionation of particles in hydrocyclones
As a model system, the fractionation of Sephadex particles in the hydrocyclone was studied. Using a 20% underflow it was found that the concentration of the Sephadex particles in the underflow from the bottom of the hydrocyclone was 28 times greater than that of the overflow. Analysis of the sizes of the particles in the underflow and overflow of the hydrocyclone revealed that, compared with the original sample (Figure 3a), the flow from the top of the hydrocyclone was significantly enriched in smaller diameter particles (Figure 3b) compared with the underflow (Figure 3c). Hence, not only were most of the particles in the underflow but also the particles present in the overflow of the

hydrocyclone were smaller.

When yeast cells were passed through the hydrocyclone using the standard conditions with a 10% underflow it was found that the concentration of yeast cells in the underflow was two-fold greater than that in the overflow. In addition, compared with the original culture (Figure 4a) there was a greater proportion of large clumps of cells in the underflow as compared with the overflow (Figures 4b and 4c).

In the case of red blood cells, either no enrichment was found or a slight enrichment of cells was found in the overflow even though the cells were denser than the isotonic saline in which they were suspended. Hence the combination of the small size and small density difference between the cells and the medium in which they were suspended was not enough to allow fractionation of cells in this hydrocyclone under the experimental conditions used.

CONCLUSIONS

It is clear from this work that using the current conditions of flow and pressure in the test hydrocyclone, as a general rule cells, even animal cells, are compatible with separations in hydrocyclones in terms of being able to resist the shear forces generated in the pump and hydrocyclone. Furthermore, the results using the model Sephadex particles and the yeast cells for dense separations clearly show that some degree of separation can be achieved using this non-idealised hydrocyclone geometry. The separation of red blood cells is marginal in the current test rig. However, the separation of red blood cells in this rig could be improved by optimising the geometry of the hydrocyclone in order to obtain a more effective separation environment. The stability shown by the biological materials tested so far also promises that further separation efficiency could be obtained by reducing the size of the separator; hydrocyclones as small as 4mm have been used for micron-sized particulate suspensions in the past (Haas *et al.*, 1957). In addition, there is a necessity to investigate the potential scope of light separations for the efficient separation of biological materials in hydrocyclones.

REFERENCES

Flanigan, D.A., Skilbeck, F., Stolhard, J.E. and Shimoda, E. (1989) SPE 19743. 64th Ann. Tech. Conf. and Ex. of Soc. Petroleum Engrs. (San Antonio, Texas).

Ford, T.C. and Rickwood, D. (1982) Analyt. Biochem. 124, 293-298.

Haas, P.A., Nurmi, E.O., Whatley, M.E. and Engel, J.R. (1957) Chem. Eng. Prog. 53, 203-207.

Johnson, R.A., Gibson, W.E. and Libby, D.R. (1976) Ind. Eng. Chem. Fundam. 15, 110-115.

Smyth, I.C. (1989) Ph.D. Thesis Southampton University.

Smyth, I.C., Thew, M.T. and Colman, D.A. (1984) in Proceedings of the Second Conference on Hydrocyclones, Bath. BHRA, Cranfield.

TABLES

Table 1. Properties of Model Particles.

Particle type	Size (μm)	Density (g/ml)	Shape	Comments
Sephadex	10-130	1.19	spherical	polysaccharide particles insensitive to shear
Yeast	3-33	1.13	ovoid	grows in clumps of 2-6 cells cells have thick cell walls
Blood cells	3-5	1.09	discoid	thin cell membrane

Table 2. Analysis of red blood cell lysis

	Optical densities of samples at 410nm		
		Treatment of cells	
	Original sample	pump	pump+hydrocyclone
Total haemoglobin	0.598	0.345	0.425
Soluble haemoglobin*	0.010	0.011	0.006
Percentage lysis	1.7	3.2	1.4

*haemoglobin not pelleted by centrifugation at 2000g

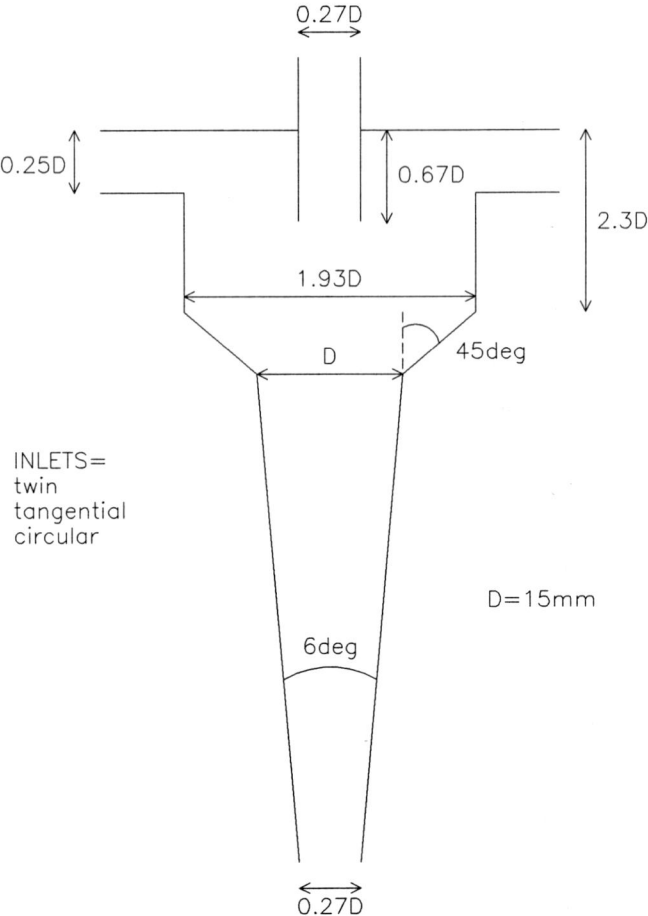

Figure 1. Geometry of the test hydrocyclone.

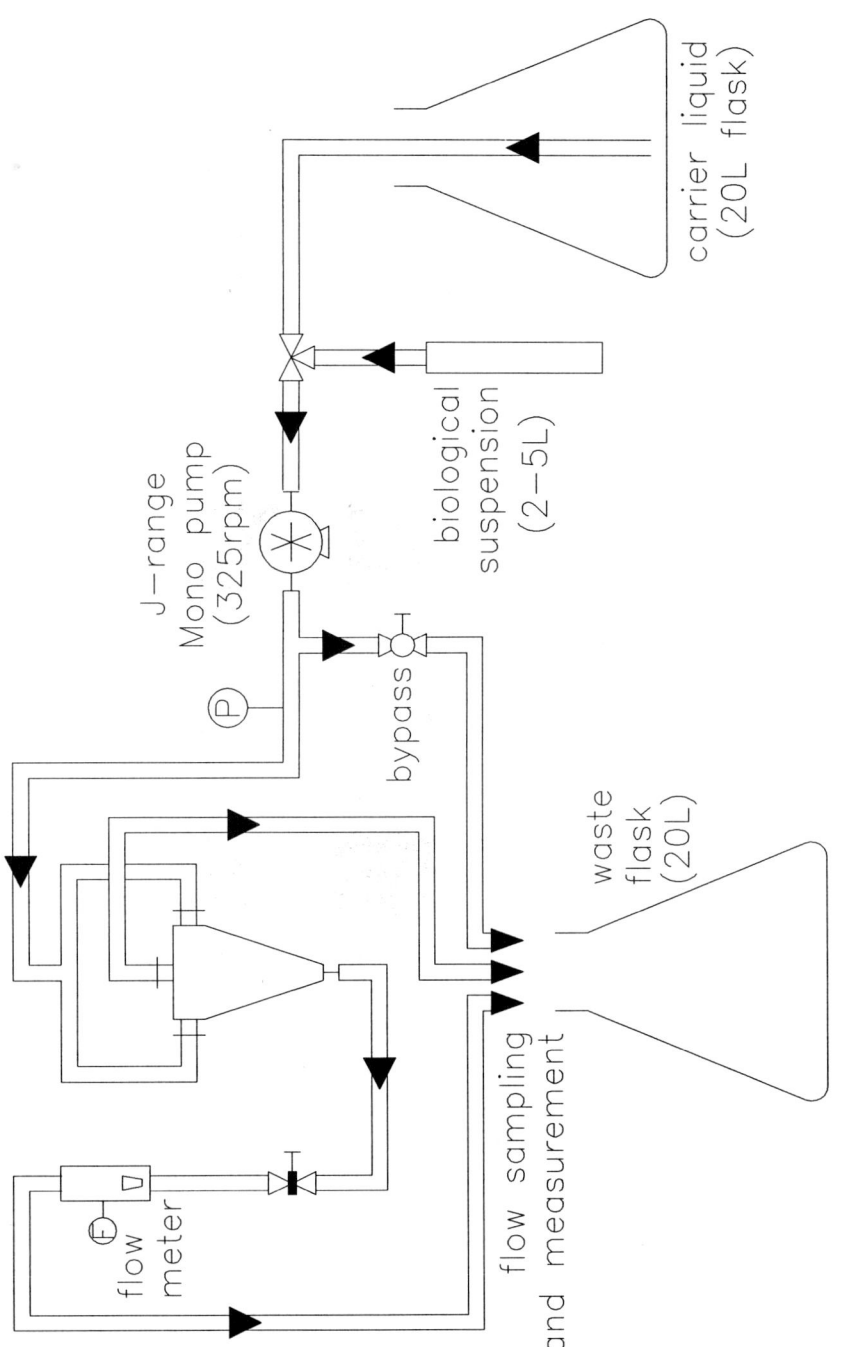

Figure 2. Hydrocyclone test rig.

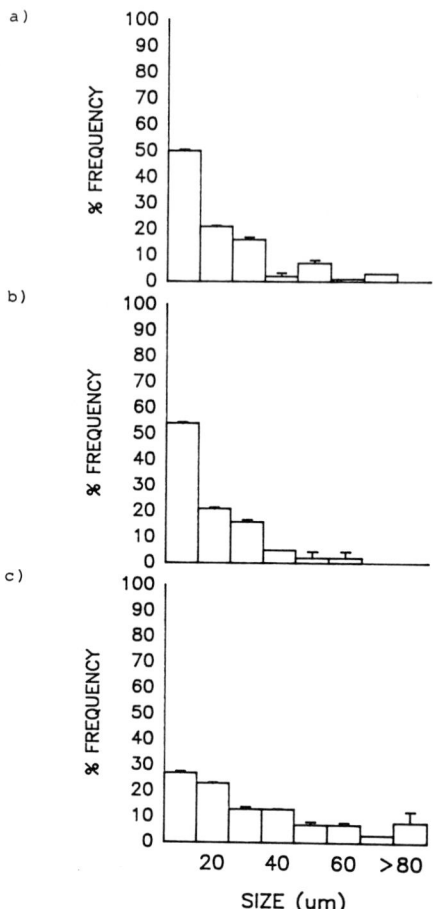

Figure 3. Separation of Sephadex polysaccharide particles. Histograms showing the size distribution in: (a) original sample; (b) particles in the overflow and (c) particles in the underflow.

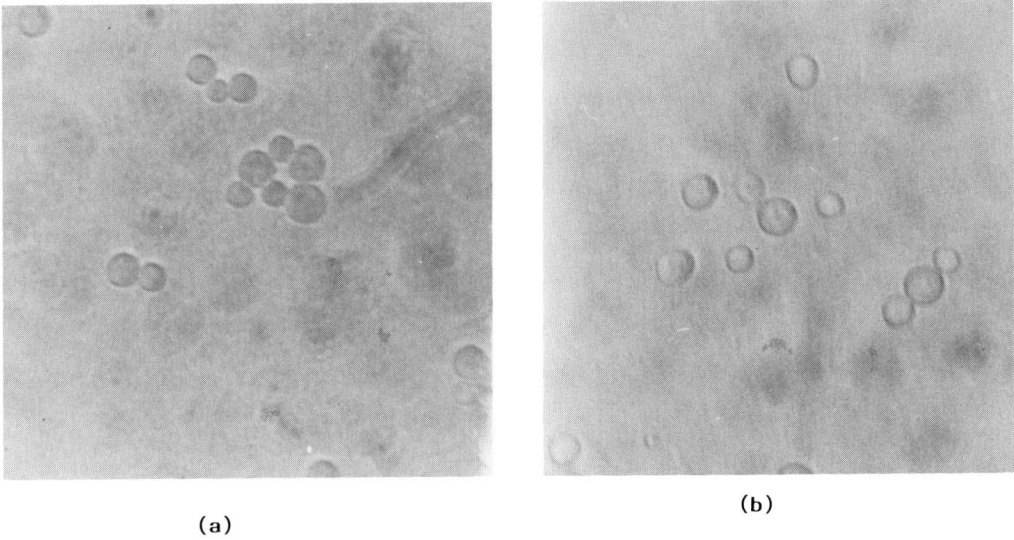

Figure 4. Comparison of yeast cells in the overflow and underflow fractions of the hydrocyclone. (a) original suspension of yeast cells; (b) yeast cells present in the overflow; (c) yeast cells present in the underflow.

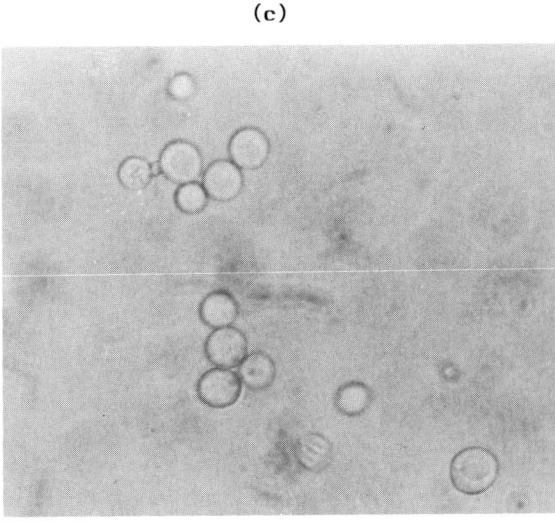

A PARAMETRIC EVALUATION OF THE HYDROCYCLONE SEPARATION OF DRILLING MUD FROM DRILLED ROCK CHIPPINGS

K.J. WALKER, T.J. VEASEY, I.P.T. MOORE

School of Chemical Engineering,
University of Birmingham, Edgbaston, Birmingham B15 2TT.

ABSTRACT

Hydrocyclones are used in the oil industry to separate both oil and water based drilling muds from drilled rock chippings. However, as yet no successful attempt has been made to optimise this process. This project has been undertaken with a view to gaining a more thorough understanding of the behaviour of drilling muds within a hydrocyclone. The aim of the work is to produce a simulated system which will predict the separation characteristics of a drilling mud in a hydrocyclone.

The paper describes results obtained using Laser Doppler Velocimetry of velocity profiles within a hydrocyclone with fluids of varying rheological characteristics. Both mean axial and tangential velocities were measured within the main body of the hydrocyclone as were Reynold Stresses.

INTRODUCTION

Drilling muds are a complex mixture of liquids, solids and chemicals used to maintain the quality of drilled holes primarily by carrying drilled rock cuttings to the surface. In order to avoid an undesirable build-up of chippings, these must be removed at the surface.

The current work is concerned with the final stage of solids separation using the hydrocyclone, which is a continuously operating classifying device which utilises centrifugal force to accelerate the settling rates of particles.

One of the most important considerations in the experimentation was to simulate the most important properties of a drilling mud since for both flow visualisation and L.D.V. work a transparent fluid was required. Therefore a standard drilling mud bentonite suspension would not be suitable. Although many drilling muds in use are Newtonian in nature the majority of operational muds have non-Newtonian rheological characteristics. A short introduction to the differences in properties of these fluids is given to aid the understanding of the results obtained.

Newtonian Fluids
In a Newtonian fluid, the shear stress is directly proportional to the shear rate and the rheological equation is given as:

$$T = \mu \dot{y} \qquad (1)$$

where μ is the dynamic viscosity and $\dot{y}$ is the shear rate.

The plot of shear stress against shear rate is a straight line passing through the origin; the fluid begins to move as soon as a non-zero force is applied, examples of such fluids are water or petrol.

For a Newtonian fluid, the ratio for the apparent viscosity, μ_a:

$$\mu_a = \frac{T}{\dot{y}} \qquad (2)$$

is a constant at constant temperature and pressure [1].

Non-Newtonian Fluids
A non-Newtonian fluid is defined as a substance which does not conform to Newtonian equations, such as most drilling muds. Non-Newtonian fluids can exhibit a variety of different rheological behaviour, with some even behaving in a solid-like manner. As a result there are a number of equations to describe such fluids.

Time -independent fluids are characterised purely by the rate of shear at any point being some function of the shearing stress at that particular point. These fluids can be sub-divided into three groups, according to their differing relations between shear viscosity and shear rate.

Carboxy-methylcellulose (C.M.C.) is classified as a time independent fluid of the pseudoplastic or shear-thinning type, where the apparent viscosity of the fluid decreases with increasing shear rate [2]. The experimental measure of shear stress and shear rate for such fluids shows a linear relationship on a log-log plot over most of the flow range. An empirical relationship has been developed, known as the "Power Law" equation, to describe such fluids, which is mathematically expressed as:

$$T = K\dot{y}^n \qquad (3)$$

In terms of shear viscosity, this equation becomes:

$$u = K\dot{y}^{n-1} \qquad (4)$$

Where K and n are fixed parameters for a particular fluid. K is a measure of the fluid consistency and n is a simple measure of the degree of departure from Newtonian behaviour. The reasonable accuracy and simple form of this equation has led to a wide application for engineering purposes.

Although this model is commonly used, it is limited in accuracy, it cannot describe the rheological behaviour for both large and small shear rates. Another drawback is the confusion caused in the dimension involved in both K and n, which changes with varying K or n.

However, for the purposes of defining the characteristics of a simulated drilling mud, this equation is appropriate, as the Power Law model is often used in drilling mud applications [3]. In these experiments the importance is placed on the viscosity and shear thinning properties of the fluid, rather than a complete match to a bentonite suspension. If an accurate simulation were to be attempted, the rheology would be complicated by the addition of another variable, yield stress. The yield point only applies to homogeneous fluids and suspensions and is present when the internal structure is broken down at a certain shear rate. Thus the fluid behaves in the same way as a solid until its yield point has been reached.

The highly shear thinning fluid C.M.C., although exhibiting the same characteristics as a fluid with a yield point, is a polymer. The long polymer chains are stretched rather than being broken down so there is no true yield stress for C.M.C. solutions. However, for the purposes of this work the simulation achieved is more than adequate, as shown by Figure 1.

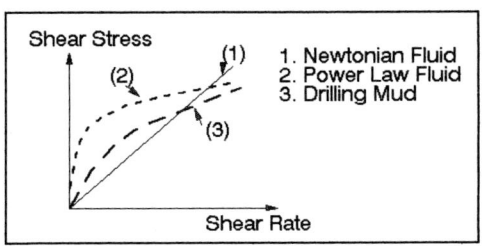

Figure 1. A plot of shear stress against shear rate for varying rheological characteristics.

EXPERIMENTAL TECHNIQUES

Laser Doppler Velocimetry (L.D.V.) is a velocity measuring technique which utilises a frequency shift caused by motion commonly known as the Doppler effect [4], [5].

L.D.V. Optics

Figure 2 shows a basic L.D.V. system, where the laser beam is split into two parallel beams and focused by a lens. The point of intersection of the two beams is known as the measuring volume, this region is critical to the success of any measurements made. Therefore, the optics which produce this region and what is inside the region itself are extremely important.

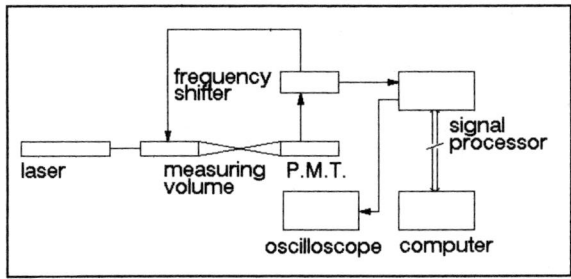

Figure 2. A basic L.D.V. system.

L.D.V. Equipment

The equipment necessary for the velocity measurement consisted of a glass hydrocyclone encased in a glass box, a Mono pump, a sump, a L.D.V. and a three dimensional traverse assembly. A detailed drawing of the hydrocyclone assembly is shown by Figure 3. The hydrocyclone comprised a blown glass conical section, of dimensions similar to a standard Mozley 2" hydrocyclone, attached to a standard Mozley polyurethane vortex finder cap. A machined nylon gasket was used between the glass body and the polyurethane cap and the whole assembly was bolted together between two metal plates top and bottom. A surrounding box of 4mm thick glass plate was attached to the lower support plate and filled with water to reduce the optical refraction of the laser beams. The glass cyclone body was joined to the surrounding box through a ground glass joint and a section of flexible tubing. The flexible tubing served to reduce stresses in the glass cyclone body and the ground glass joint. The cyclone assembly was bolted through the lower support plate to a stand capable of moving vertically with horizontal movement provided by the mountings of the laser itself. The feed, underflow and overflow pipework used was all high pressure flexible tubing in an attempt to reduce any vibration which may be transmitted from the pump.

The laser which was used was a 100mW dual beam argon laser with lenses fitted of focal distance 122mm, operated in backscatter mode. Operation in the back scatter mode was necessary since the scattered light of the laser beams will not transmit through the liquid/air interface present in the cyclone [6].

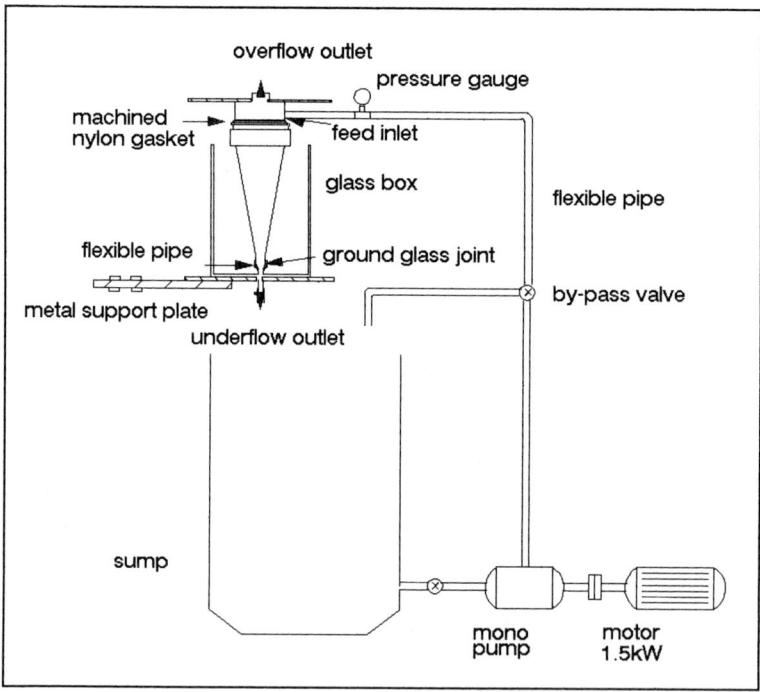

Figure 3. Schematic diagram of the L.D.V. equipment.

EXPERIMENTATION

Preliminary experiments were conducted to study the hydrocyclone separation of solids from a simulated drilling mud. The polymer C.M.C. was used to provide the non-Newtonian flow characteristics and thus simulate the bentonite clay suspension of a real mud. Tests were conducted to determine the efficiency of solids separation by, firstly, varying the fluid viscosity, secondly, changing the feed pressure, thirdly, changing the cyclone geometry and finally, using solids in the fluid to simulate the drilled cuttings within a drilling mud, the solids used being silica sand.

The results of this work suggested that the non-Newtonian characteristics of the fluids being used were affecting the flow patterns within the hydrocyclone. Therefore, flow visualisation work was undertaken to assess both the extent and nature of any changes to the flow patterns. A glass 2" hydrocyclone was made for this purpose, similar to the Mozley hydrocyclone used in the earlier tests and the operating conditions employed were determined by the results of the preliminary sand separation studies.

The polymer C.M.C. was again used in a range of concentrations: 0.3%; 0.5%; 0.7% and 1.0% C.M.C.. Table 1 shows the values of K, n and the apparent viscosities of these concentrations of C.M.C. measured at a shear rate of $1020s^{-1}$, using a variable speed Fann viscometer. Visualisation of the flow patterns was achieved by using different coloured plastic beads of densities 1.14 and 0.86 kgm^{-3} in the solutions.

Table 1
Rheological characteristics of varying concentrations of C.M.C.

% C.M.C.	Apparent viscosity (cp)	K	n
0.3%	23	1.6	0.49
0.5%	40	4.1	0.43
0.7%	56	6.1	0.42
1.0%	106	8.6	0.35

The extent of the shear degradation of the polymer was also measured by viscosity testing both before and after the hydrocyclone operation, using a variable speed Fann viscometer.

The further flow visualisation work conducted included quantifying the velocity profiles of the Non-Newtonian liquids within the simulated muds using the technique of Laser Doppler Velocimetry.

Both mean axial and tangential velocities were measured using the system described, at 15mm vertical intervals and 2mm horizontal intervals, as shown by Figure 4. The other measurements taken at these locations were of velocities at a 45° angle to both the axial and tangential velocities. These measurements were made to allow the calculation of the Reynold stresses throughout the hydrocyclone. In each case the feed pressure to the hydrocyclone was 10 psi, as a result of L.D.V. constraints.

DISCUSSION OF RESULTS

The flow visualisation work has produced interesting results for the behaviour of the non-Newtonian fluids with distinct differences in the flow patterns being noted. An important observation was the absence of an air core in the hydrocyclone when it was operated at higher fluid viscosities, 0.7% and 1.0% C.M.C..

The results of the preliminary separation work established that a higher feed pressure caused a sharper separation in addition to lowering the cut point size. It was also found that a small apex (spigot) diameter caused a poor separation due to an increased "hindered settling" effect. An increase in viscosity decreased the efficiency of both size classification and the bulk separation of sand. Tables 2 and 3 give the measured d_{50} sizes at different hydocyclone geometries, circulating fluid viscosities and feed pressures.

The L.D.V. data presented here was obtained using a 2" Mozley hydrocyclone operating at a feed inlet pressure of 10 psi, with an 8mm vortex finder and a 12mm spigot size. These conditions were employed to allow the most effective use of the L.D.V. system available.

Table 2
Measured d_{50} sizes (in microns) with different hydrocyclone geometries with 0.5%
C.M.C. as the circulating fluid

Feed Pressure (psi)	Spigot Size	Vortex Finder Size	
		8 mm	14 mm
10	6.4 mm	48	104
10	9.4 mm	29	37
20	6.4 mm	31	33
20	9.4 mm	18	32
30	6.4 mm	26	28
30	9.4 mm	16	28

Table 3
Measured d_{50} size (in microns) using different concentrations of C.M.C. with an 8 mm vortex finder and a 9.4 mm spigot size.

% C.M.C.	Feed Inlet Pressure		
	10 psi	20 psi	30 psi
0.3%	18	15	12
0.5%	29	18	16
0.7%	47	38	31

Although it is appreciated that 4" hydrocyclones of a different geometry are used in the oil industry, the flow patterns are fundamentally different when operating with a shear thinning non-Newtonian fluid. Therefore, an investigation into the extent of these differences is required before any attempt can be made to optimise the efficiency of separation. A 2" Mozley hydrocyclone was chosen for several reasons: the operating curves suggest that a finer d_{50} size is possible; the flow patterns are accentuated due to the smaller diameter of the hydrocyclone and the physical constraints of using the L.D.V. system on a much larger hydrocyclone would have been impractical.

The first flow measurement in a hydrocyclone was conducted by Kelsall [7] and was supplemented by the flow visualisation study by Bradley and Pulling [8]. The results of this work for liquid velocities in a hydrocyclone operating with water as the circulating fluid are now generally accepted.

Axial Velocity Results
The results of the 0.3% axial velocities, shown by Figure 5, show downward velocities in the region close to the hydrocyclone wall, with a fluid flow reversal giving upward velocities further towards the centre of the cyclone. The Locus of Zero Vertical Velocity (LZVV) in this case is similar to that which would be expected from a Newtonian fluid [7] with a cylindrical shaped LZVV observed in the upper regions of the hydrocyclone and a more conical shape shown in the lower regions. Any further fluid flow reversals in the central air core region are most likely to have been caused by turbulence effects.

In the case of the 0.5% C.M.C. results, shown by Figure 6, downward velocities are also noticed along the cyclone walls. However, there is a significant difference in the fluid flow reversal with only slight upward velocities being evident in the upper regions of the hydrocyclone.

The results for the 0.7% C.M.C., shown by Figure 7, show this trend continuing with an increase in viscosity, the only upward velocities present being in the very uppermost areas of the cyclone body. In this case the LZVV is conical in shape in the upper regions of the cyclone with only downward velocities being present towards the underflow outlet.

Tangential Velocity Results
The increase in viscosity had a more dramatic effect on the tangential velocity results, as shown by Figure 8. The 0.3% C.M.C. results showed a maximum velocity towards the centre of the hydrocyclone in the upper regions with the profiles flattening further down and decreasing towards the central region as the underflow outlet was reached.

In the 0.5% C.M.C. the velocities, shown by Figure 9, in the upper regions of the hydrocyclone were of similar proportions to those obtained with 0.3%C.M.C., however, rather than increasing towards the centre of the hydrocyclone the velocities decreased towards the air core. In the lower regions of the cyclone body the magnitude of the velocities decreased while the profiles remained similar to those of the less viscous 0.3% C.M.C.

This trend continued in the most viscous fluid, 0.7% C.M.C., with the velocities decreasing almost to zero towards the centre of the cyclone in the upper regions, shown by Figure 10. The magnitude of the velocities decreased towards the underflow outlet.

Reynold Stress Results
The turbulence effects are shown in the form of Reynold Stresses calculated according to the method given by Tropea [9]. The results, Figures 11 and 12, show there is considerably more turbulence in the 0.3% C.M.C. than in the 0.5% although in both cases the only turbulence in the main flow occurs close to the cyclone walls in the upper regions of the cyclone body. The predominant difference is found in the area along the air core. Inside the air core itself is a region of turbulent fluctuations which in shear thinning fluids are transmitted to the surrounding area. The air core in the 0.5% C.M.C. does not transmit fluctuations to the surrounding fluid which is of a higher viscosity.

There is evidence of laminar eddy currents present at levels 3 and 4 in both the 0.3% and 0.5% C.M.C. There is no significant turbulence shown on the Reynold Stress plots but axial mean flow reversals are observed in this region. these eddys were also observed in the flow visualisation.

The limitations to this work are that the velocities were only measured from the cyclone wall to the air core and assumed to be symmetrical, the inlet pressure although within the recommended operational limits was low and so may have resulted in any flow patterns obtained being accentuated. the other important variable which will be investigated is whether the flow patterns shown were as a result of the viscosity of the fluids or their shear thinning nature.

CONCLUSIONS

1. The axial results show that the effective length of a hydrocyclone is reduced with increasingly shear thinning fluids.

2. The tangential velocity results show that the swirling motion of a fluid is reduced by increasing the shear thinning nature of the circulating fluids.

3. An increase in viscosity of a fluid dampens any turbulent flow around the air core region of a hydrocyclone.

FURTHER WORK

Although the results obtained show dramatic differences to velocities measured under similar conditions with the circulating fluid being water the reasons for these differences require further investigation. Attempts to use a viscous Newtonian fluid as a comparison have to date been unsuccessful. The problems encountered using such fluids need to be overcome before any relevant attempt to model the fluid flow of drilling muds can be achieved. It is also intended to extend the L.D.V. work by operating a larger hydrocyclone and increasing the feed pressures.

The results of both the Newtonian and non-Newtonian velocity measurements will be related to the separation characteristics of solutions containing a particulate phase. With the aid of a computational fluid dynamic model an overall picture of the flows within a hydrocyclone can be produced.

A more extensive study is planned in terms of investigating the separation efficiencies of a range of sizes of Mozley hydrocyclones and also a conventional 4" Oiltools hydrocyclone, as used in the oil industry. The testwork will be an extension of the preliminary work with sand and magnetite being used to determine the effect of the presence of different density solids.

ACKNOWLEDGEMENTS

The author would like to thank both the S.E.R.C. and the B.P. Research Centre, Sunbury for their financial support of this project; Richard Mozley Ltd for their technical support and the technical staff of the School of Chemical Engineering for their help and advice.

REFERENCES

1. Fredrickson A.G., Principles and applications of rheology. Prentice Hall, 1964.

2. Bertrand J. and Crouderc J.P., Numerical and experimental study of flow induced by an anchor in viscous non-Newtonian and pseudoplastic fluids. International Chemical Engineer, Volume 28 (2), April, 257-270.

3. Editions Technip, Drilling mud and cement slurry rheology manual. Editions Technip, Paris, 1982.

4. Polyakov A.F., Thermo and Laser Anemometry, Hemisphere, 1989.

5. Watrasiewicz B.M. and Rudd M.J., Laser Doppler measurements. Butterworths, 1976.

6. Hsieh K.T. and Rajamani R.K., Mathematical model of the hydrocyclone based on physics of fluid flow. American Institute of Chemical Engineering Journal, Volume 37, May 1991, No.5, 735-746.

7. Kelsall D.F., A study of the motion of solid particles in a hydraulic cyclone. Transactions of the Institute of Chemical Engineers, 1952, Volume 30, 87-104.

8. Bradley D. and Pulling D.J., Flow patterns in the hydraulic cyclone and their interpretation in terms of performance. Transactions of the Institute of Chemical Engineers, Volume 37, 34-45.

9. Tropea C., A note concerning the use of one-component LDA to measure shear stress terms. Exp. in Fluids, Volume 1, 209.

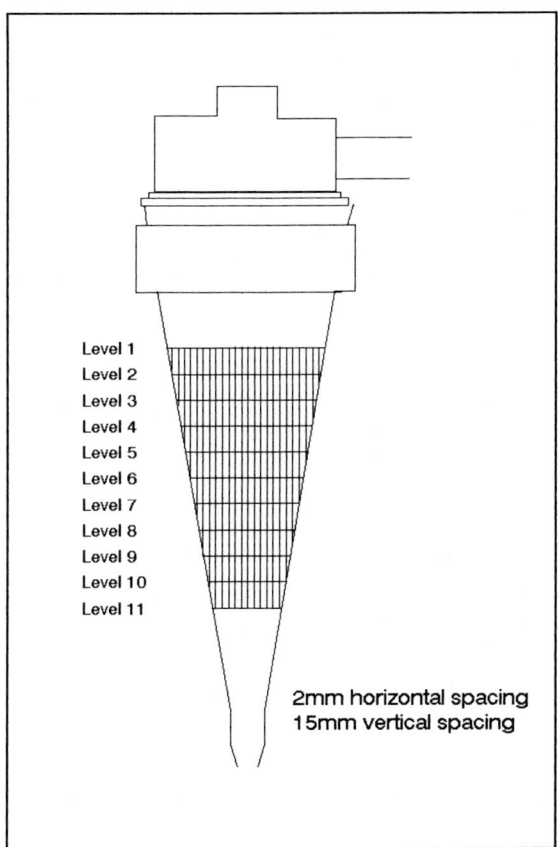

Figure 4. Positions of velocity measurements.

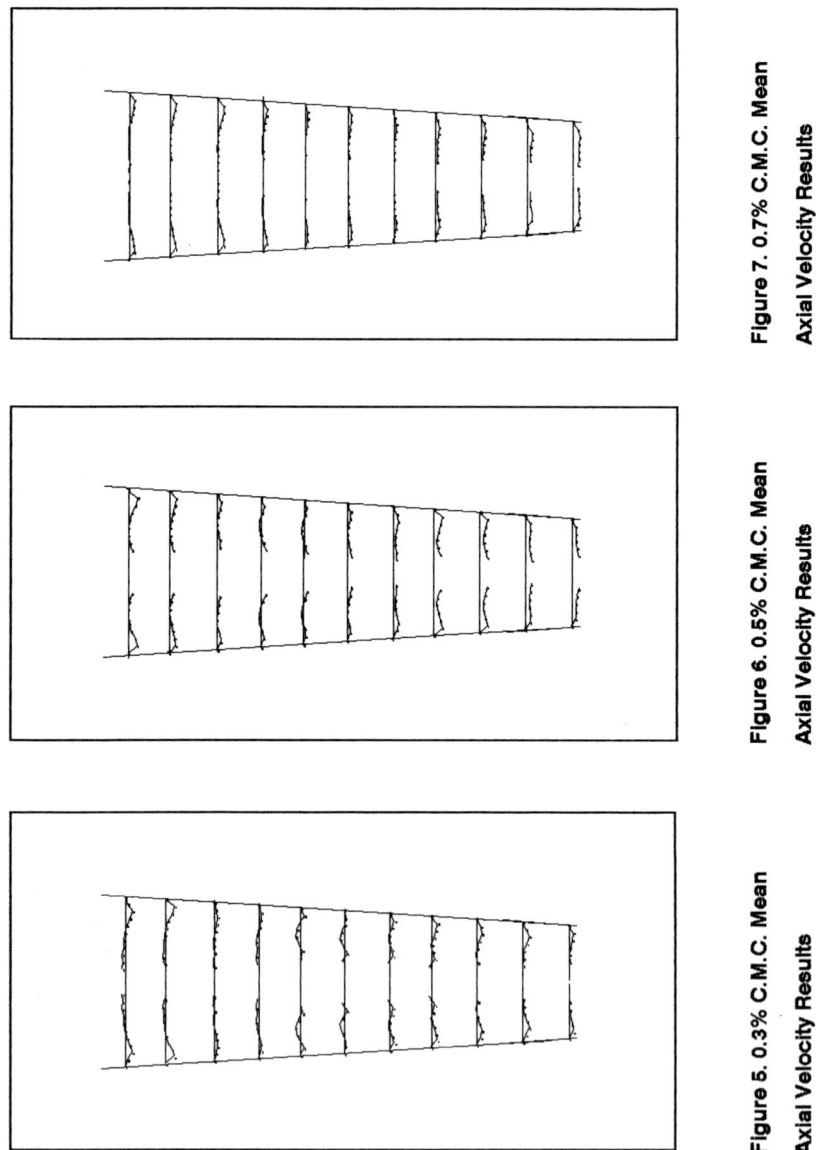

Figure 7. 0.7% C.M.C. Mean Axial Velocity Results

Figure 6. 0.5% C.M.C. Mean Axial Velocity Results

Figure 5. 0.3% C.M.C. Mean Axial Velocity Results

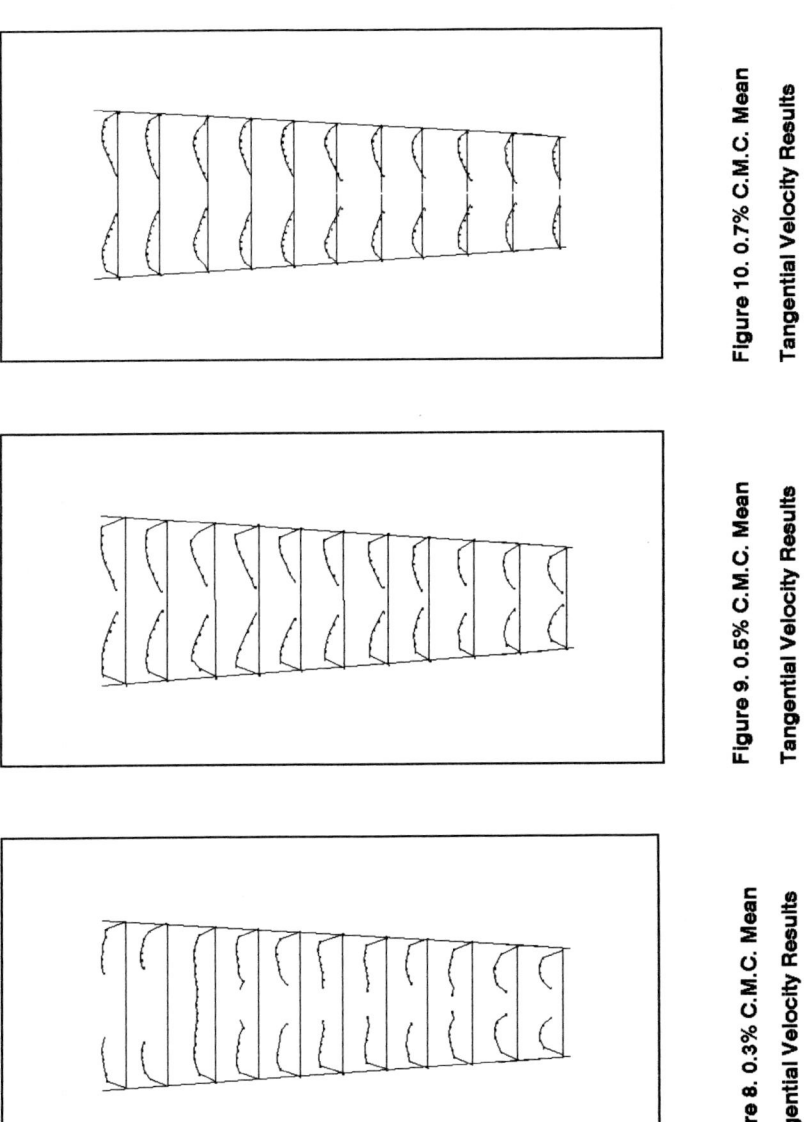

Figure 10. 0.7% C.M.C. Mean Tangential Velocity Results

Figure 9. 0.5% C.M.C. Mean Tangential Velocity Results

Figure 8. 0.3% C.M.C. Mean Tangential Velocity Results

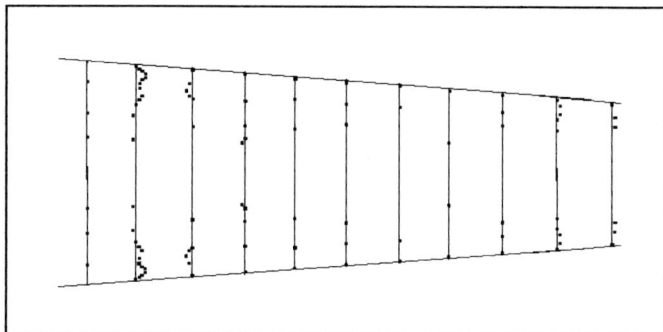

Figure 12. Reynold Stresses in 0.5% C.M.C.

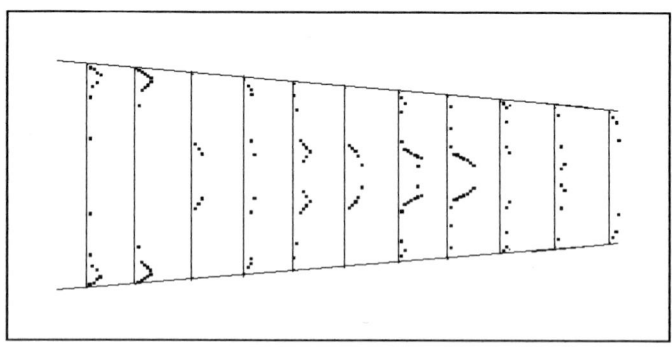

Figure 11. Reynold Stresses in 0.3% C.M.C.

DESIGN AND OPERATION

A NEW METHOD OF TESTING HYDROCYCLONE GRADE EFFICIENCIES

Ladislav Svarovsky
Department of Chemical Engineering
University of Bradford
Bradford BD7 1DP

Jan Svarovsky
Bradford Grammar School
Bradford BD9 4JP

ABSTRACT

This paper describes a new and simple experimental method for obtaining the reduced cut size and the rest of the reduced grade efficiency curve of an operating hydrocyclone. The method relies on feeding a known and fully-characterised slurry to the hydrocyclone under test, and on measuring only two solids concentrations (in the feed and in the overflow), one static pressure differential and the slurry temperature. These measurements are best done and logged by a personal computer, and have to be repeated at two different pressure settings.
The new method eliminates the need for sampling, particle size determinations (except that of the feed suspension but this is only done once for a whole string of experiments) or flowrate measurements, thus making the tests simpler than the conventional test methods and also capable of being performed by simple, on-line instrumentation and a computer.
The underlying theory is fully developed in the paper, together with its application to a specific case of testing a small diameter hydrocyclone for monitoring of very fine particle size in industrial slurries.

NOMENCLATURE

c is mass concentration of solids in the feed

c_o is mass concentration of solids in the overflow

D is the internal diameter of a hydrocyclone

E_T is the total coarse efficiency (or recovery)

E_T' is the reduced total efficiency defined in eqn.5
Eu is Euler number as defined in eqn.12
$F(x)$ is the cumulative percentage undersize in the feed
$G(x)$ is the actual grade efficiency function (curve)
$G'(x)$ is the reduced grade efficiency function, eqn.1
R_f is the underflow-to-throughput ratio (by volume)
Q is the volumetric flowrate of the feed
Stk_{50}' is Stokes number as defined in eqn.13
v is characteristic velocity as defined in eqn.14
x is particle size as a variable
x_g is the mass median of the feed solids [$F(x_g) = 0.50$]
x_{50}' is a the reduced cut size [$G'(x_{50}') = 0.50$]
Δp is static pressure drop across the hydrocyclone
μ is liquid viscosity
ρ is liquid density
ρ_s is solids density
σ_g is the geometric standard deviation of $F(x)$
σ_s is the geometric standard deviation of $G'(x)$

INTRODUCTION

Testing the grade efficiencies of hydrocyclones is a tedious and time-consuming task. It is still very much needed, however, particularly when investigating the effect of feed concentration of solids where theory still requires experimentally-determined constants.

The conventional test method requires sampling of at least two of the material streams involved, followed by full particle size analyses (and concentration measurements) of these samples and, also, by flowrate measurements for the necessary determination of the total solids recovery to underflow. The total solids recovery may also be evaluated from the particle size distribution of the third stream.

The grade efficiency curves evaluated from such conventional tests are subject to large random variations because all errors of sampling, particle size measurements and concentration measurements are propagated into the final result.

In our on-going search for new methods of on-line particle size analysis and in developing one based on a hydrocyclone, we have discovered an elegant way of simplifying the grade efficiency testing which also, at the same time, gives more reproducible results than the conventional tests. The theory and an example of use of the test method are given in the following.

GRADE EFFICIENCY OF A HYDROCYCLONE

For any separator with a size-dependent performance, such as a hydrocyclone, the grade efficiency varies with particle size, and a graphical representation of this is called the grade efficiency curve. As the value of the grade efficiency has the character of probability, it is sometimes referred to as the partition probability; the curve then becomes the partition probability curve or Tromp curve.

In practice, the grade efficiency curve is a continuous function of particle size x - see Figure 1 which gives a typical grade efficiency curve for a hydrocyclone.

The effect of flow splitting (or "dead flux") in applications with appreciable and dilute underflow, as is common with hydrocyclones, is to modify the shape of the grade efficiency curve making it appear as if the performance of the hydrocyclone were better than it actually is. An example is given in Figure 1 where a typical grade efficiency curve of a hydrocyclone is plotted. The curve does not start from the origin (as it should for inertial separation) but has an intercept, the value of which is usually equal to the underflow-to-thoughput ratio R_f. This is because the very fine particles simply follow the flow and are split between the underflow and the overflow in the same ratio as the fluid. The R_f ratio is defined as the fraction of the volumetric feed rate which turns up in the underflow, i.e. the underflow rate, divided by the feed rate.

In order to remove the effect of flow splitting from the efficiency definition so that it describes only the true "centrifugal efficiency", the grade efficiency is "reduced" by the following equation (1,2):

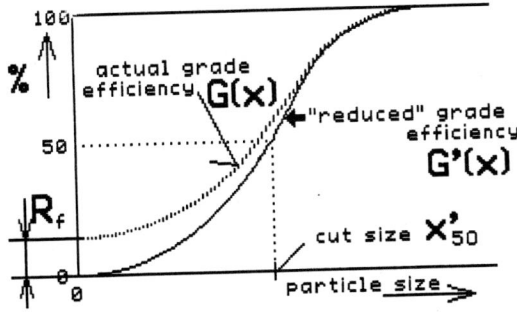

Figure 1. Grade efficiency curves for a hydrocyclone

$$G'(x) = \frac{G(x) - R_f}{1 - R_f} \qquad (1)$$

This forces the curve to pass through the origin as indicated by the second curve, G'(x), in Fig.1. The reduced grade efficiency curve can, for some separators, be approximated by an analytical expression such as the one used in this method - see eqn.2 given in the following section.

The grade efficiency curve, while it provides the most comprehensive description of separation efficiency, is rather clumsy for correlation with operating variables or for simple equipment comparisons. Such applications call for a single number, independent of the size of the feed solids, as a measure of efficiency. This is available in the form of the "reduced cut size" which is defined as the size corresponding to 50% on the reduced grade efficiency curve G'(x) - see Fig.1. Most mathematical descriptions of the performance of hydrocyclones are in terms of the reduced cut size.

THE THEORY

The reduced grade efficiency curves of hydrocyclones can be fitted by a cumulative log-normal function in the following form (3):

$$G'(x) = 0.5 + 0.5 \, \text{erf} \left[\frac{\ln x - \ln x'_{50}}{\sqrt{2} \ln \sigma_s} \right] \qquad (2)$$

where the erf function is defined as:

$$\text{erf}(z) = \frac{2}{\sqrt{\pi}} \int_0^z e^{-t^2} dt \qquad (3)$$

and erf(z) can be evaluated using tables, series or analytical approximations. Note that x'_{50} and σ_s can be determined from a plot of the grade efficiency curve in a log-probability graph paper or by using a curve-fitting package.

The preferred definition of particle size in grade efficiency testing of hydrocyclones is the Stokes' diameter as measured by laboratory methods involving gravitational or centrifugal sedimentation. This incorporates the effects of particle density, shape and fluid viscosity on particle-fluid interaction, within the validity of Stokes' Law.

Once the grade efficiency curve is known for a given set of operating conditions, the total efficiency E_T (recovery of solids into the underflow) expected with a particular feed of a cumulative particle size distribution F(x) can be predicted using the following relationship :

$$E_T = \int_0^1 G(x) \, dF \qquad (4)$$

$$E'_T = \frac{E_T - R_f}{1 - R_f} \qquad (5)$$

Note that eqn.4 can also be used to relate the reduced efficiencies, i.e.

$$E'_T = \int_0^1 G'(x)\, dF \qquad (6)$$

According to the present method, the particle size distribution of the feed solids must be capable of being closely approximated by the log-normal distribution in the following form, analogous to eqn.2 (as cumulative fraction undersize):

$$F(x) = 0.5 + 0.5\, \text{erf}\left[\frac{\ln x - \ln x_g}{\sqrt{2}\, \ln \sigma_g}\right] \qquad (7)$$

where the erf function is again defined in eqn.3.

The total reduced efficiency E'_T can then be predicted from $G'(x)$ and $F(x)$ by integration (eqn.6) and this, using eqns 2 and 7 leads to the following formula (1):

$$E'_T = 0.5 + 0.5\, \text{erf}\left[\frac{\ln x_g - \ln x'_{50}}{\sqrt{2}\, \sqrt{(\ln^2 \sigma_g + \ln^2 \sigma_s)}}\right] \qquad (8)$$

The total reduced efficiency E'_T can also be evaluated directly from the feed solids concentration c and the solids concentration in the overflow c_o using another formula (1):

$$E'_T = 1 - \frac{c_o}{c} \qquad (9)$$

Note that, conveniently, eqn.9 does not require the knowledge of the flow ratio R_f.

It is the combination of equations 8 and 9, by eliminating E'_T, which forms the foundation of the present method:

$$1 - \frac{c_o}{c} = 0.5 + 0.5\, \text{erf}\left[\frac{\ln x_g - \ln x'_{50}}{\sqrt{2}\, \sqrt{(\ln^2 \sigma_g + \ln^2 \sigma_s)}}\right] \qquad (10)$$

In the above equation, the response of the separator to the operating conditions, in terms of the cut size x'_{50} and the standard geometric deviation of the reduced grade efficiency σ_s, can be found experimentally by using a feed of a known particle size distribution described by eqn.7 above and by monitoring the two concentrations c and c_o.

As one equation is not enough for calculating the two parameters, another equation has to be generated. This is done by changing the operating conditions whilst feeding the separator with the same slurry. With hydrocyclones, for example, eqn. 10 can be duplicated by taking measurements at two different pressure drops but, as the cut size changes with pressure drop, a fundamental model describing such change also has to be available. It is assumed here, however, that the value of σ_s does not change with a small change in pressure drop.

PRACTICAL USE OF THE METHOD

The sequence of the measurement, preferably controlled by a computer, is to take the readings of c and c_o at one pressure drop (result set 1), then switch to another pressure drop and repeat the measurement (set 2). Fig.2 shows a schematic diagram of the measurement positions, with all of the readings being capable of being taken and logged by a computer. The concentration readings are from two separate and suitably calibrated densitometers such as gamma gauges, vibrating U-tube density meters or ultrasonic devices.

It is also possible to use just one densitometer and re-route periodically and alternately the feed and overflow streams through it. This is best done by solenoid or motorised valves, controlled by the computer (see an example in Fig.3 described below).

The evaluation of results is done with a computer which uses a mathematical model of the separator function and the following simple algorithm for the evaluation of the two parameters x'_{50} and σ_s:
1. Assume σ_s and evaluate $(x'_{50})_1$ from eqn.10 and the result set 1,
2. change Δp and calculate $(x'_{50})_2$ from the result set 2,
3. use a fundamental model which should link the two values of x'_{50} and if it does not, keep changing σ_s and re-calculating the two values of x'_{50} until it does.
4. print or display the two final values of x'_{50}, each corresponding to a different pressure drop, and the one common value of σ_s.

Given the simplicity of the above calculations and the speed of personal computers, the calculations are done virtually instantaneously. The only delay in the measurement is through the necessity of switching to another pressure drop and the need to establish steady-state operation after the change. This is, however, quite short (a few seconds) in hydrocyclones.

According to the present test method, the feed slurry containing the test solids is recirculated through a specially built and instrumented test rig (containing the hydrocyclone under test) such as the one shown schematically in Fig.3 and further described in the following section. The small step change in pressure drop required can be easily introduced by using a solenoid or motorised valve, or by changing the speed of the supply pump, with the change initiated by the controlling computer.

The only the particle size analysis required by the present method is that of the test solids in the feed stream and this is done off-line and repeated many times to reduce errors, using laboratory analytical equipment.

THE EXPERIMENTAL APPARATUS

In the experimental rig shown in Fig.3, which we used to test a 10mm, stainless steel "Doxie" hydrocyclone from Dorr-Oliver, the change from one operating pressure to another is achieved by switching on or off the solenoid valve (sv) in the by-pass line line 1. This is, however, only possible when the test solids contain only very fine particles which do not separate significantly in the T-junction 2 which splits the pumped flow into the hydrocyclone feed line 3 and the by-pass line 1.

In the case of coarser slurries used in testing larger hydrocyclones, this arrangement would not be acceptable and the change in the operating pressure drop might better be achieved by changing the speed of the supply pump.

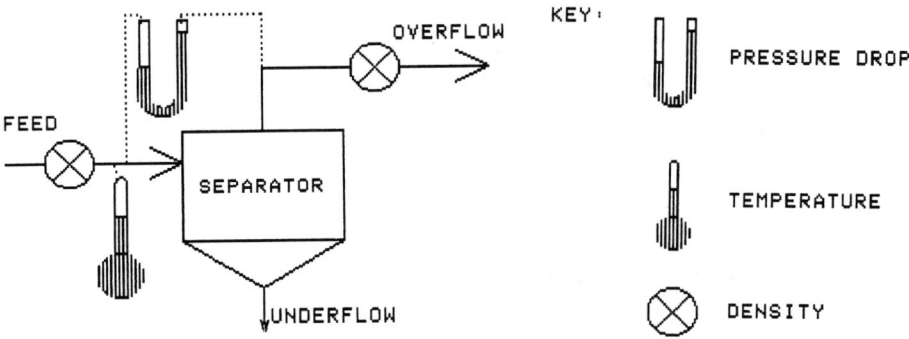

Figure 2. A schematic diagram of the measurement positions

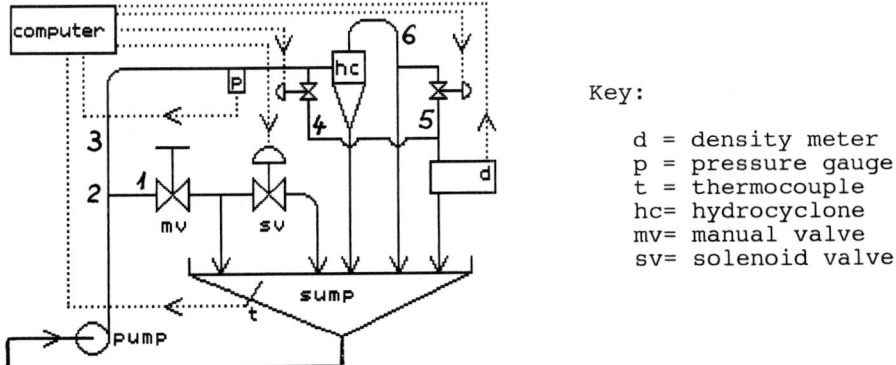

Figure 3. A schematic diagram of the experimental setup

It may also be noted from Fig.3 that, in this example, only one densitometer is used and the two streams to be measured, the sample stream 4 from the feed stream 3 and the sample stream 5 from the overflow 6, are switched through it alternately, with a delay in between to allow each sample stream to reach the sensor head of the instrument. A vibrating U-tube density meter is used in the example. Note also that both of the streams taken through the densitometer have to be well de-aerated and this is achieved by venting the lines and routing them in such a way that de-aeration takes place by gravity.

EXAMPLE OF TEST RESULTS

Fig. 4 shows the particle size distribution, by mass, of the solids used in the tests (chalk), as obtained with the Ladal Pipette Centrifuge and the Andreasen Pipette Method. As can be seen from the log-probability plot, the distribution is very nearly log-normal and thus suitable for testing the performance of the hydrocyclone. Furthermore, the medium size of the chalk (3.9 microns) is within the range of cut sizes expected from the hydrocyclone (2 to 4 microns) which is a requirement for effective separator testing.

The most important effect to be tested in hydrocyclones is that of the feed solids concentration. The range to be covered was from 0 to 20 % by volume. An IBM-compatible microcomputer was used to control the rig and log the data, using a specially-developed program which included the necessary algorithms mentioned previously.

Besides the feed solids concentration c, there is a whole host of variables affecting hydrocyclone performance and these are conveniently grouped together in dimensionless groups. A model based on the use of such groups has been published before (1) and the present method is based on an adaptation of this model to small diameter hydrocyclones.

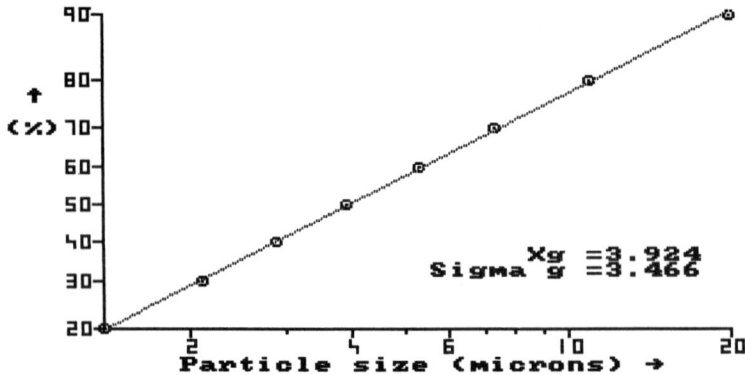

Figure 4. Mass distribution of the test powder (chalk).

Whilst with larger hydrocyclones the resistance coefficient called Euler number Eu depends of Reynolds number, it is practically constant for for very small hydrocyclones such as the one used in this work. We felt justified, therefore, in using such combination of Eu with Stk (instead of a straight product of Stk.Eu as derived from first principles (2)) which would obviate the need to measure flowrates on the rig and, therefore, greatly simplify the testing procedure.

The main equation relating the relevant dimensionless groups is as follows:

$$Stk'_{50} \sqrt{Eu} = k_1 \exp(k_2 c) + k_3 \qquad (11)$$

where the dimensionless groups are defined as follows:

The Euler number Eu is a pressure loss factor based on the static pressure drop across the cyclone:

$$Eu = \Delta p / (\rho v^2 / 2) \qquad (12)$$

Stk'_{50} is the Stokes number (for reduced cut size x'_{50}) defined as:

$$Stk'_{50} = x'^2_{50} (\rho_s - \rho) v / (18 \mu D) \qquad (13)$$

where ρ_s and ρ are densities of the solids and of the liquid respectively, μ is liquid viscosity and D is cyclone diameter, x'_{50} is the reduced cut size (see the Nomenclature for the definitions).

Both of the above equations use the superficial velocity in the cyclone body as the characteristic velocity, i.e.

$$v = 4Q / (\pi D^2) \qquad (14)$$

The leftt-hand term in eqn.11 is a dimensionless group that has its theoretical basis in the turbulent two-phase flow theory for hydrocyclones (4) whilst the right-hand side is a semi-empirical expression of the effect of feed concentration, which has its roots in the theory of hindered settling (5).

The test rig described in previously was run with the chalk slurry at several solids concentrations ranging up to 20% by volume, with duplicated measurements at each concentration. The range of reduced cut sizes measured was from 2 microns at low concentrations to 5 microns at 20%. The geometric standard deviation of the reduced grade effciency curve varied between 1.8 and 2.5. The above results agree with those obtained with the conventional test methods (6,7).

Fig.5 shows the results in dimensionless form and the best fit curve obtained by the minimum sum of squares method. This provided the constants for equation 11, with the following values (if c is in %): $k_1 = 0.083419 \times 10^{-3}$, $k_2 = 0.22359$ and $k_3 = 1.1335 \times 10^{-3}$.

Eqn.11 then represents the performance of the hydrocyclone within the range of operating conditions used in its testing, and can be employed, for example, as a model in particle size measurement of unknown slurries in the same rig as used in testing the hydrocyclone.

CONCLUSIONS

The advantages of the new method for testing hydrocyclones can be summarised as follows:

1. testing of hydrocyclones is greatly simplified and sped up, compared to conventional methods, and the method can be used in research and development.
2. a personal computer can be used to automatically control the measurement, to take multiple readings of each variable (for improved accuracy) and to log and evaluate the data,
3. errors of sampling, concentration measurement, flowrate measurement and particle size measurement normally present in conventional testing are either reduced or completely eliminated,
4. the same equipment and method can be used for on-line monitoring of the median particle size and of the spread of the distribution in the feed slurry.

Any suitable feed solids material may be used for the testing but it has to be one with a mono-modal particle size distribution that approximately follows the log-normal law. This requirement is not too limiting because the test material is a matter of choice anyway.

The new test method was recently used by the authors in the course of the development of a new, hydrocyclone-based, on-line monitor of particle size in fine suspensions.

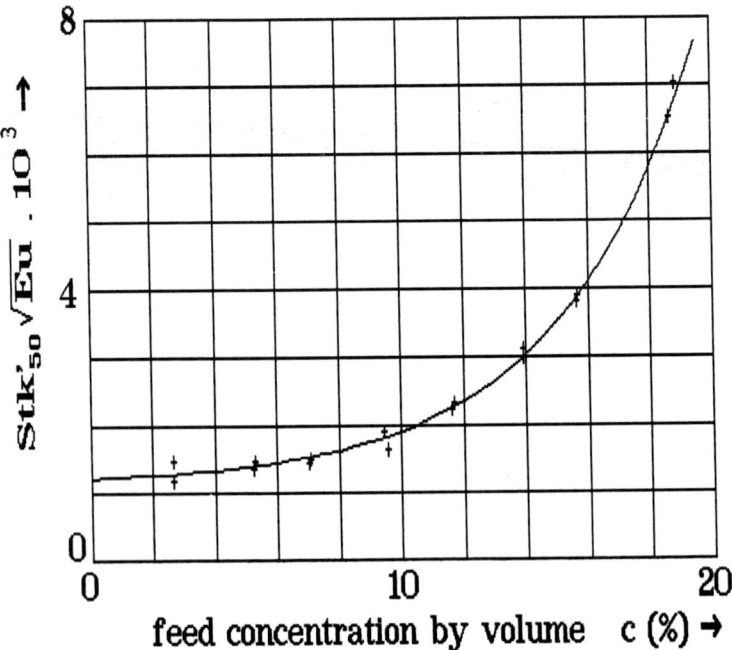

Figure 5 Test data obtained with a 10 mm hydrocyclone

REFERENCES

1. Svarovsky, L., (ed), 'Solid-liquid Separation', 3rd edition, Butterworths, London, 1990

2. Svarovsky, L., 'Hydrocyclones', Holt Rinehart and Winston, London, 1984

3. Gibson, K., "Large scale tests on sedimenting centrifuges and hydrocyclones for mathematical modelling of efficiency", pp 1-10, in Proc. Symp. on Solid-liquid Separation Practice, Yorkshire Branch of the I. Chem. E., Leeds, 27-29 March 1979.

4. Schubert, H., and Neesse, T., "A hydrocyclone separation model in consideration of the turbulent multi-phase flow", Paper 3, Proc. Int. Conf. on Hydrocyclones, Cambridge, 1-3 October, 1980, BHRA Fluid Engineering, Cranfield, 1980, pp 23 - 36

5. Davies, L., Dollimore, D., and McBridge, G.B., Powder Technology, 1977, 16, 45

6. Svarovsky, L., Evaluation of clarification performance of a two-stage hydrocyclone system, 2nd International Conf. on Hydrocyclones, Bath, 19-21 Sept.1984, Paper J2, Proceedings published by BHRA, Cranfield, 1984

7. Svarovsky, L., Evaluation of small diameter hydrocyclones, I.Chem.E. Symposium on Solid/Liquids Separation Practice and the Influence of New Techniques, Leeds, 3-5 April, 1984, Paper 24, pp 193-205, Inst. of Chemical Engineers, Yorkshire Branch, 1984

EFFECT OF SOLIDS FEED GRADE ON THE SEPARATION OF SLURRIES IN HYDROCYCLONES

E. ORTEGA-RIVAS
Department of Chemical Sciences; University of Chihuahua
Apdo. 1542-C; Chihuahua, Chih.; MEXICO.

L. SVAROVSKY
Department of Chemical Engineering; University of Bradford
Bradford, West Yorkshire BD7 1DP; ENGLAND

ABSTRACT

In developing a dimensionless scale-up model for small diameter hydrocyclones the influence of the particle size distribution was studied. In order to do that a batch of dolomitic limestone was divided to obtain three different grades of the same material. Suspensions with concentrations of 15% (v/v) and 20% (v/v) for each fraction of powder were tested in three geometrically similar hydrocyclones of 22mm, 44mm and 88mm in diameter. A dimensionless expression was developed to account for the particle size distribution of feed, which presented appropriate goodness of fit and showed good agreement with correlations available in the literature.

NOMENCLATURE

(Dimensions given in terms of mass M, length L, time T and temperature θ)

- a: constant
- A: settling area (L^2)
- A_p: projected area of particles (L^2)
- b: partial regression coefficient
- C: feed solids concentration by volume
- C_D: drag coefficient
- D_c: hydrocyclone body diameter (L)

D_o: overflow diameter (L)
D_u: underflow diameter (L)
Eu: Euler number
F_D: drag force (MLT^{-2})
g: gravity acceleration (LT^{-2})
K: constant
k_1, k_2: constants
k_d: correction factor
k_p: empirical constant for a family of geometrially similar cyclones
K': fluid consistency index (MT^nL^{-2})
l: vortex finder length of cyclone (L)
L: overall length of cyclone body (L)
m: exponent of correction factor depending on D_c
n: flow behaviour index for power-law fluids, empirical exponent
n_p: empirical constant for a family of geometrically similar cyclones
n': flow behaviour index
Q: volumetric flow rate of feed suspensions (L^3T^{-1})
r: correlation coefficient
R: multiple correlation coefficient
Re: Reynolds number
Re^*: Reynolds number for power-law fluids
Re_p^*: particle Reynolds number for power-law fluids
Rf: underflow-to-throughput ratio
Stk_{50}: Stokes number
$Stk_{50}(r)$: reduced Stokes number based on reduced cut size
$Stk^*_{50}(r)$: reduced Stokes number for power-law fluids
u: particle-fluid relative velocity (LT^{-1})
u_g: terminal settling velocity under gravity (LT^{-1})
v: superficial velocity in the cyclone body (LT^{-1})
x: independent variable
X: particle size (L)
X_g: mass median size (L)
X_{50}: cut size (L)
$X_{50}(r)$: reduced cut size (L)

y: dependent variable
γ: viscosity coefficient for power-law fluids $(ML^{-1}T^{n-2})$
$\dot{\gamma}$: shear rate (T^{-1})
ΔP: pressure drop $(ML^{-1}T^{-2})$
μ: viscosity $(ML^{-1}T^{-1})$
μ_a: apparent viscosity $(ML^{-1}T^{-1})$
ρ: liquid density (ML^{-3})
ρ_m: suspension density (ML^{-3})
ρ_s: solids density (ML^{-3})
τ: shear stress $(ML^{-1}T^{-2})$

INTRODUCCION

According to several authors [1,2], the particle size distribution of the powder in suspension fed to a hydrocyclone is an important variable in the separation process, which has a direct effect on the reduced cut size. Schubert and Neesse [1] reported a theoretical equation, with an empirical correction factor, based on the turbulent multi-phase flow theory. The effect of the particle size distribution of feed is included in the above mentioned correction factor, which also bears an exponent taking different values according to the hydrocyclone body size, when it falls below 0.1m in diameter. The study of the turbulent multi-phase flow theory is only one of four known approaches which try to describe the behaviour of particles in a hydrocyclone. A brief description of this theory will be presented in the following.

Turbulent two-phase flow theory
This theory states the importance of the effect caused by turbulent flow in hydrocylones. According to this, it is proposed that separation takes place under the influence of the centrifugal field and the turbulent transport in the two-phase flow. The influence of turbulence can be described by an eddy diffusion coefficient equal to, or smaller than, the

diffusion of the fluid particles. The assessment of the influence of such diffusion on separation is very difficult in practice. Bloor and Ingham [3,4] have contributed to the theoretical background of this theory. They derived an analytical form of the radial velocity and found an analytical solution to the equation of motion for the tangential speed. They also developed a more elaborated and physically realistic model for the eddy viscosity based on the Prandtl mixing theory, which greatly simplified the equation of motion.

Mueller, Neesse and Schubert [5] proposed a set of equations and derived two models for turbulent cross-flow wet classification: the suspension-partition model and the suspension-tapping model. In the former the flow is divided into overflow and underflow without changes in the total cross section, whereas in the latter overflow and underflow are "tapped" from the main flow through small outlet openings. According to them, the suspension-tapping model is analogously applicable to the separation in hydrocyclones. For performance predictions on the basis of the theory discussed above Schubert, Bohme, Neesse and Espig [6] derived the following relationship for hydrocyclones

$$X_{50} = K \left[\frac{\mu D_c \ln (D_o/D_u)^3}{(\rho_s-\rho)(1-C)^{4.65} (\Delta P/\rho_m)^{\frac{1}{2}}} \right]^{\frac{1}{2}} \qquad (1)$$

Where X_{50} is the cut size (i.e. the size which has equal probability of being separated with the underflow or the overflow); μ is the medium viscosity; D_c, D_o and D_u are hydrocylone diameter, overflow diameter and underflow diameter respectively; ρ_s is the solids density; ρ is the liquid density; C is the volume percent of solids; ΔP is the pressure drop and ρ_m is the suspension density. The value of K equals 0.12.

Equation (1) appears to have included the previously mentioned correction factor in the value of K. This factor has been reported [1] as

$$k_d = \left[0.00022 X_g [(\rho_s-\rho)/D_c]^{\frac{1}{2}}\right]^m \qquad (2)$$

Where X_g is the mass median size of the feed material and the exponent m depends on D_c.

Dimensionless scale-up approach for hydrocylones

A scale-up procedure to select hydrocylones has been recently proposed [7,8]. The method consists of selecting a known design of hydrocyclone and, by using a series of dimensionless relations, making the necessary predictions as changes in fixed variables are introduced. As the performance characteristics of hydrocylones involve a great number of variables, the use of dimensionless groups in this method is an obvious advantage. This approach has the merit of using families of geometrically similar hydrocyclones so that all design variables are omitted from the scale-up correlations, except the size of the underflow orifice which should be variable and is considered to be an operating variable.

For low solids concentrations, dimensional analysis coupled with two theories of separation in hydrocyclones gives two basic relationships between three dimensionless groups

$$Stk_{50} \, Eu = constant \qquad (3)$$

$$Eu = k_p (Re)^{np} \qquad (4)$$

Where the Reynolds number is defined as

$$Re = \frac{D_c v \rho}{\mu} \qquad (5)$$

The Euler number is the well known pressure loss factor, i.e.

$$Eu = \frac{2\Delta P}{\rho v^2} \qquad (6)$$

Where ΔP is the pressure drop across the cyclone and the Stokes number is defined as

$$Stk_{50} = \frac{x^2_{50}(\rho_s-\rho)v}{18\mu D_c} \qquad (7)$$

Equations (5), (6) and (7) use the superficial velocity in the cyclone body as the characteristic one, i.e.

$$v = \frac{4Q}{\pi D_c^2} \qquad (8)$$

Where Q is the suspension flow rate and all the remaining symbols in last equations have been previously defined.

At higher concentrations, the feed concentration as a fraction by volume, C, has to be included as an additional dimensionless group. Svarovsky and Marasinghe [9] have developed the following expression for the effect of high feed solids contents

$$Stk_{50}(r) = k_1(1-Rf)\exp(k_2C) \qquad (9)$$

Where $Stk_{50}(r)$ as defined in equation (7) includes the reduced cut size, i.e. taking into account the "dead flux" effect of very fine particles simply following the flow and split in the same ratio as the liquid. k_1 and k_2 are empirical constants which depend on the hydrocyclone configuration and Rf is the underflow-to-throughput ratio.

For concentrations higher than 10% solids by volume, many practical slurries show non-Newtonian behaviour and it can be shown [10] that Reynolds and Stokes numbers can be re-expressed to consider this non-Newtoniancy. For the case of the Stokes number, it can be derived as follows.

If a particle moves relative to the fluid in which it is suspended, the force opposing the motion is known as the drag force. Conventionally, the drag force F_D is expressed according to Newton

$$F_D = C_D A_p \frac{\rho u^2}{2} \quad (10)$$

Where C_D is a coefficient of proportionality known as the drag coefficient, A_p is the area of the particle projected in the direction of the motion, ρ and u are the fluid density and the particle-fluid relative velocity respectively.

The drag force on a sphere for creeping flow can be determined theoretically from Navier-Stokes equations and the solution is known as Stokes law

$$F_D = 3\pi\mu u X \quad (11)$$

Where μ is the medium viscosity and X is the particle size.

For a Newtonian fluid the viscosity can be obtained from the characteristic function.

$$\tau = \mu \dot{\gamma} \quad (12)$$

Where τ is the shear stress and $\dot{\gamma}$ the shear rate.

As distinct from Newtonian fluids, in most non-Newtonians the shear stress is not proportional to the shear rate, but to its nth power; hence the name power-law fluids. The equation of flow is

$$\tau = K(\dot{\gamma})^n \quad (13)$$

Clearly, in logarithmic coordinates the flow curve is a straight line, in which n represents its slope and K is given by the intersection with the axis at $\dot{\gamma}=1$.

An apparent viscosity substitution, using equation (13), can be expressed as:

$$\mu_a = K'(\dot{\gamma})^{n'-1} \quad (14)$$

Where μ_a is the apparent viscosity, K' is the fluid consistency index and n' the flow behaviour index. For Newtonian fluids n'=1, while for others, the greater the divergence from unity the more non-Newtonian is the fluid in question. On the other hand as the name of the fluid consistency index suggests, the larger its value the thicker or more viscous the fluid.

Rotational viscometers can be used for evaluation of K and n by means of equation (13). It has been established that K and n can be related to K' and n' only when these properties are constant over a reasonable range of shear stresses [11]. In such a case n'=n and K'=f (K,n).

By combining equations (10), (11) and (14), the drag coefficient C_D for steady-state creeping flow past a sphere, may be obtained as

$$C_D = \frac{24}{Re_p^*} \qquad (15)$$

The particle Reynolds number Re_p^* is given by

$$Re_p^* = \frac{X^{n'} \rho u^{2-n'}}{K'(3)^{n'-1}} = \frac{X^{n'} \rho u^{2-n'}}{\gamma} \qquad (16)$$

In deriving equation (15) an average shear rate equal to 3u/X and the fluid consistency for flow between two infinite parallel plates, as suggested elsewhere [12], were employed.

For a spherical particle in a power-law model fluid, the drag coefficient C_D is also given by Stokes law as

$$C_D = \frac{4}{3} \left(\frac{(\rho_s - \rho)}{\rho} \right) \frac{gX}{u^2} \qquad (17)$$

Where g is the acceleration due to gravity. Equations (15), (16) and (17) can be combined to derive a particle settling velocity as

$$u_g = \left[\frac{(\rho_s-\rho)gx^{n+1}}{18\gamma}\right]^{1/n} \quad Re_p^* < 0.1 \quad (18)$$

Where u_g is the terminal settling velocity under gravity. From simple sedimentation theory it follows than $u_g = Q/A$. Substituting in equation (18) and re-arranging terms the reduced Stokes number for power-law fluids is obtained, i.e.

$$Stk^*_{50}(r) = \frac{[x_{50}^{n'+1}(r)](\rho_s-\rho)v^{2-n'}}{18\gamma D_c} \quad (19)$$

MATERIALS AND METHODS

Test materials and equipments

A batch of dolomitic limestone was divided in a centrifugal laboratory air classifier, the batch was split by its median size in order to obtain volumes approximately similar for each fraction. By doing this three parts were obtained: a fine, a medium (the original one) and a coarse. As can be seen in figure 1, the median sizes are approximately 6μm, 14μm and 32μm for fine, medium and coarse fractions respectively. The densities of the three powders, determined by air pycnometry, were 2774kg/m³ for fines, 2885kg/m³ for medium and 2832kg/m³ for coarse.

Suspensions of 15% and 20% solids contents by volume for each powder were tested in three geometrically similar hydrocylones of 22mm, 44mm and 88mm body diameters. All design proportions were according to Rietema's optimum design [13], as shown in table 1 below

TABLE 1

Rietema's optimum hydrocylone proportions (from [13])

D_i/D_c	D_o/D_c	D_u/D_c	l/D_c	L/D_c	cone angle (°)
0.28	0.34	0.20	0.40	5	20

Three pressure drops for each unit were tested. The pressures were as follows: small cyclone 151700Pa, 303400Pa and 455100Pa; medium cyclone 110300Pa, 220600Pa and 330900Pa and large cyclone 68900Pa, 137800Pa and 206700Pa.

The fine fraction was treated in the small and medium units, the medium fraction was tested in all units whereas the coarse one was employed in the medium and large cyclones.

The set of experiments necessary to develop the model was arranged in a two-way classification manner, so all the possible runs over the whole range were performed. Such an arrangement, considering all the variables involved described above, resulted in a total of 96 experiments including a replicate for each one.

The rig illustrated in figure 2 was used for the experiments. As can be seen, it consists of a large vessel equipped with baffles in order to facilitate mixing as it is promoted by a variable speed agitator. The prime mover is a rubber lined centrifugal pump. The undeflows of all units are conveyed by pipes direct to the feed tank whereas the overflows are discharged by smoothly bent pipes to a v-notch type weir-box, which is used for their flow rate measurement.

Rheograms of the feed suspensions were obtained by a rotational viscometer, the BRABENDER Rheotron, in which the outside cylinder impresses a shear flow onto the slurry and the resultant shear force is measured at the inside cylinder. The suspension state was aided using sodium-hexa-meta

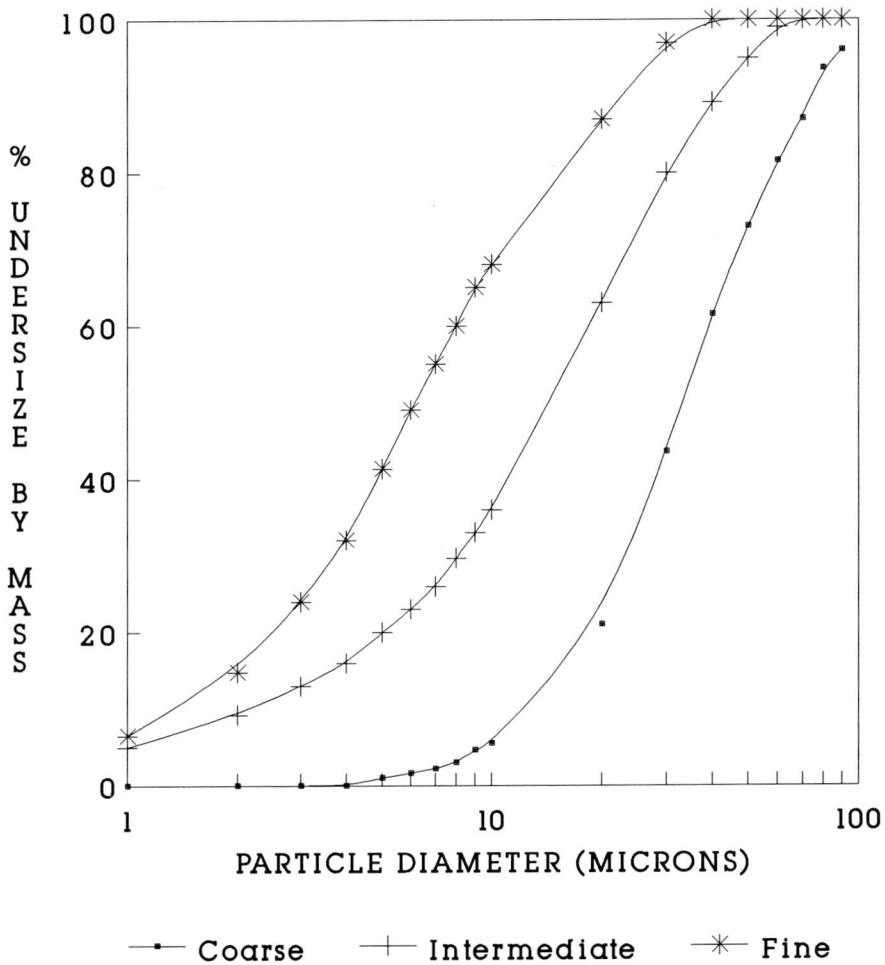

Figure 1. Particle size distributions of the powders used.

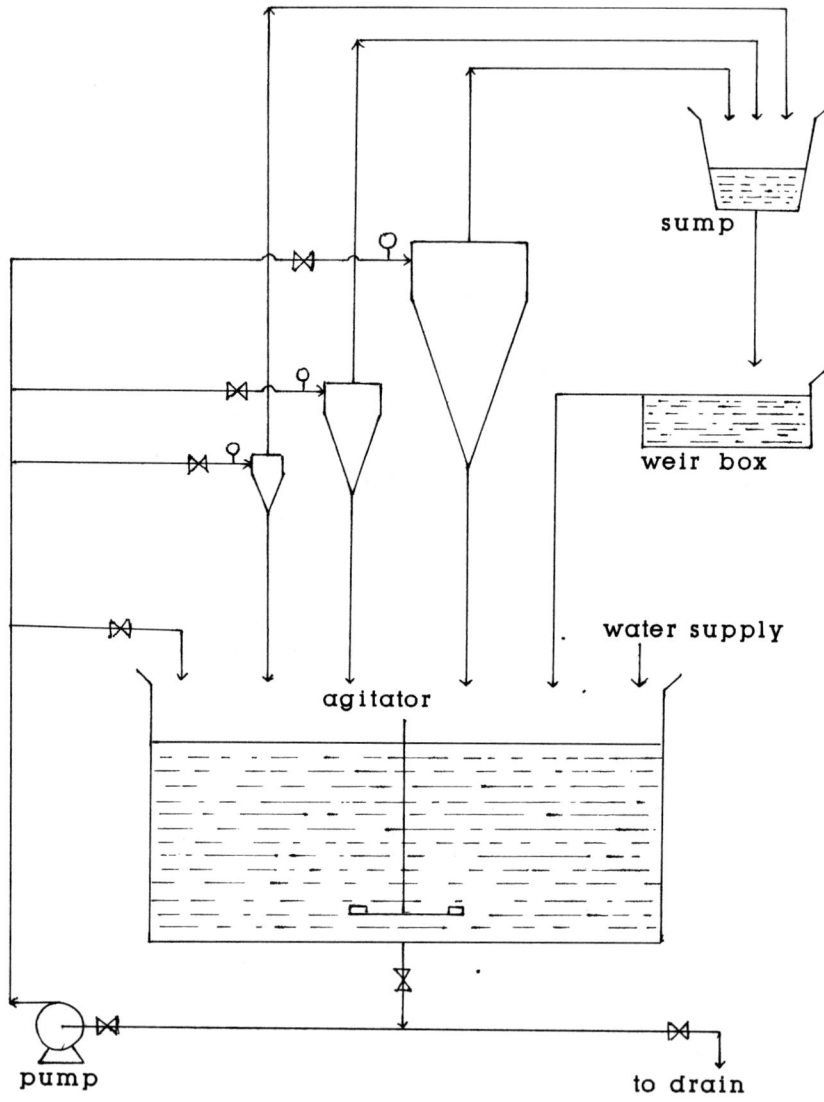

Figure 2. Flow diagram of the test rig used in experiments.

phosphate as dispersant and the readings were rapidly taken to avoid errors due to settling of particles.

The particle size analyses of both leaving streams were carried out by a combined Andreasen-Pipette Centrifuge method. This way was chosen in order to obtain directly the Stokes diameter, considering that the size range covered was quite wide.

Handling and processing of data

In trying to obtain relationships from raw data in a case such as this the approaches suggested in the literature [14,15] normally involve plotting of one variable against another so a "scattegram" is obtained. By looking at this, it may be determined whether or not a trend in the variables is evident. If some recognisable pattern is identified, the analytical representation of the function can be obtained by means of regression analysis.

To verify the adequacy of the model a correlation coefficient can be derived. Such a coefficient r would give a value equal to one for perfect prediction. Furthermore, the validity of a regression model can be checked by inspecting the residuals, which are the difference between the observed data and the values predicted by the equation. It has been stated [14,15] that the residuals are normally distributed for an appropriate model and two useful graphs are residuals versus the predicted y-value, as well as normal probability plot of residuals. The choice of a right model will give no defined pattern in the former and an approximate straight line in the latter.

In many practical cases a variable may depend on more than one independent variable. If these independent variables tend to vary according to some pattern, simple regression leads to misleading results, and multiple regression represents a better choice. In this case a multiple linear regression is recommended and the fitted equation may be represented by

$$y = a + b_1 x_1 + b_2 x_2 + b_3 x_3 + \ldots b_n x_n \qquad (20)$$

Where y is the dependent variable; x_1, x_2, x_3,...x_n are the independent variables and a, b_1, b_2, b_3,...b_n are the estimated parameter of the model or partial regression coefficients.

When the relationships between the variables is clearly non-linear a transformation by means of logarithms may be used. For example, in trying to fit data to the model

$$y = a(x_1)^{b_1} (x_2)^{b_2} (x_3)^{b_3} \ldots (x_n)^{b_n} \qquad (21)$$

To obtain a linear relationship we take logarithms of both sides to give

$$\ln y = \ln a + b_1 \ln x_1 + b_2 \ln x_2 + b_3 \ln x_3 + \ldots b_n \ln x_n \qquad (22)$$

The adequacy of the fitted model can be verified by means of the multiple correlation coefficient R, which also takes a value close to unity when predicted and observed values have a high degree of correlation. The value of R^2 gives a measure of the proportion of variation in the dependent variable explained by the regression on the explanatory variables. The closer the value of R^2 is to one the most valuable is the fitted equation for predicting future values of y. Further details are given elsewhere [16].

A sophisticated computerised method recently described by Svarovsky [17] was employed to obtain grade efficiency curves (Tromp curves) from data collected by the combined technique mentioned above and thus, cut sizes and Stokes numbers were evaluated in order to derive correlations relevant to the study.

RESULTS

As an extensive review of the literature has shown [10], at high concentrations the most correlated variables in hydrocyclone operation are the $Stk_{50}Eu$ group, the underflow-to-throughput ratio and the suspension concentration. Also, as has been previously mentioned, the particle size distribution of the feed powder has a direct effect in the reduced cut size [1,2].

Taking into account the above considerations, by combining equations (1),(2) and (19) a dimensionless groups relation may be derived, i.e.

$$Stk^*_{50}(r) Eu^{\frac{1}{2}} = 8.8 \times 10^{-4} (k_d)^{n+1} \left[\frac{(1.27)^{2-n} \ln(D_o/D_u)^3 D_c^{2n-2}}{(1-c)^{4.65} Q^{n-1}} \right] \quad (23)$$

It can be shown that the above relation can be re-expressed as

$$Stk^*_{50}(r) Eu^{\frac{1}{2}} = 8.8 \times 10^{-4} (k_d)^{n+1} \left[\frac{(1.27)^{2-n} \ln(1-Rf/Rf) D_c^{2n-2}}{(1-c)^{4.65} Q^{n-1}} \right] \quad (24)$$

The correction factor, k_d, is in this case

$$(k_d)^{n+1} = \left[220 X_g [(\rho_s - \rho)/D_c]^{\frac{1}{2}} \right]^m \quad (24b)$$

Equation (24) can take the following form

$$Stk^*_{50}(r) Eu^{\frac{1}{2}} = a \left[(k_d)^{n+1} \right]^{b_1} \left[\frac{(1.27)^{2-n} \ln(1-Rf/Rf) D_c^{2n-2}}{(1-c)^{4.65} Q^{n-1}} \right]^{b_2} \quad (25)$$

So a model like that represented by equation (21) can be used to fit the data, where the sought parameters would be a, b_1 and b_2.

A multiple regression analysis was performed and the results are given in Table 2 below

TABLE 2
Results of regression analysis

Cyclone diam(mm)	Number of experiments	a	b_1	b_2	R^2
22	28	0.0029	-0.373	0.133	0.752
44	40	0.0001	-0.813	1.990	0.911
88	28	0.0007	-2.190	0.443	0.940
All		0.00045	$-20(D_c)$	0.94	0.810

As can be seen in Table 2, a trend in the parameter b_1 (exponent m) can be noticed and, hence, a generalised expression can be represented by

$$Stk^*_{50}(r) Eu^{\frac{1}{2}} = 0.00045 (k_d)^{n+1} \left[\frac{(1.27)^{2-n} \ln(1-Rf/Rf) D_c^{2n-2}}{(1-C)^{4.65} Q^{n-1}} \right]^{0.94} \quad (26)$$

Where

$$(kd)^{n+1} = \left[220 X_g [(\rho_s - \rho)/D_c]^{\frac{1}{2}} \right]^m \quad (26b)$$

And $m = -20 D_c$

The above expression presented a goodness of fit R^2 of 0.810 and its adequacy can be checked further by means of validity tests as those described in the previous section. Figures 3 and 4 show the standardised residuals and normal probability plots respectively for equation (26). As can be seen, the residuals are randomly scattered around the mean value of zero with no systematic pattern indicating that the model is acceptable.

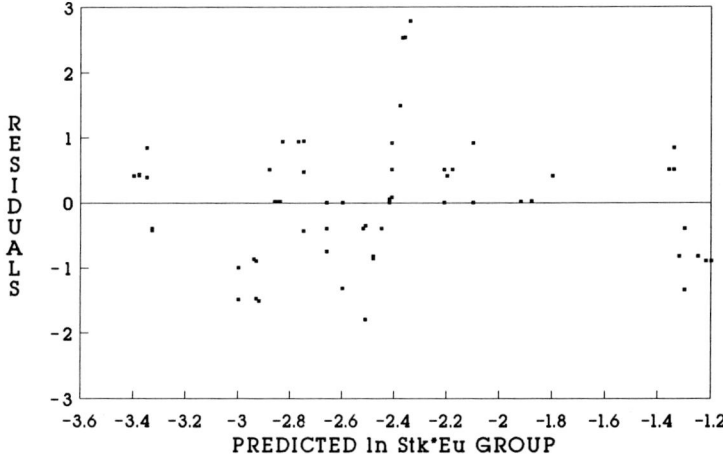

Figure 3. Standardised residuals plot for equation (26)

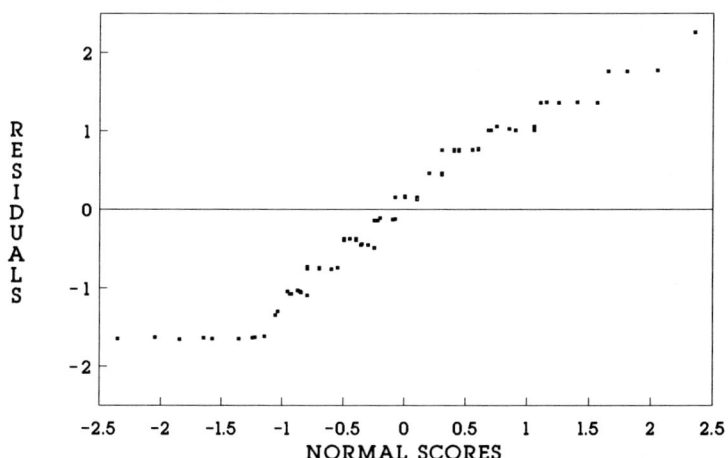

Figure 4. Normal prob. plot of residuals for eq. (26)

In trying to compare equation (26) with other correlations available in the literature, the key component of the $Stk^*_{50}(r)Eu^{\frac{1}{2}}$ group (i.e. $X_{50}(r)$) should be made explicit because dimensionless groups have not been generally employed. Furthermore, apparently, the parameters of characteristion of non-Newtonian slurries have not been used in a context such as ours. Assumptions will be, hence, necessary in order to test the validity of the proposed relation in terms of agreement with other researchers' findings.

The first stage would consist in giving a fixed value to the flow behaviour index n, which was determined by fitting the power law to the experimentally determined rheograms. As can be seen in figure 5 the behaviour exhibited by the curves is dilatant. The goodness of fit of the rheograms was appropriate with r values always over the 0.99 figure. The values of n ranged from approximately 1 to 2 so an average of n equal to 1.5 was taken for the purpose of comparison. Also, a proper equivalent for the viscosity μ, would be the generalised viscosity coefficient γ. Thus, considering that the exponent 0.94 in equation (26) is close enough to one for ease of transposition, by combination of equations (4), (19) and (26) the following expression was derived

$$X_{50}(r) = 0.13(k_d) \left[\frac{\gamma D_c \ln(1-Rf/Rf)}{(1-C)^{4.65}(\rho_s - \rho)(\Delta P/\rho_m)^{\frac{1}{2}}} \right]^{0.4} \quad (27)$$

Where, again, the power of k_d equals the value of $-20D_c$.

To test the adequacy of equation (27) the plot illustrated in figure 6 was obtained. Although the scatter did not improve enough, considering the validity tests performed in equation (26), it can be said that the model was reasonably fitted.

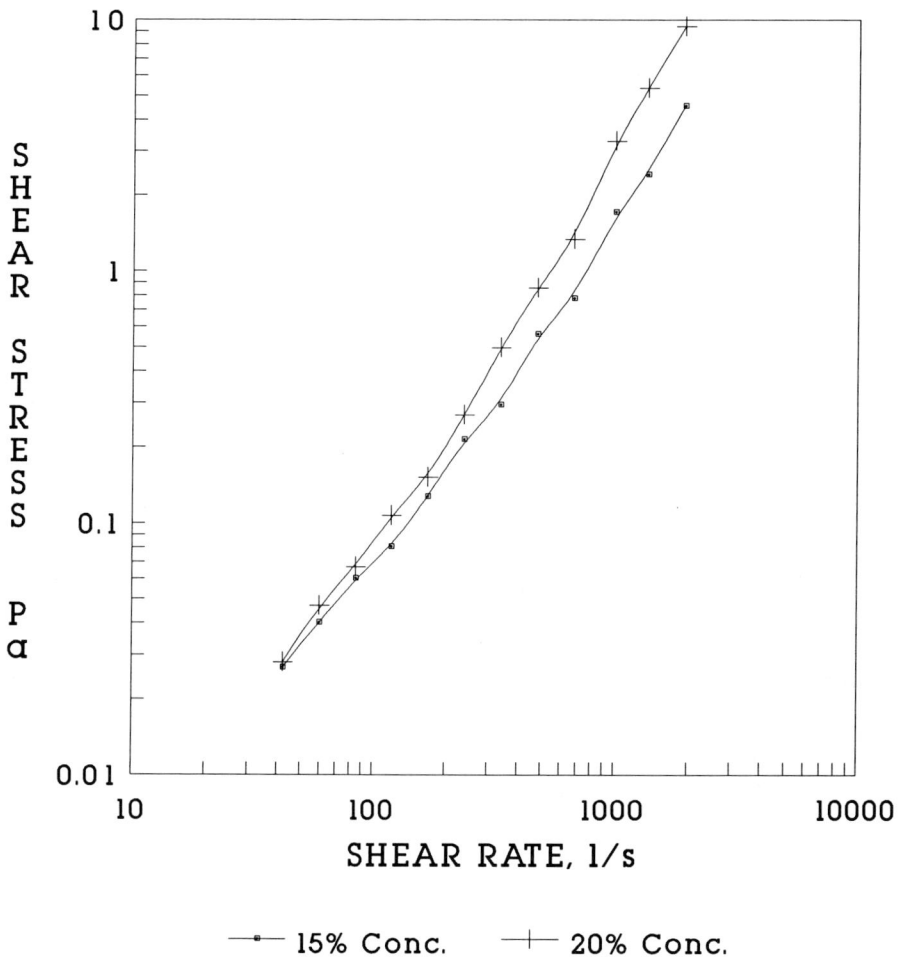

Figure 5. Rheological characterisation of feed slurry.

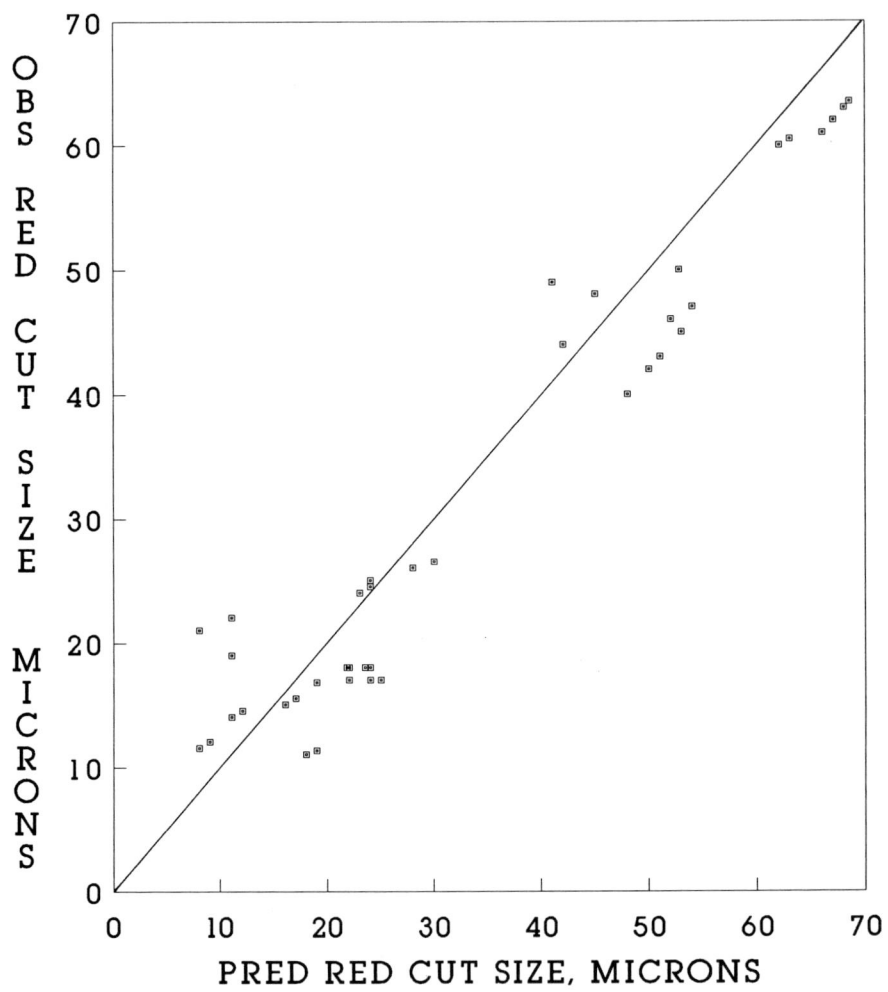

Figure 6. Observed vs predicted value of X50(r) for eq. (27)

DISCUSSION

The agreement between equation (27) and the theoretical model of Schubert, Bohme, Neesse and Espig [6] (equation (1)) is evident. Whereas the value of 0.13 for the constant is virtually the same reported by them, the exponent of 0.4 in this mentioned equation is close to the 0.5 power due to the square root of the term in the original relation of the authors above named. In spite of the claimed agreement, the set of equations developed in the previous section does not completely fit in a scale-up model becuase it presents a dimension. An attempt of modifying equation (27), as a way of matching it to the scale-up model, is given in the following.

The literature presents a number of correlations for cut size as a function of fluid properties [13,18,19,20]. Bradley [21] reduced a series of equations to a common expression including the term $[\mu D_c^3/Q(\rho_s-\rho)]^{\frac{1}{2}}$. Svarovsky [22] proposed later an equation with a concentration correction term for concentrated suspensions, which also includes the above factor obtained by Bradley. Thus, whether the concentration is considered or not, it can be seen that in terms of fluid properties $x_{50} \propto [\mu/(\rho_s-\rho)]^n$.

For the case of this work, taking into account concentration and non-Newtonian behaviour effects, it was found that $X_{50}(r) \propto [\mathscr{I}\rho m/(\rho_s-\rho)]^{0.43}$. Under some assumptions, this factor was compared with the above mentioned factors due to Bradley and Svarovsky, and the results were appropriate [10].

Thus, it is possible to express the influence of the feed material on the cut size as

$$X_{50}(r) \propto \left[\frac{\rho_m}{\rho_s-\rho}\right]^n \qquad (28)$$

Where n is an empirical factor. Also, according to equation (27).

$$X_{50}(r) \propto \left[X_g [(\rho_s-\rho)/D_c]^{\frac{1}{2}} \right]^m \qquad (29)$$

A sort of compromise between the effects due to solids density and particle size distribution may be reached by combining equations (28) and (29) as follows: introducing a correlation constant and equating them, the relation below is obtained

$$\frac{X_g^m \rho_m^n}{D_c^{0.5m}(\rho_s-\rho)^{n-0.5m}} = \text{constant} \qquad (30)$$

Since the values of the powers in equation (30) have been derived empirically, they are able to change in seeking a better fit. Equation (30) indicates that there should be an ideal combination of the ratios shown, which could be used as an alternative to the correction factor, k_d, to account for feed material effects. The presence of the hydrocyclone diameter D_c is important due to its well known direct influence on the reduced cut size. Furthermore, its inclusion keeps the dimensionless feature of this proposed factor represented above. Taking into account all these considerations, equation (30) may be re-arranged to give

$$\left[\frac{X_g}{D_c} [\rho_m/(\rho_s-\rho)]^{\frac{1}{2}} \right]^n = \text{constant} \qquad (31)$$

Where n is an empirically derived exponent.

In order to obtain a dimensionless relation, equation (31) could substitute k_d in equation (26). Also, since our slurries were slightly non-Newtonian, the ratio D_c^{2n-2}/Q^{n-1}, will approximate unity and the factor $(1.27)^{2-n}$, will incorporate to the constant term.

Under all the assumption above considered, equation (26) may be re-expressed as

$$\text{Stk}^*_{50}(r) E_u^{\frac{1}{2}} = a \left[\frac{X_g}{D_c} (\rho_m/\rho_s - \rho)^{\frac{1}{2}} \right]^{b_1} [\ln(1/Rf)]^{b_2} (1-C)^{b_3} \quad (32)$$

Where the term $(\ln(1/Rf))$ is closely related to $(1-Rf/Rf)$

Again, a model like the represented by equation (21) was fitted to our experimental data and the parameters obtained were $a = 2.8 \times 10^{-6}$, $b_1 = -0.55$, $b_2 = 2.44$ and $b_3 = -5.15$ with a goodnes of fit $R^2 = 0.816$. Consequently, the resultant equation is

$$\text{Stk}_{50}(r)^* Eu^{\frac{1}{2}} = 2.8 \times 10^{-6} \left[\frac{X_g}{D_c} \left(\frac{\rho_m}{\rho_s - \rho} \right)^{\frac{1}{2}} \right]^{-0.55} [\ln(1/Rf)]^{2.44} (1-C)^{-5.15} \quad (33)$$

The same test performed in equation (26) to check the validity of the model were carried out in equation (33). Figures 7 and 8 show the patterns obtained, which follow the trends considered normal, so the model may be applicable.

Finally, transposing for the cut size to be explicit in equation (33), and giving the resultant equation a structure resembling equation (1), it can be shown that [10]

$$X_{50}(r) = K(k_d) \left[\frac{\eta D_c [\ln(1/Rf)]^{2.44}}{(\rho_s - \rho)(1-C)^{5.15} (\Delta P/\rho_m)^{\frac{1}{2}}} \right]^{0.4} \quad (34)$$

$$k_d = \left[\frac{X_g}{D_c} [\rho_m/(\rho_s - \rho)]^{\frac{1}{2}} \right]^{-0.22} \quad (34b)$$

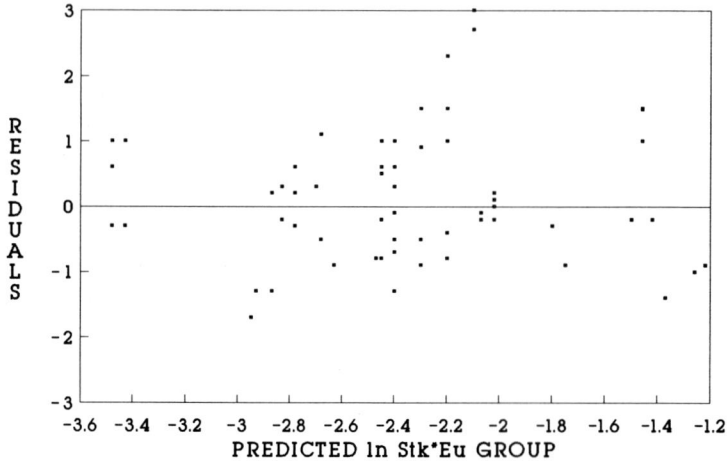

Figure 7. Standardised residuals plot for equation (33)

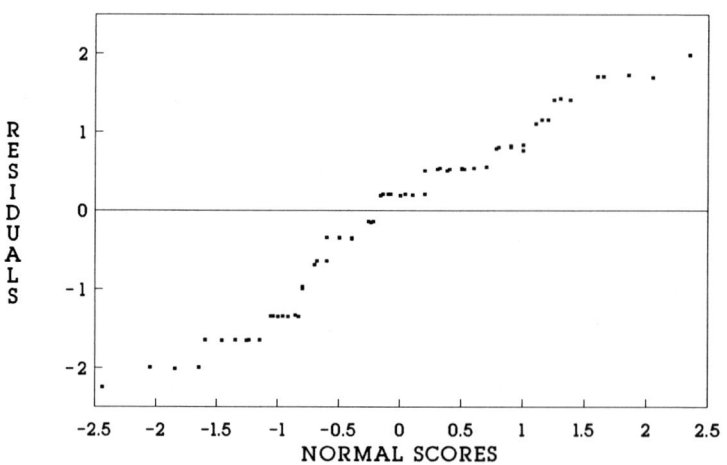

Figure 8. Normal prob. plot of residuals for eq. (33)

Since Rf and D_o/D_u are related (i.e. $(D_o/D_u)^3 = 1-Rf/Rf$), it can be stated that reasonable agreement remains with the model proposed by Schubert, Bohme, Neesse and Espig [6]. The value of the exponent of equation (34) compares well with that of equation (1), while the power of the concentration correction factor remains close enough to the 4.65 in this same equation (1).

Finally, to test further the validity of equation (34) an observed against predicted plot of $X_{50}(r)$ was obtained. This plot is given in figure 9. Since some scatter is still noticeable, the adequacy of the model can be, again, justified in terms of the patterns shown for the tests carried out to equation (33) illustrated in figures 7 and 8.

CONCLUSIONS

The apparent applicability of the equations developed in this work, in relation with the turbulent two-phase flow theory, is quite reasonable. Comparison was only possible with the model proposed by Schubert, Bohme, Neesse and Espig [6], mainly because this topic has been generally neglected in the literature. Even these mentioned authors centre their studies in modelling classification in turbulent two-phase flows and not in the effect of particle size distribution in separation.

The effect of feed particle size on classification has been longely recognised [2] and has been attributed to the turbulence damping effect of the solids [1]. Classification may be considered as the separation of particles according to differences in their settling rates which they follow in a fluid under the action of field forces, drag forces and inertial forces. Since turbulent flow conditions prevail in hydrocyclones turbulent transport phenomena are superimposed which result in a mixing effect impairing the separation efficiency. At high concentrations of coarser particles the crowding of particles in the conical part of the hydrocyclone will result in a sediment formation. In an extreme case this

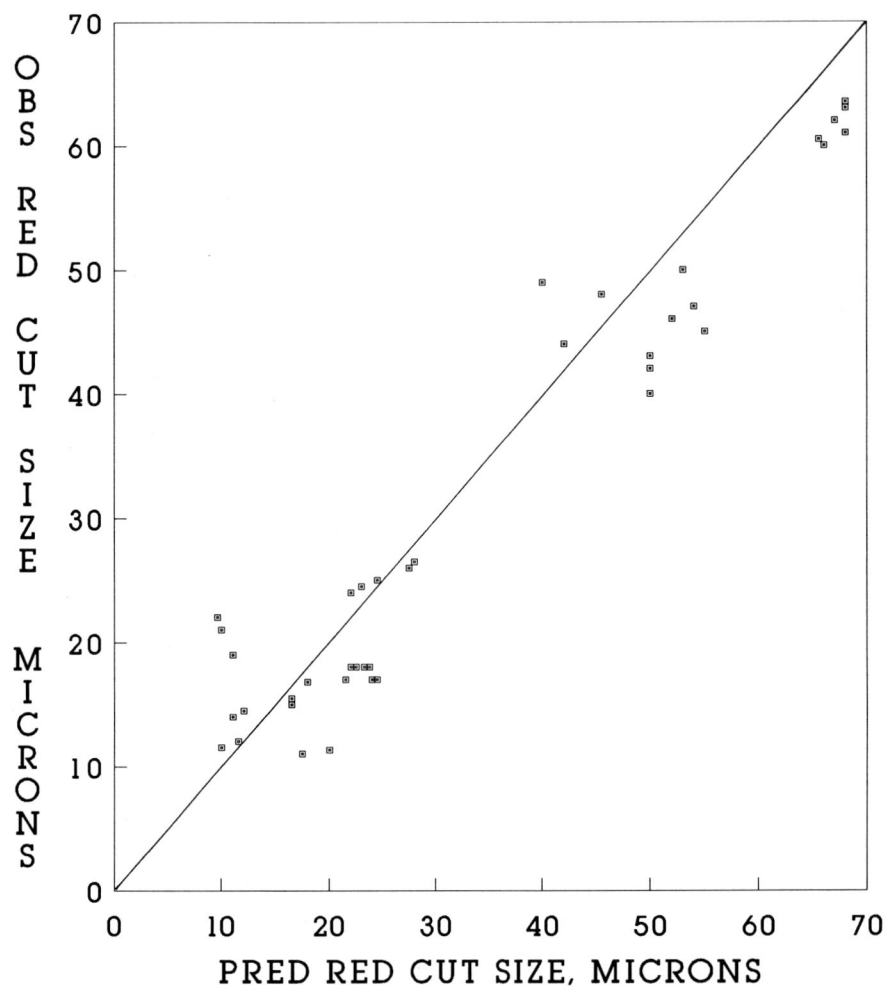

Figure 9. Observed vs predicted value of X50(r) for eq. (34)

will cause a rope discharge and a partial redispersion of sediment.

Considering the above mentioned reasons, it is obvious that a very important role play the design variable D_c as well as the operating variable Rf, combined with the particle size distribution of feed, in classification. Equations (33) and (34), clearly show this influence. The relation of the power of the correction factor with the hydrocyclone diameter was well tested in this work. The effect of turbulent flow and crowding of particles should, therefore, be more drastic in small units with high pressure drops. This situation may be worsen if the feed particles are coarse in general terms. This might be the reason why, for our project, a better fit was obtained in the large hydrocyclone (see Table 2).

The results obtained appear to suggest that the cut size is a function of slurry characteristics, particulary its solids grade, combined with the main geometric feature of the hydrocyclone, its body diameter. For this reason, it is believed that the set of equations presented may be very useful for industrial duties in order to predict with confidence the cut size based on the above mentioned parameters.

Since, as pointed out earlier, there is a dearth of information in the subject, more work might be necessary to test further the usefulness of the model. The main suggestion would be to carefully select the grade of feed powder in relation with the diameter of the cyclone to be tested.

ACKNOWLEDGEMENTS

The work reported here formed part of a PhD project carried out af Bradford University. Thanks are given to its Chemical Engineering Department for the generous use of its facilities, as well as for assistance provided by administrative, technical and academical staff. Project funding in the form of a research grant obtained from the National Council of

Science and Technology (CONACYT, México), is gratefully acknowledged.

The help given by J. Vinicio Torres in preparing some plots and the typing performed by Carmelita Gonzalez are truly appreciated.

REFERENCES

1. Schubert, H. and Neesse, T., A hydrocyclone separation model in consideration of the turbulent multi-phase flow. Part. Sci. and Tech., 1985, **3**, 1-13.

2. Fahlstrom, P.H., Studies of the hydrocyclone as a classifier, Proc. 6th Int. Congress in Mineral Process., Cannes 1963.

3. Bloor, M.I.G. and Ingham, D.B., Theoretical investigation of the flow in a conical hydrocyclone. Trans. Instn. Chem. Engns., 1973, **51**, 36-41.

4. Bloor, M.I.G. and Ingham, D.B., Turbulent spin in a cyclone. Trans. Instn. Chem. Engns., 1975, **53**, 1-6.

5. Mueller, B., Neesse, T. and Schubert, H., Design of a hydrocyclone based on a turbulent model. Freiberg. Forschungsh. A., 1975, **A544**, 31-43.

6. Schubert, H., Bohme, S., Neesse, T. and Espig, D., Classification in turbulent two-phase flows. Aufbereitungs-Technik, 1986, **Nr.6**, 295-306.

7. Svarovsky, L., Hydrocyclones, Holt Saunders, Eastbourne, 1984.

8. Svarovsky, L., Hydrocyclone selection and scale-up. Filtration and Separation, 1981, **18**, 551-554.

9. Svarovsky, L. and Marasinghe, B.S., Performance of hydrocyclones at high feed solids concentrations. BHRA 1st Int. Conference on Hydrocyclones, Cambridge 1980.

10. Ortega-Rivas, E., Dimensionless scale-up of hydrocyclones for separation of concentrated suspensions, PhD Thesis, University of Bradford, Bradford 1989.

11. Metzner, A.B., and Reed, J.C., Flow of non-Newtonian fluids-correlation of the laminar, transition and turbulent-flow regions. A.I.Ch.E. Journal, 1955, **1**(4),434-440.

12. Daneshy, A.A., Numerical solution of sand transport in hydraulic fracturing. *Journal of Petroleum Technology*, 1978, **January 1978**, 132-140.

13. Rietema, K., Performance and design of hydrocyclones. Parts I to IV. *Chem. Eng. Sci.*, 1961, **15**, 298-325.

14. Daniel, C. and Wood, F.S., *Fitting Equations to Data*, 2nd ed., John Wiley & Sons, New York, 1980.

15. Stoodley, K.D.C., Lewis, T. and Stainton, C.L.S., *Applied Statistical Techniques*. Ellis Horwood Ltd., Chichester, 1980.

16. Draper, N.R. and Smith, H., *Applied Regression Analysis*, John Wiley and Sons, New York, 1966.

17. Svarovsky, L., Evaluation of grade efficiency using multiple function curve fittings of particle size distributions. In *Particle Size Analysis 1985*, ed. P.J. Lloyd, John Wiley & Sons, Chichester, 1985.

18. Bradley, D. and Pulling, D.J., Flow patterns in the hydraulic cyclone and their interpretation in terms of performance. *Trans. Instn. Chem. Engrs.*, 1959, **37**, 34-44.

19. Yoshioka, N. and Hotta, Y., Liquid cyclone as a hydraulic classifier. *Chem. Eng. Japan.*, 1955, **19**(12),632-641.

20. Lilge, E.O., Hydrocyclone fundamentals, *Trans. Inst. Min. and Metall.*, 1962, **71**, 285-337.

21. Bradley, D., *The Hydrocyclone*, Pergamon Press, Oxford, 1965.

22. Svarovsky, L., (ed.), *Solid-Liquid Separation*, 2nd. ed., Butterworths, London, 1981.

A CYLINDRICAL HYDROCYCLONE

HUIXIN YUAN
Department of Mechanical Engineering
Wuxi Institute of Light Industry
Wuxi, Jiangsu, P.R. China

ABSTRACT

Hydrocyclones have been evaluated and designed mainly in terms of the "Separation Efficiency" or the "Cyclone Number". As a result, the hydrocyclone is less effective in thickening.

In this paper, an efficiency criterion is found suitable to the hydrocyclone for thickening. In terms of the efficiency criterion, an optimum cyclone thickener with a cylindrical body has been abtained on the basis of experiment. Compared with a commercial conical hydrocyclone, the cylindrical cyclone thickener has high capacity in thickening with low pressure drop across the cyclone body.

NOMENCLATRURE

b	width of inlet
C	mass fraction of solids (by volume)
Dc	cylone diameter
d	orifice diameter
Ek	Centrifugal Efficiency
Et	Total Efficiency
l	cyclone lenghth
lo	lenghsth of votex finder
p	pressure
Q	volumetric flow rate
Rf	volumetric rate of underflow to feed
W	weight

Suffix
c	clarification
i	inlet
o	overflow
t	thickening
u	underflow

INTRODUCTION

A hydrocyclone is a very effective separator. It can find wide application in industries, such as clarification, classification, thickening, washing, etc. But the hydrocyclone, like other kinds of separators, is not universal for every kinds of separation duties. In other words, the hydrocyclone can not perform every duty effectively. Therefore, there is possibly no point in claiming the high efficiency or optimum dimensions of a hydrocyclone regardless of the duties required. In fact, there is an efficiency criterion or optimum dimension of hydrocyclones for each duty. As discussed in appendixes, the definition of efficiency of hydrocyclones used for thickening is quite different from that for clarification.

Hydrocyclones have been designed mainly in terms of the "Separation Efficiency" [1,2] or the "Cyclone Number" Cy50 [3], which, in fact, concern the clarification efficiency. The hydrocyclone, therefore, is a clarifier which can cause rather clean overflow with very wet underflow.

There is often the demand for high content of solids recovered from suspensions in industries. In this case, A thickener is needed to thicken suspensions. For example, to get dry products, a thickener could be used in association with a centrifuge. The thickener can feed the centrifuge with thick suspensions so that the number of centrifuges or the size of centrifuges could be reduced. Even for clarifying of suspensions, the efficiency of a clarifier could be increased by deluting the suspension first with the thickener. In the case where both clean liquid and high content of solids are required, the clarifier and the thickener could be employed in association because both clean overflow and dry underflow can not be got from solid/liquid suspensions with only a clarifier or a thickener.

In industries, a hydrocyclone designed in terms of the "Separation Efficiency" or the "Cyclone Number" is often used for thickening. The operation of thickening, of course, can not be performed satisfactorily as a result of that the hydrocyclone is a clarifier rather than a thickener. This has been recorgnised by young[4]. In order to design a thickener, he used a definition of efficiency which is silica solids removed by the hydrocyclone devided by silica solids at the feed, together with the content of solids in underflow, to evaluate the thickener. The efficiency, in fact, is the "Total Efficiency"[5] which does not represent the net separation effect of hydrocylones. In other words, the dryness of underflow may not be increased by increasing the "Total Efficiency".

The aims of the research are as follows:
1. to find a suitable evaluation of efficiency for a thickener.
2. the cyclone design to have high capacity of thickening.

SEPARATION CRITERIA

To design a hydrocyclone, a criterion is needed. The efficiency criterion is one of the most important criteria to design a hydrocyclone besides the pressure drop across the hydrocyclone.

As detailed in appendixes, the clarification efficiency could be evaluated with Equation (A-2):

$$Ekc = \frac{Ci - Co}{Ci}$$

The efficiency criterion used to evaluate a thickener in this research is defined as:

$$Ekt = \frac{Cu - Ci}{1 - Ci}$$

MATERIALS AND METHODS

Hydrocyclones

A miniature hydrocyclone shown in Figure 1 was explored in this paper. To reduce the cost of manufacture, the bore of the cyclone body was set at 10 mm. Large cyclones designed to avoid the plugging due to large size of solids in suspensions could be abtained by the way of scaling. The inlet enters the cyclone tangentially and is rectangular in shape with the lenghth parallelling to the axis of the cyclone body and the width b being half the lenghth. The hydrocyclone could be assembled interchangebly with pieces machined out of plexiglass in a range of dimensions. The high value and the low value for each of design variables were tested as follows:

$$b/Dc = 0.12, 0.3$$
$$do/Dc = 0.25, 0.4$$
$$lo/Dc = 0.40, 2.5$$
$$l/Dc = 4.00, 7.5$$

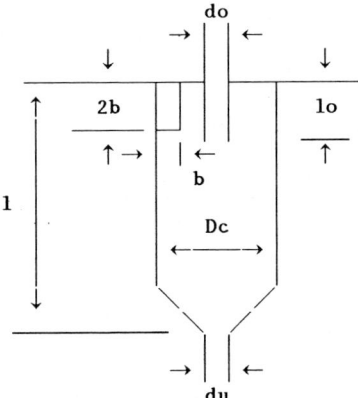

Figure 1. A Experimental hydrocyclone

In addition, a commercial conical hydrocyclone (10mm) was used for comparision.

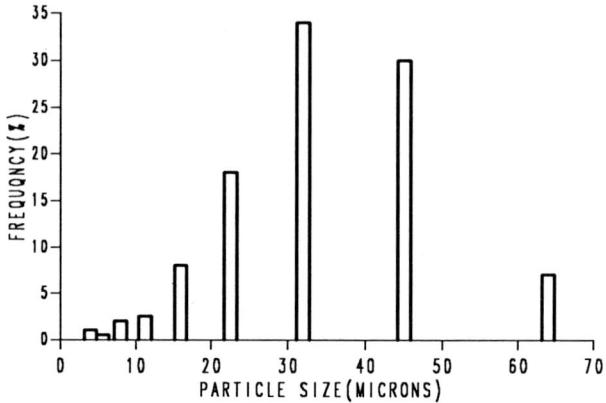

Figure 2. Size distribution of starch particle

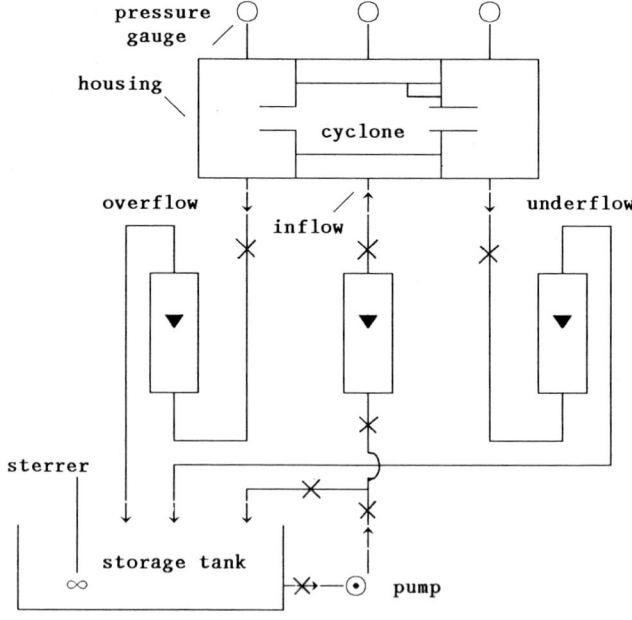

Figure 3. Flow diagram of test rig

Materials

Starch/water suspensions were used in the test. The starch is made out of sweet potatoes and its size distribution is shown in Figure 2. The concentration of the suspension ranges from 1.77% to 2.4%.

Test Rig

As shown in Figure 3, a closed circuit system was used to allow sufficient time for test without very large storage tanks. Hydrocyclones were mounted in a housing which is divided into three champs: inflow champ, overflow champ and underflow champ. The suspension in a storage tank was pumped into the inflow champ where the suspension was fed into the hydrocyclone tangentially and the pressure of feed was measured by a pressure gauge. The overflow of the cyclone went out axially and into the overflow champ where the overflow pressure was monitored. Finally, the overflow was freely discharged into the storage tank. The underflow of the cyclone was freely discharged into the tank too via the underflow champ. The flowrate of the three streams were all measured by rotator flowrators. With the help of a sterrer and the blending effect of return of overflow and underflow, the concentration of the feed in the storage tank could be kept homogeneous and constant. Samples were taken from the outflows of the hydrocyclone for every test condition. The starch content was measured using geometric determination as detailed in appendixes. The results discussed later are based solely on measurements of the starch content of samples taken during the tests.

RESULTS AND DISCUSSION

To find the physical significant of E_{kc} and E_{kt} further, simple performance tests were run respectively on a conical hydrocyclone and a cylindrical hydrocyclone. As shown in Figure 4 through Figure 7, E_{kc} corresponds to the concentration of overflow, C_o, and E_{kt} to the concentration of underflow, C_u. In other words, E_{kc} is clarification efficiency and E_{kt} thickening efficiency. The thickness of underflow could be reflected by the thickening efficiency E_{kt} rather than the clarification efficiency E_{kc}.

In order to find an optimum design of the thickener, a orthogonal test was run. The results of the test are shown in Table 1. According to the range analysis (Table 2) of the results, the importance and selection of design variables in terms of E_{kt} are quite different from that in terms of E_{kc} and E_t.

Firstly, the width of inlet, b, is the most important in all design variables in the cases of both thickening and clarifying. As expected, the smaller the width, the higher the clarification efficiency E_{kc} and the "Total Efficiency" E_t. Consibering the pressure drop across hydrocyclones, $b/d_c = 0.2$ is favourable for a clarifier . In the case of thickening, there are two possible selections of b/d_c: 0.12 and 0.20. The pressure drop, when $b/d_c = 0.20$, is only about one third of that when $b/d_c = 0.12$. So $b/d_c = 0.20$ is favourable for a thickener.

The diameter of the votex finier is secondary of importance in design variables for both clarification and thickening. The effect of thickening, of course, depends mainly on the ratio of overflow orifice d_o to underflow orifice d_u besides the inlet orifice. The large ratio of d_o to d_u may result in high concentration of underflow. The ratio could be increased by

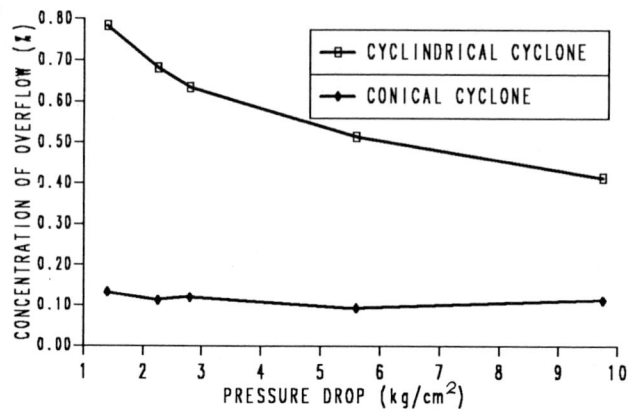

Figure 4. Concentration of overflow vs pressure drop

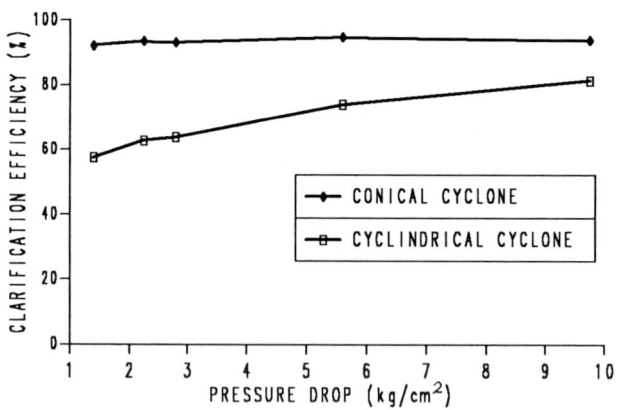

Figure 5. Clarification efficiency vs pressure drop

183

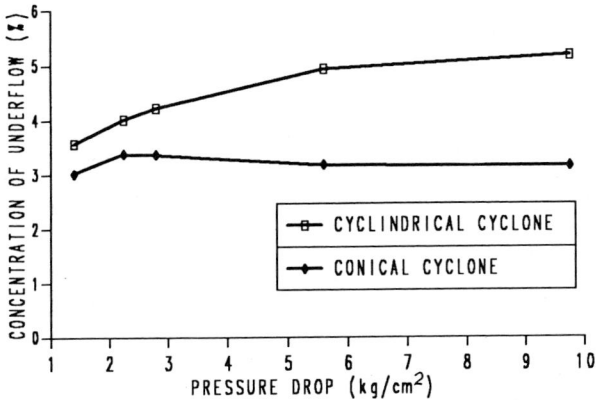

Figure 6. Concentration of underflow vs pressure drop

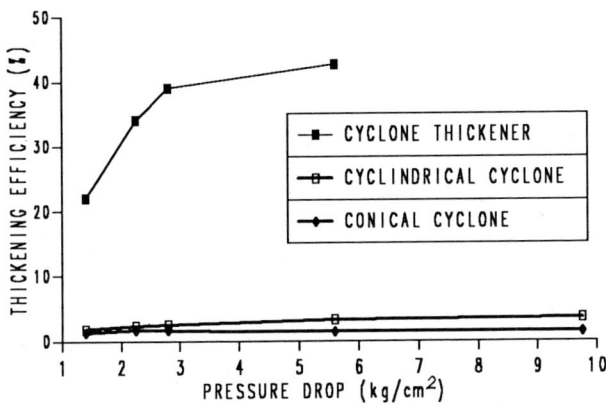

Figure 7. Thickening efficiency vs pressure drop

TABLE 1

Results of the Orthogonal Test

Exp. No.	b (mm)	do (mm)	l (mm)	lo (mm)	Qi (1/h)	Qo (1/h)	Qu (1/h)	Rf	Co (%)	Cu (%)	$\triangle$p (kg/cm^2)	Et (%)	Ekc (%)	Ekt (%)
1	1.2	2.5	40	4	99	70	29	.29	.22	7.5	1.75	91.5	90.8	5.2
2	1.2	3.0	50	8	138	110	28	.20	.17	11.2	3.00	95.1	92.9	9.0
3	1.2	3.5	60	15	198	176	22	.11	.33	19.9	4.80	92.3	86.2	17.9
4	1.2	4.0	70	25	257	244	13	.05	.73	37.0	5.30	78.3	69.5	35.5
5	1.5	2.5	50	15	257	180	77	.30	.19	7.9	6.50	98.5	92.1	5.6
6	1.5	3.0	40	25	198	165	33	.17	.27	13.3	4.45	92.5	88.7	11.2
7	1.5	3.5	70	4	138	115	23	.17	.73	9.2	.80	64.4	69.5	7.0
8	1.5	4.0	60	8	99	80	19	.19	.78	9.5	.50	76.5	67.4	7.3
9	2.0	2.5	60	25	138	95	43	.31	.59	6.2	1.10	78.0	75.3	3.9
10	2.0	3.0	70	15	99	65	34	.34	.57	5.0	.35	71.7	76.2	2.7
11	2.0	3.5	40	8	257	245	12	.05	.59	40.1	2.90	78.4	75.3	38.6
12	2.0	4.0	50	4	198	180	18	.09	.43	24.1	1.35	91.7	82.0	22.2
13	3.0	2.5	70	8	198	120	78	.39	.73	5.2	1.10	85.7	69.5	2.9
14	3.0	3.0	60	4	257	190	67	.26	.71	7.1	2.00	77.1	70.3	4.8
15	3.0	3.5	50	25	99	74	25	.25	1.9	4.1	.15	43.0	18.8	1.7
16	3.0	4.0	40	15	138	120	18	.13	1.2	8.1	.30	44.4	49.0	5.9

TABLE 2

Range analysis of the orthogonal test's results

Efficiency	factors	b	do	l	lo	Qi	Selection and importance of factors major ⟶ minor
Et	I	357.2	353.7	306.7	324.7	282.7	
	II	331.9	336.4	328.3	335.7	281.9	
	III	319.8	278.3	323.9	306.9	362.2	b1, Qi3, do1, lo1, l2
	IV	250.2	290.9	300.1	291.8	332.3	
	R	107.0	75.4	28.2	43.9	80.0	
Ekc	I	339.4	327.7	303.8	312.6	253.2	
	II	317.7	328.1	285.5	305.1	286.7	
	III	308.8	249.8	299.2	303.5	326.4	b1, do1/do2, Qi3, lo1, l1
	IV	207.6	267.9	284.7	252.3	307.2	
	R	131.8	78.3	19.1	60.3	73.2	
Ekt	I	67.6	17.6	60.9	39.2	16.9	
	II	31.1	27.7	38.5	57.8	25.8	
	III	67.4	65.2	33.9	32.1	54.2	Qi4, do4, b1/b3, l1, lo2
	IV	15.3	70.9	48.1	52.3	84.5	
	R	52.3	53.3	27.0	25.7	67.7	

R = Range.

decreasing du. But this may take the risk of plugging the underflow orifice. The ratio of do/du depends on do for the underflow orifice is fixed at 2 mm in diameter in this test.

The small overflow orifice could lead to a fairy recovery of solids in underflow. As a result, both high total efficiency and high clarification efficiency could be achieved. As shown in Table 2, both Ekc and Et reach maximum when do/Dc = 0.25 and do/Dc = 0.20 respectively. It is hardly surprising that a small overflow orifice (do/Dc = 0.25) was got for thickening by Young. As a thickening criterion, the concentration of underflow might have not been weighed in his work.

On the other hand, a large overflow orifice could cause high concentration of underflow at the price of loss in cleanness of overflow. As expected, the thickening efficiency Ekt reaches a maximum when do/Dc = 0.4. It seems that the larger the overflow orifice, the better the thickening. The underflow orifice, however, may be plugged by dry underflow. So the fluidity of the underflow through the outlet must be considered. 40% of starch content in the underflow was still practical under the test condition here. But it may be difficult for the underflow to go out with further increase in the concentration. The value of 0.4 for do/Dc, therefore, is decided as an optimum overflow orifice.

The lenghth of cyclone and the lenghth of the votex finder are less important for both clarification and thickening. A long body of hydrocyclones seems to be unnecessary. The reason for this is probably that there has been sufficient time for most number of solids (above 11.28 microns) to sediment in the short cyclone (l/Dc = 4.0). It is uneconomical for fine solids to be separated from suspensions in thickening. And short cyclone could reduce the costs of construction and the installation space required. A bit longer votex finder (lo/Dc = 0.8) appears favourable for a thickener while a short votex finder (lo/Dc = 0.4) for a clarifier.

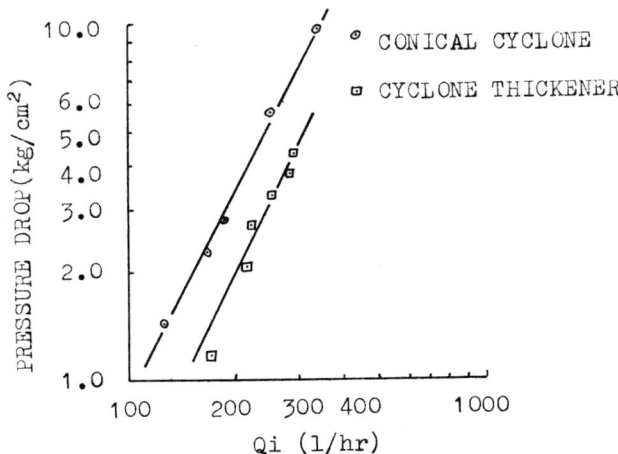

Figure 8. Pressure Drop vs flowrate

As shown in Figure 7, the thickening efficiency increases sharply with increased pressure drop across the cyclindrical hydrocyclone up to about 3 kg/cm². Above the head, pressure drop begins to absorb appreciable power and the efficiency increases slowly. If the pressure supplied is too high, the underflow orifice of the thickener could be plugged with thick underflow and the seal of pumps could be destroied frequently. The food industry has being used 2 kg/cm² to 3 kg/cm² as feed pressure supplied to hygrocyclones.

In summary, an optimum cylindrical clarifier is $b/Dc = 0.2$, $do/Dc = 0.25$, $l/Dc = 5$, $lo/Dc = 1.5$, an optimum cyclone thickener $b/Dc = 0.2$, $do/Dc = 0.4$, $l/Dc = 4.0$ and $lo/Dc = 0.8$.

The cylindrical clarifier could not work as effectively as the conical cyclone in terms of the clarification efficiency (see Figure 5). But the conical cyclone is not qualified to thicken because its thickening efficiency is much lower than the cylindrical thickener's (Figure 7). Besides being higher in thickening efficiency, the thickener is less in power consumption than the conical cyclone (see Figure 8).This optimum experimental cyclone will have a flow rate of 260l/hr at 3 kg/cm² of head. Its thickening efficiency could reach about 40% but the conical cyclone's below 2%.

CONCLUSIONS

The criterion for best thickening performance should be defined as:

$$Ekt = \frac{Cu - Ci}{1 - Ci}$$

rather than the efficiency Ekc, which is the criterion for best clarification performance.

The results of the test work show that a cyclone thickener of the following design provides optimum thickening:

$$b/Dc = 0.2$$
$$do/Dc = 0.4$$
$$l/Dc = 4.0$$
$$lo/Dc = 0.8$$

The optimum 10 mm cyclone thickener has a flow rate of 260l/hr at 3 kg/cm² of head.

ACKNOWLEDGEMENTS

The author would like to thank Shijazhuang Pharmaceutical Factory, P.R. China, for providing the commercial conical hydrocyclone and also Ms Bin Feng for her help with the experiments.

REFERENCES

1. Kimber, G.R. and Thew, M.T., Experiments on oil/water separation with hydrocyclones. Paper E1, 1st European Conf. on Mixing & Centrifugal Sepn, Cambidge, Sept. 1974.

2. Svarovsky, L., Selection of hydrocyclone design and operation using dimensionless groups. Paper A1, 3rd Int. conf. on Hydrocyclones, Oxford, Sept. 1987. Pub. Elsevier for BHRA, Cranfield (1987).

3. Rietema, K., Performance and design of hydrocyclones. Chem. Engng. Sci., 1961, 151, 298-325.

4. Young, G., An experimental investigation of the dimensional and operating parameters of a hydrocyclone in a drilling mud. Paper J1, 3rd Int. Conf. on Hydrocyclones, Oxford, Sept. 1987. Pub. Elsevier for BHRA, Cranfield (1987).

5. Svarovsky, L., Solid-Liquid Separation, 2nd ed., Butterworth, 1981.

6. Kelsall, D.F., A further study of the hydraulic cyclone, Chem. Engng. Sci., 1953, 2, 254.

APPENDIXES

Efficiency Criteria

The "Centrifugal Efficiency" E_k given by Kelsall [6] is defined as the fraction which is separated of the feed material which presents itself for classification. Thus:

$$E_k = \frac{E_t - R_f}{1 - R_f}$$

where

$$E_t = \frac{C_u Q_u}{C_i Q_i}, \text{ i.e. the "Total Efficiency" or "Gross Efficiency"}.$$

R_f is the volumetric ratio of underflow to feed flow.
In terms of the "Centrifugal Efficiency", the efficiency of a clarifier, which is used to deplet solids from suspensions, can be defined as:

$$E_{kc} = \frac{\dfrac{C_u Q_u}{C_i Q_i} - \dfrac{Q_u}{Q_i}}{1 - \dfrac{Q_u}{Q_i}} = \frac{C_u - C_i}{C_i} \cdot \frac{R_f}{1 - R_f} \quad (A-1)$$

or, alternatively:

or

$$Ek_c = \frac{C_i - C_o}{C_i} \quad (A-2)$$

or

$$Ek_c = \frac{C_u - C_o}{C_u - C_o + C_o/R_f} \quad (A-3)$$

Equation (A-2) is known as the "Clarification Number" [5] or the "Separation Efficiency"[1]. And if the content of solids is measured by volume,

$$Ek_c \leq 1 - C_o \quad (A-4)$$

$Ek_c = 1$ provided $C_o = 0$, that is, the overflow is clean liquid, regardless of the amount of the clean liquid. In other words, separation of one drop of pure liquid from suspensions could mean complete separation. It is unreasonable. Nevertheless, the "Separation Efficiency" could be used to evaluate a clarifier because the overflow will never reach pure in the normal operation of the claifier.

Similarily, the efficiency of a thickener, which is used to deplet liquid from suspensions, could be defined as:

$$Ek_t = \frac{\frac{(1-C_o)Q_o}{(1-C_i)Q_i} - \frac{Q_o}{Q_i}}{1 - \frac{Q_o}{Q_i}} = \frac{C_i - C_o}{1 - C_i} \cdot \frac{1 - R_f}{R_f} \quad (A-5)$$

or, alternatively:

$$Ek_t = \frac{C_u - C_i}{1 - C_i} \quad (A-6)$$

or

$$Ek_t = \frac{1 - R_f}{\frac{1 - C_o}{C_u - C_o} - R_f} \quad (A-7)$$

Certainly,

$$Ek_t \leq C_u \quad (A-8)$$

$Ek_t = 1$ when $C_u = 1$, that is, the underflow is completely dry. Like the definition of clarification efficiency, there is a drawback of these definitions, that is, $Ek_t = 1$ so long as the underflow is dry regardless of the amount of the underflow. However, completely dry underflow will never appear in normal operation of a cyclone thickener. The definition of thickening efficiency, therefore, could be used as an efficiency criterion of the thickener.

The thickening efficiency Ekt, theoritically, can be caculated with any one of Equation (A-5) through Equation (A-7). Unfortunately, the value of Ekt caculated from test data with each of these equations appears different as a result of errors from tests and measurement. The Ekt value caculated with Equation (A-7) lies between that with Equation (A-5) and Equation (A-6). Which value of Ekt from these equations is the most reasonable depends on the purpose of the separation. In the case of thickening, the concentration of underflow, Cu, is the most concerned. In order to reflect the extent of thicken thickening, the value of thickening efficiency, therefore, should be caculated with Equation (A-6) rather than Equation(A-5) or Equation (A-7). Otherwise, the error of the concentration of overflow may enlarge the error of the Ekt value even though the O-U approach could reduce the effect of the error of Co to some extent. Similarily, in the case of clarifying, the value of clarification efficiency Ekt should be caculated with Equation (A-2).

Starch Content Determination

All the starch content are measured with a precision analytical balance. The starch content of samples can be caculated from the weight of empty containers, We, the weight of wet samples with container, Ww, and the weight of dry samples with container, Wd. Thus:

$$C' = \frac{Wd - We}{Ww - We} \quad \text{(by weight)}$$

Content C on a volumetric basis can be abtained by converting C' to volumetric basis:

$$C = \frac{1}{1 + \frac{ds}{dw}\left(\frac{1}{C'} - 1\right)} \quad \text{(by volume)}$$

where
    ds is the density of starch
    dw is the density of water.

PREDICTION OF UNDERFLOW MEDIUM DENSITY IN DENSE MEDIUM CYCLONES

T.J. NAPIER-MUNN
I.A. SCOTT[**]
R. TUTEJA[*]
J.J. DAVIS
T. KOJOVIC

Julius Kruttschnitt Mineral Research Centre
Isles Road, Indooroopilly, 4068
Brisbane, Australia

ABSTRACT

The prediction of the underflow medium density, ρ_u, in a dense medium cyclone is important, both because many empirical and phenomenological models of DMC performance incorporate ρ_u as an independent variable, and for plant design purposes. This paper reviews the published correlations for ρ_u, and presents some recent work on the subject. Three approaches have been used: empirical regression equations, correlations based on dimensional reasoning, and methods based on modelling the classification of the medium in the cyclone. The latter method has been successful in some cases, but suffers from a sensitivity to the accurate determination of certain parameters, particularly water recovery. Correlations based on dimensional reasoning give good results within a particular dataset, and show some extrapolative power, but more comprehensive datasets will be required to provide correlations of more general applicability.

[**] Now with AMC Mineral Sands, Perth, Western Australia

[*] Now at the W.A. School of Mines, Kalgoorlie, Western Australia

INTRODUCTION

It is well known that density separations in dense medium cyclones (DMCs) are controlled by the behaviour of the medium. Some of the earliest workers in the field dealt with the performance of DMCs in terms of the medium behaviour. Herkenhoff [1] for example, in studying iron ore separations, noted that the medium classified in the cylone and showed that finer media resulted in lower medium density differentials[1] and increased ore yields to underflow (implying reduced δ_{50}s). Tarjan [2] was the first to consider theoretically the performance of a DMC in terms of the classification of the medium. He defined the density of separation (δ_{50}) as the density of the medium prevailing at the locus of zero axial velocity, and used classification theory to predict the medium density profiles in the cyclone. Medium differential and the density separation were shown to be a function of medium size distribution, pressure drop and cyclone geometry. Lilge [3] also correlated product medium densities with medium size distribution, cyclone operating conditions, and separating performance, and in other papers [4,5] developed a classification theory for the medium and ore which related measurements of shear in the cyclone with the rheological properties of the medium.

Schubert [6] reported data of other workers demonstrating the classification and consequent thickening of a magnetite medium (of density 1500 kg m^{-3}, in a 75mm x 20° cyclone), and cited this as the reason for his observation that $\delta_{50} > \rho_m$. Classification of a magnetite medium was also reported by Khaidakin [7], who observed that the classification size of the magnetite increased substantially as the proportion of fine coal added to the medium increased, leading to reduced segregation of the medium and consequently a change in the separating density of the coal. Sokaski and Geer [8] evaluated the performance of a 254mm x 20° cyclone in the treatment of a number of coals using magnetite media. They found that fine media gave sharper separations (less misplaced material) than coarse media, although cut-points did not vary.

Since DMCs operate at high total solids loading (particularly high density operations using ferrosilicon media), one might expect the separation to be influenced by the solids-handling capacity of the apex, a concept used to model classifying hydrocyclones by Fahlstrom [9]. Cohen and Isherwood [10] and Upadrashta and Venkateswarlu [11] both invoked mechanisms of this kind for DMCs.

Three groups of workers [11-13] have reported data which suggest an equivalence between the separating density, δ_{50}, and the underflow medium density, ρ_u:

---

[1] The differential reflects the instability of the suspensoid medium. Two common definitions are recognised: $\Delta_{u/f} = \rho_u - \rho_f$ $\Delta_{u/o} = \rho_u - \rho_o$

$$\delta_{50} = \rho_u \tag{1}$$

Collins et al [13] noted that this relation only applied under 'normal' or non-segregating conditions, in which the underflow/overflow density differential, $\Delta_{u/o}$, remained in the range 200-500 kg m^{-3}, with an optimum at 400 kg m^{-3}.

Data obtained by Napier-Munn [14,15] on diamondiferous ores with a 610 mm DMC of Dutch State Mines (DSM) design and ferrosilicon media under production conditions did not confirm eqn (1); in this case the ore yields to underflow were generally low. He developed an arbitrary regression model [15] to describe the separation in terms of medium and ore characteristics and operating conditions. Subsequent data obtained by Napier-Munn with ferrosilicon in a 104mm cyclone using density tracers (i.e. at very low 'ore' loadings) were found to be well described by the expression [14]:

$$\delta_{50} = a + b \rho_f + c \rho_u \tag{2}$$

in which the constants a, b and c are determined empirically. The same equation was used by Davis [27] and by Davis and Napier-Munn [16] to describe tracer separations at lower density using magnetite media, and Wood [17] used a similar expression (with ρ_f being replaced by ρ_o) to describe full-scale coal separations:

$$\delta_{50(4\times2)} = a + b \rho_o + c \rho_u \tag{3}$$

Scott [18, 19] has developed a more comprehensive model of DMCs in terms of the 'pivot' phenomenon, in which size-by-size partition curves are observed to pass through a single point defined by two parameters, the pivot density, ρ_p, and the pivot partition number, Y_p, which are characteristic of the separation. Scott developed regression models for these and two other separation parameters [15, 18, 19] in terms of cyclone geometry, medium properties, operating conditions and $\Delta_{u/o}$.

The phenomenological or entirely empirical models offered by all these workers require a knowledge of one or both product medium densities in order to predict the parameters of the density separation. There arises therefore a need for models to predict the medium behaviour under a particular set of operating conditions. This paper reviews some previously published correlations, and introduces two new procedures developed recently in DMC research at the JKMRC.

EXISTING MODELS FOR MEDIUM UNDERFLOW DENSITY

Background

The medium underflow density is determined by the mass flows of medium solids and carrying fluid (water plus ultrafine contaminants) reporting to the cyclone spigot. These in turn are a function of the vortex and axial flows developed in the cyclone and the dimensions of the air core and spigot, together with the segregation of the (fine) medium solids in the carrying fluid and the nature of the boundary layer flow developing adjacent to the wall of the conical section of the cyclone. These flows can be disturbed and modified by the presence of significant quantities of (coarse) ore or coal reporting to the sink product, which may be thought of as displacing a proportion of those medium particles having a lower settling velocity in the fluid, the proportion increasing with the ratio of respective settling velocities.

Whilst the application of the methods of computational fluid dynamics to the analysis of this problem is likely to yield useful results in the long term, the attempts to date have served mainly to emphasise the complexity of this approach, and the limitations of some of the assumptions required. Particular problems include the preferred need to operate in three dimensions (though two may be acceptable), the need to handle 3 phases (liquid, fine medium solids and coarse ore/coal solids), and the characterisation of medium viscosity and its prediction in different parts of the cyclone.

Accordingly, the limited number of predictive methods published to date have been empirical or phenomenological. There are three main approaches:

- Prediction of the classification or sedimentation of the medium
- Correlations based on dimensional reasoning
- Entirely empirical.

The published correlations will now be reviewed.

Published correlations for underflow medium density, ρ_u

Napier-Munn [15,20] reported an empirical correlation for underflow density for a constant geometry 610mm cyclone using ferrosilicon media with feed densities in the range 2600-2950 kg m^{-3}. Based on 43 tests at a range of feedrates, pressure heads and ferrosilicon size distributions, the following regression equation was obtained:

$$\rho_u = -0.350 + 0.992 \, \rho_f + 0.000675 \, F + 0.222 \, (H/\rho_f) \\ - 0.0078 \, (M_A)^{(1/M_B)} \quad (4)$$

($R^2 = 0.904$; n = 43)

The goodness of fit is illustrated in Figure 1.

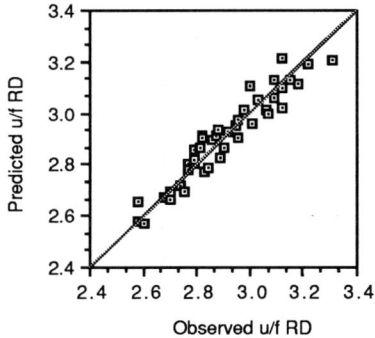

Figure 1. Observed vs Predicted u/f RD for eqn 4 (Napier-Munn [15,20])

The equation form is arbitrary and valid only over the range of variables investigated. However, within this range ρ_u is seen to vary linearly with feed density and head. The dependence on feedrate and M_A is small, but the dependence on the gradient of the medium size distribution, M_B, is strong up to a value of $M_B \approx 2$, whereafter ρ_u is relatively constant. Thus a wider size distribution reduces the underflow density.

Upadrashta and Venkateswarlu [11] published an empirical expression for ρ_u, based on 28 tests with a 100mm cyclone operating on magnetite ore with a ferrosilicon medium:

$$\rho_u = 4.32 - 10.4\,(D_u/D_o) + 1.88\,\rho_f + 1.16\,P_i + 0.48\,(D_u/D_o)\,\rho_f$$
$$-0.22\,(D_u/D_o)\,P_i - 0.10\,\rho_f P_i + 4.74\,(D_u/D_o)^2$$
$$-0.31\,\rho_f^2 - 0.18\,P_i^2 \tag{5}$$

The goodness of fit is shown in Figure 2.

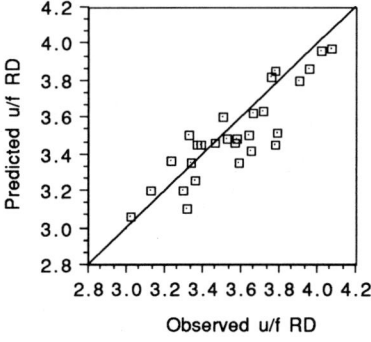

Figure 2. Observed vs Predicted u/f RD for eqn 5 (Upadrashta & Venkateswarlu [11])

Response surface analysis suggested that ρ_u increases with feed density and inlet pressure (as for eqn 4), but at low values of the ratio D_u/D_o, ρ_u is high and relatively unaffected by feed density and pressure. As D_u/D_o increases, ρ_u declines until again it is constant and independent of ρ_f and P_i.

Wood's empirical model of DSM-type production scale dense medium cyclones in coal washing applications [17,21] includes an expression for the underflow medium density, the latest version of which is [21]:

$$\rho_u = 0.459 \, \rho_f \left(\frac{Q_u}{Q_f}\right)^{(0.194(\rho_f - 2.04))} M_A^{0.17} \, H/D_c^{0.082} \, D_c^{-0.10} \qquad (6)$$

The quality of fit to the data is demonstrated in Figure 3.

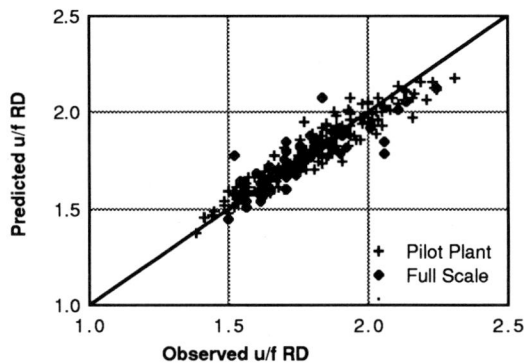

Figure 3. Observed vs Predicted u/f RD for eqn 6 (Wood [21])

Again, ρ_u is seen to increase with feed medium density, but the dependence on head is complicated by the fact that Q_u is inversely related to head.

Equations 4-6 are empirical in nature, regression equations incorporating those operating and design variables thought to be important in determining ρ_u. Napier-Munn [14] suggested the following relation from dimensional reasoning:

$$\frac{\rho_u}{\rho_f} = K \, (1-C_v)^\alpha \, (Re_i)^\beta \, L^\gamma \qquad (7)$$

where   L = $\dfrac{P_i}{0.5 \, \rho_f \, v_i^2}$ (pressure loss coefficient, or the pressure drop, P_i, expressed as number of inlet velocity heads)

and Re_i = $\dfrac{\rho_f \, v_i \, D_i}{\eta_{a(min)}}$ (inlet Reynolds number)

For a 104 mm cyclone using milled ferrosilicon media in the density range 2043 - 3053 kg m^{-3}, the values of the constants were found to be:

K = 0.804
β = 0.115
α = 1.735
γ = -0.372
(R^2 = 0.950, n = 24)

The goodness of fit is illustrated in Figure 4.

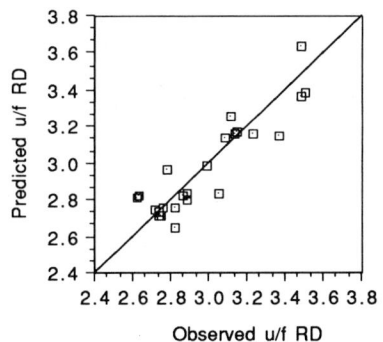

Figure 4. Observed vs Predicted u/f RD for eqn 7 (Napier-Munn [14])

The normal pressure-flowrate relationships [22] substituted into eqn 7 indicate that the thickening dependency on flowrate (or more particularly on the inlet velocity, v_i) is relatively small. Eqn 7 also reveals a dependency of ρ_u on medium viscosity, the underflow density falling as viscosity increases (other things being equal).

In the same study, Napier-Munn [14,23] showed that the bulk hydrocyclone classification model of Holland-Batt [24] could be adapted to predict the concentration of medium solids at the cyclone overflow and hence, knowing the recovery of medium to underflow, the underflow medium density:

$$(\rho_u - \rho_l) = (\rho_f - \rho_l) \left[\dfrac{1}{R_m} + \left(1 - \dfrac{1}{R_m}\right) \cdot \exp\left[A - \dfrac{hz}{Q_f} B\right] \right] \qquad (8)$$

Unfortunately the model could not be tested against data as the sedimentation property, z, was not measured, but back-calculating z from eqn 8 (other terms being known) gave plausible results based on other measurements of ferrosilicon media sedimentation rates. Other authors, however, have not been able to apply eqn 8 successfully [18,30].

Interestingly, both eqns 7 and 8 can predict negative differentials ($\rho_u/\rho_f < 1$), and no special assumptions need to be invoked. This phenomenon has been observed experimentally [23,25]. It arises as a consequence of an unusual combination of a low flowrate, low medium sedimentation rate (fine particles, high concentration), high medium viscosity, and appropriate cyclone geometry (including a wide cone angle), and leads to unusually shaped partition curves for the density separation.

As noted earlier, several authors have considered the medium behaviour in terms of its classification in the cyclone. Napier-Munn's use of the Holland-Batt model utilises this approach, even though the quality of the classification observed with the ferrosilicon media at high solids concentration was generally poor [14], probably because of the high solids concentrations prevailing (typically 25-35% v/v). Indeed the Holland-Batt model can be interpreted in this context as describing a bulk sedimentation mechanism. Certainly Napier-Munn's data showed no evidence of the correspondence between water and fines recovery found in conventional hydrocyclones, and it was concluded that this phenomenon was peculiar to dense medium separations using dense solids (eg ferrosilicon) [14]. In such separations high density underflows are the norm, and it is likely that crowding mechanisms prevail and entrainment of all sizes in the underflow product is possible.

In coal separations, however, in which a lower density solid (magnetite) is used to make up lower density media, there is evidence that the medium classifies as in conventional hydrocyclones. Recently Vanangamudi *et al* [26] showed that magnetite was classified normally in a coal-cleaning Vorsyl separator, as long as the vortex finder length was not too short. Water flowrate to overflow was a linear function of feed water flowrate, and the conventional concept of a corrected (and reduced) efficiency curve was found to be applicable to the magnetite classification. However the authors did not utilise this observation to predict the density of the underflow medium.

Earlier, Davis [27] had studied the segregation of magnetite media in a 200 mm DM cyclone of DSM design, without coal feed. He showed that the magnetite did classify, and the resulting efficiency curve could be fitted to the Whiten classification function [28]:

$$E = C\left(\frac{e^\alpha - 1}{e^{\alpha d/d_{50c}} + e^\alpha - 2}\right) \qquad (9)$$

Typical efficiency curves are shown in Figure 5. Davis noted that highly segregated medium (producing a large density differential) was generally correlated with distinct classification curves and correspondingly high values

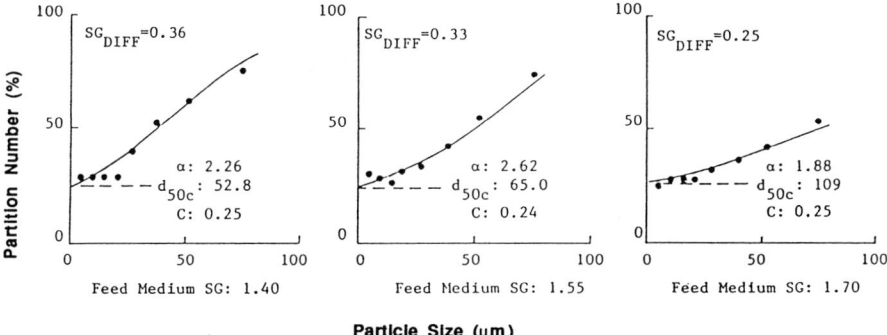

Figure 5. Typical efficiency curves for classification of magnetite in a 200 mm DSM cyclone (Davis [27])

of the efficiency parameter, α. In attempting to model the classification, he observed that the water split to overflow, C, must be determined to a high degree of precision to predict satisfactorily the resulting medium underflow density. For example, with a water split of 90% and an underflow relative density of 1.80, C must be determined to ± 0.6% relative to restrict the error in estimating ρ_u to ± 0.03 RD units. Unfortunately, in hydrocyclone classification C is usually the least well defined classification parameter, and Davis found that it was not possible to develop a regression equation for C with the necessary precision, which he attributed to "severe non-linearities in the data with apex diameter and feed medium specific gravity." Accordingly he utilised an empirical expression to predict underflow density directly from the feed density, feed pressure and magnetite size (normalised to the cyclone diameter):

$$\rho_u = \rho_f + K_0 + K_1 (D_{50}/D_c) + K_2 H/D_c - K_3 \rho_f \qquad (10)$$

where $K_0 ... K_3$ are constants to be determined from the data. Different sets of constants were estimated by regression analysis for four different apex sizes, in the range:

$$\begin{aligned}
D_u &= 0.25\, D_c - 0.40\, D_c \\
K_0 &= -0.102 - 0.316 \\
K_1 &= 1790 - 5055 \\
K_2 &= 0.01 - 0.038 \\
K_3 &= 0.054 - 0.452.
\end{aligned}$$

The quality of fit for all four data sets is illustrated in Figure 6.

Napier-Munn [14], Davis [27] and Scott [18] have all observed that the medium classification responded as would be expected to the influence both of solids concentration and liquid viscosity, the d_{50} increasing with increase in these variables, and the classification efficiency declining.

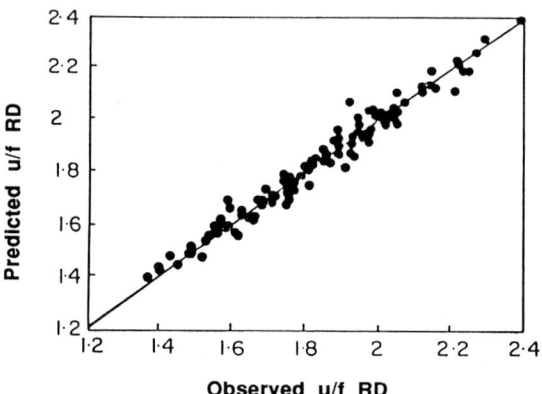

Figure 6. Observed vs Predicted u/f RD for eqn 10 (4 data sets plotted) (Davis [27])

Davis and Napier-Munn [16] obtained an expression for the underflow medium density for magnetite media in Davis' 200 mm cyclone in terms of the medium viscosity, which was varied both by adjusting the medium density and by the addition of clay:

$$\rho_u = 0.196 \, \rho_f^{1.131} \, (\eta_a / \rho_f)^{-0.093} \, (D_u/D_c)^{-0.325} \qquad (11)$$

$$(R^2 = 0.902, n = 12)$$

where η_a = apparent equivalent Newtonian viscosity (Nsm^{-2} x 10^3) in the experimental range 5-70 Nsm^{-2} x 10^3, measured by a Debex on-line rotational viscometer [29]. The goodness of fit is shown in Figure 7.

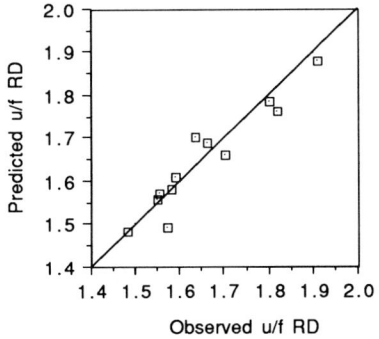

Figure 7. Observed vs Predicted u/f RD for eqn 11 (Davis & Napier-Munn [16])

This equation decouples the effects of medium density and viscosity (which are strongly correlated), and demonstrates that the underflow density falls as the kinematic viscosity increases. ρ_u also falls as the apex diameter increases, diverting more water to the underflow.

There is thus some consensus amongst the published correlations as to the influence of design and operating variables on the underflow medium density. Underflow density increases with feed density and medium flowrate (usually expressed as head of medium), and is also influenced by the size distribution of the medium; a coarser media with a narrow size distribution increases the value of ρ_u. The density falls as the ratio D_u/D_o increases. An important conclusion is that a high medium viscosity decreases ρ_u. The dependency of ρ_u on medium density and viscosity and on apex diameter is illustrated in Figure 8 (from Davis and Napier-Munn [16]). It should be noted that in either limit there is a tendency for ρ_u to reach a maximum or a minimum value. The maximum is attained when the solids-handling capacity of the apex is reached, which can occur for example even at low feed densities when the segregation effect is strong [13]. When medium viscosity is high, the apex is large and/or the medium solids are fine, the underflow density and overflow density both approach the value of the feed density, i.e. the differential approaches zero, and no segregation occurs. Either extreme is known to be deleterious to the quality of the associated density separation, and in practice there is agreement that for conventional separations in which the split of ore or coal between overflow and underflow is relatively even, the optimum range of the differential $\Delta_{u/o}$ is 200-500 kgm^{-3} [13, 27].

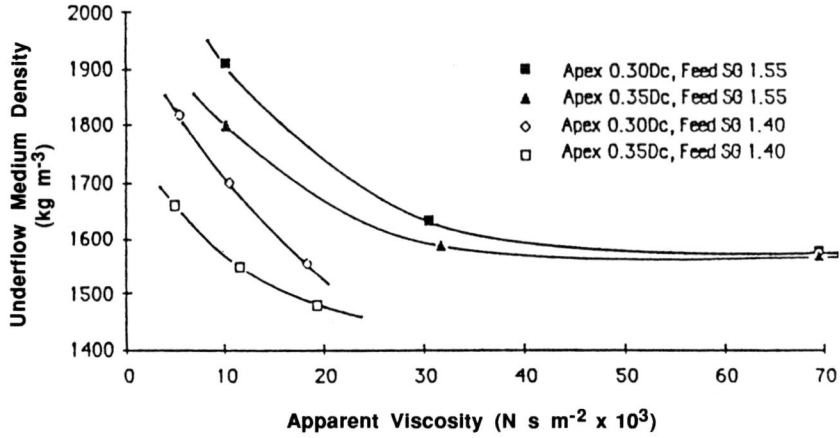

Figure 8. Dependency of u/f density on apex diameter and medium density and viscosity for magnetite in 200 mm cyclone (Davis & Napier-Munn [16])

NEW CORRELATIONS FOR UNDERFLOW DENSITY

Scott [18] evaluated several approaches to predicting underflow density. He found that Holland-Batt's modified bulk classification model (eqn 8), utilising a medium sedimentation rate determined by the Debex stability meter [29] was not successful in predicting ρ_u. Nor was an empirical expression, also incorporating sedimentation rate as an independent variable characterising the tendency of the medium to segregate. He therefore considered two alternatives: a classification model, and another empirical model in terms of pulp viscosity, feed density and cyclone dimensions.

The classification model was developed on the basis of 10 data sets obtained from a 200 mm gravity-fed DSM cyclone using ferrosilicon media, with molasses added to vary the liquid viscosity. The 'ore' loading was negligible, and the apex diameter was fixed at 0.35 D_c. The static head was 3.2 m (or 16 diameters), and feed medium densities ranged from 1400 to 2700 kg m^{-3}. The classification curve partition numbers were determined from mass balanced medium size distributions (obtained with a Microtrac laser-dispersion particle size analyser) and mass flows, and the resulting efficiency curves are shown in Figure 9. The partition data were fitted to the Whiten function (eqn 9) but in terms of recovery to underflow, i.e. with C replaced by (1-C) (water recovery to underflow).

The parameters α and d_{50c} were then determined by a weighted least squares non-linear parameter estimation procedure, using values of (1-C) obtained from mass balanced experimental data rather than by fitting, owing to the uncertainties in determining the very fine size partition numbers; values of

(1-C) were in the range 20-31%. The fitted curves are also shown in Figure 9. A summary of the data is given in Appendix 1.

No systematic variation in the efficiency parameter, α, was evident in the data. The mean value was 2.75 (s.d. 0.67) and all but one result lay within two standard deviations of the mean. α was therefore assumed to be constant at 2.75, and correlations were sought for the other two classification parameters, (1-C) and d_{50c}. The cyclone geometry and head were constant in these experiments, and the only operating variables found to be significant were medium viscosity and volume concentration of solids. The regression equations were:

$$100\,(1-C) = 57.4\,\eta_m^{0.29}\,V_f^{-0.09} \qquad (12)$$
$$(R^2 = 0.88, n = 10)$$

$$d_{50c} = 2724\,\eta_m^{0.85}\,V_f^{-0.19} \qquad (13)$$
$$(R^2 = 0.86, n = 10)$$

Knowing α, C and d_{50c}, eqn 9 can then be used to calculate the total water and medium solids reporting to overflow and underflow, and hence the product medium densities (remembering that the density of the water was modified by the addition of molasses, and ranged from 1054 to 1315 kg m^{-3}). The underflow density predicted by this method is compared with the observed data in Figure 10. The agreement is seen to be good, the RMS prediction error being only 53 kg m^{-3}.

An alternative expression of similar predictive power was obtained for ρ_u directly, using 20 data sets from 100 mm and 200 mm DSM cyclones with varying apex diameters:

$$(VFR_u - 1) = 0.43 \times 10^{-4}\,\eta_m^{-1.52}\,V_f^{-0.35}\,(D_u/D_c)^{-2.91} \qquad (14)$$

where

$$VFR_u = \frac{V_u}{V_f} = \frac{\rho_u - \rho_l}{\rho_f - \rho_l}$$

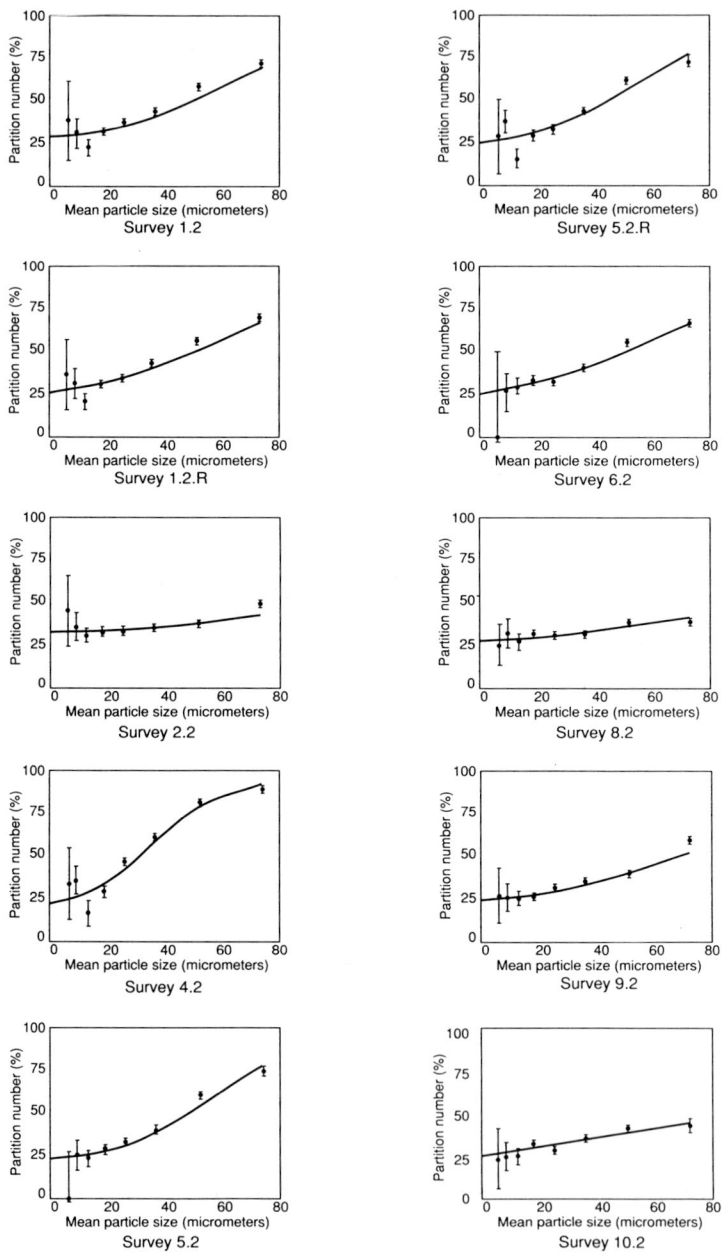

Figure 9. Classification of ferrosilicon in a 200 mm DSM cyclone (Scott [18])

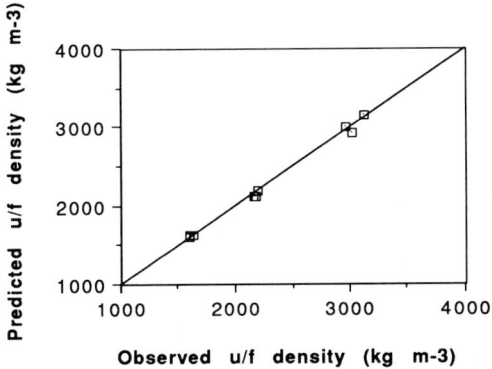

Figure 10. Observed vs Predicted u/f density for classification model, eqns 12 & 13 (Scott [18])

The quality of the fit is shown in Figure 11; the RMS error was 46 kg m^{-3}.

Tuteja [30] has recently applied the general form of eqn 14 to the 12 magnetite data sets reported by Davis and Napier-Munn [16] (previously correlated by eqn 11), and showed that the general form was appropriate, though the values of the exponents were somewhat different.

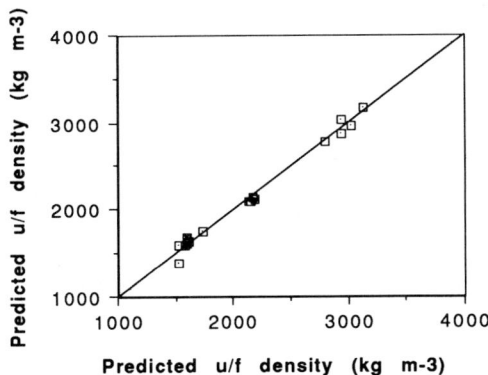

Figure 11. Observed vs Predicted u/f density for eqn 14 (Scott [18])

Tuteja also applied the hydrocyclone classification model developed by Nageswararao [31,32] to the 97 data sets of dense medium magnetite classification reported by Davis [27]. This model expresses the key performance variables (including the efficiency curve) in terms of cyclone geometry and operating conditions, in dimensionless regression equations incorporating single constants specific to the material being treated. These constants are usually determined from testwork. Tuteja found that the Nageswararao model fitted the magnetite data well, but the estimated 'constants' varied in a systematic way within Davis' experimental plan. Further work will be required to establish the cause of this variation (believed to be associated with solids concentration effects) and to account for it in the correlations.

Finally, Tuteja [30] applied a dimensionless correlation suggested by Napier-Munn to the two pilot plant data sets obtained by Davis [27] and Scott [18] described earlier in this paper, totalling 32 data sets on 100 mm and 200 mm DSM cyclones, in which medium density, medium viscosity and apex diameter were varied. The following equation resulted:

$$(\rho_u/\rho_f) = 1.872 \, Re_i^{0.0891} \, (D_u/D_c)^{-0.0872} \, (D_o/D_c)^{1.919} \tag{15}$$

where

$$Re_i = \frac{\rho_f \, v_i \, D_i}{\eta_a}$$

($R^2 = 0.95$; n = 32)

The quality of fit is shown in Figure 12.

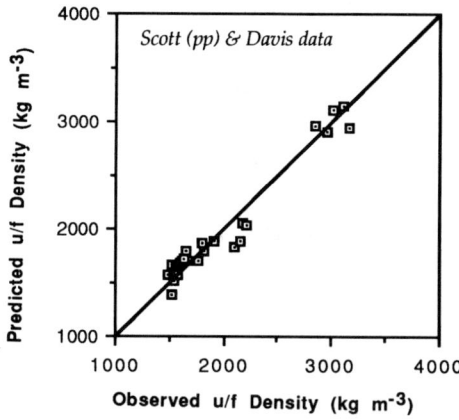

Figure 12. Observed vs Predicted u/f density for eqn 15 (Tuteja [30])

In order to test this correlation against real plant data, it was used to predict the medium underflow density for 20 data sets independent of those used to develop eqn 15: 8 data sets obtained by Tuteja from a 358 mm DSM cyclone at Hamersley Iron's Tom Price Concentrator [30] and 12 data sets obtained by Scott from 200 mm and 400 mm DSM cyclones at the Mount Isa Mines lead/zinc preconcentration plant [18]. In all these cases, medium viscosity was determined by Debex viscometer. The results are shown in Figure 13, from which it is apparent that although the trends are well predicted, there is a strong offset in predicted value which appears to be related to cyclone diameter (Figure 14). Figure 15 was prepared using the actual mean errors (shown in Figure 14) to correct the predictions of eqn 15, which suggests that if these errors can be used universally to correct eqn 15 for cyclone diameter, then this equation can provide satisfactory predictions of medium underflow density in operating plants. However validation of this hypothesis is clearly required.

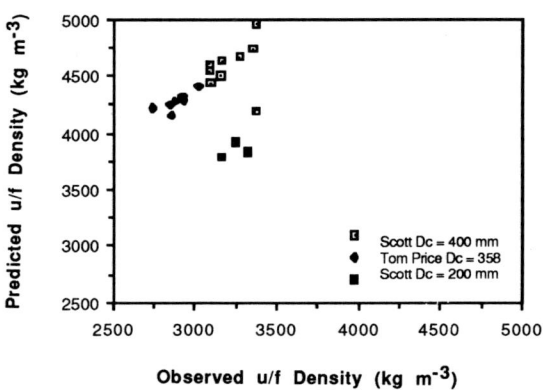

Figure 13. Observed vs Predicted u/f density for eqn 15 applied to production data

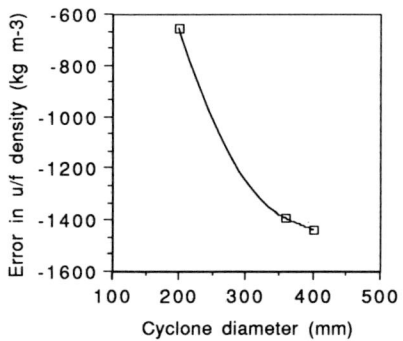

Figure 14. Predictive error in eqn 15 vs cyclone diameter

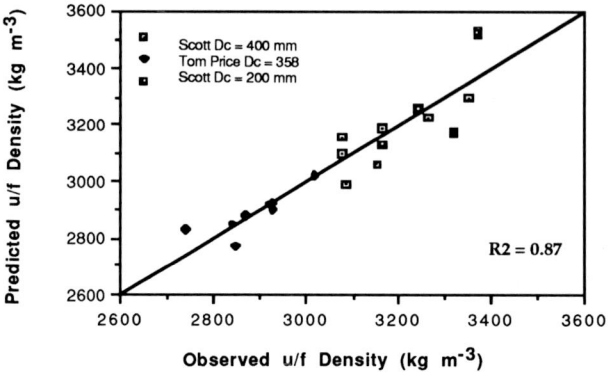

Figure 15. Observed vs Predicted u/f density for eqn 15 on production data, with correction applied

CONCLUSIONS

Three methods have been used to model the underflow medium density prevailing in a DMC under a particular set of conditions:

- empirical regression equations
- correlations based on dimensional analysis
- procedures based on predicting the classification of the medium.

Several authors have observed the classification of the medium in the cyclone and have shown that it can be treated in a similar way to classification in hydrocyclones. Many of the trends are those expected from hydrocyclone classification; for example, cut-size increases and efficiency decreases with increase in solids concentration and liquid viscosity. However some anomalies remain. For example, there is some evidence that, particularly at high solids concentrations (eg ferrosilicon media at pulp densities exceeding 2700 kg m^{-3}), the usual correspondence between water and fine particle recovery to underflow is lost, and therefore that the usual bypass mechanism is suppressed. Highly segregated medium usually corresponds to more distinct curves and more efficient classification. Certainly the fitting of conventional classification functions (eg eqn. 9) to the efficiency curves is very sensitive to the determination of water recovery to underflow. This parameter is often not estimated to a sufficient precision to obtain adequate predictions of ρ_u by using the classification model to predict solids and water splits to underflow. However Scott [18] did obtain good predictions using such a model, with the value of the efficiency curve parameters being determined empirically (eqns. 12 and 13).

Several expressions incorporating dimensionless groups have been found to correlate individual experimental datasets well. The most fundamental of these describes the thickening effect as a function of flow and energy terms, and a stability term expressed as volume concentration of solids, C_V:

$$\frac{\rho_u}{\rho_f} = K(1 - C_v)^\alpha \, Re_i^\beta \, L^\gamma \tag{7}$$

where L is the pressure loss coefficient and Re_i the inlet Reynolds number. This was shown to fit a particular dataset very well, but has not yet been applied to other datasets because of lack of information.

A similar dimensionless expression, also incorporating cyclone geometry, has been found to fit two independent pilot plant datasets satisfactorily, and also predicts full scale plant trends correctly but with a systematic offset that appears to be related to cyclone diameter:

$$\frac{\rho_u}{\rho_f} = K \, Re_i^\alpha \, (D_u/D_c)^\beta \, (D_o/D_c)^\gamma \tag{15}$$

Further work will be necessary to optimise these dimensionless correlations to permit them to handle a wider and more practical range of conditions.

The underflow density follows trends which would be expected in view of the roles which classification and sedimentation play in the segregation of medium in the cyclone. ρ_u decreases with increase in medium viscosity, and with decrease in particle size of the medium; a wider size distribution also decreases ρ_u. At high coal or ore feedrates, when apex crowding may prevail, and at high flowrates of medium, the value of ρ_u increases. ρ_u increases with increase in vortex finder diameter and with decrease in apex diameter, as expected.

A good understanding of the behaviour of underflow medium density has now been established, and thus by extension it is possible to consider the optimisation of the ore/coal concentration performance through its manipulation. However, further work is required to generalise the predictive procedures for ρ_u.

NOMENCLATURE

A, B	cyclone geometry terms
C_v	volume concentration of solids
d	particle size (µm)
d_{50}	cut-size (µm)
d_{50c}	corrected cut-size (µm)
D50	50% passing size of medium solids (µm)
D_c	cyclone diameter (m)
D_i	inlet diameter (m)
D_o	vortex finder (overflow) diameter (m)
D_u	apex (underflow) diameter (m)
E	recovery of solids to overflow (efficiency)
F	feedrate of ore (t h^{-1})
h, H	head (m)
L	pressure loss coefficient (dimensionless)
M_A	Rosin-Rammler location parameter (µm)
M_B	Rosin-Rammler scale parameter
P_i	inlet pressure (bar)
Q_f	feed flowrate (m^{3}s^{-1})
Q_u	underflow flowrate (m^{3}s^{-1})
R_i	inlet Reynolds number (dimensionless)
R_m	proportional recovery of medium to underflow (= Q_u/Q_f)
v_i	inlet velocity (ms^{-1})
V_f	volume fraction of solids in feed medium
V_u	volume fraction of solids in underflow medium
z	sedimentation rate of medium (ms^{-1})
α	classification curve efficiency parameter
δ_{50}	separating density (kg m^{-3}, or RD units)
δ_{50} (4x2)	separating density of -4 + 2mm material (kg m^{-3} or RD units)
$\Delta_{u/o}$	underflow-overflow medium density differential (kg m^{-3})
η_a	apparent viscosity (Nsm^{-2})
$\eta_{a(min)}$	minimum apparent viscosity (Nsm^{-2}) (capillary viscometer)
η_m	apparent viscosity of medium (Nsm^{-2})
ρ_f	feed medium density (kg m^{-3}, or RD units)
ρ_l	liquid density (kg m^{-3}, or RD units)
ρ_o	overflow medium density (kg m^{-3}, or RD units)
ρ_u	underflow medium density (kg m^{-3}, or RD units)

REFERENCES

1. Herkenhoff E.C. Factors affecting ore separations in a 4" DSM cyclone separator. *Eng. and Min. J.*, Aug/Sept 1953, **154**, No. 8, 88-91, and No. 9, 95-99 and 206.

2. Tarjan G. Some theoretical questions on classifying and separating hydrocyclones. *Act. Tech. Acad. Sci. Hung.*, 1961, **32**, 3/4, 357-388.

3. Lilge E.O. The operating variables of the Driessen Cone heavy media concentrator. *Trans. Can. Inst. Min. Met.*, 1956, **LIX**, 404-409.

4. Lilge E.O., Fregren T.E. and Purdy G.R. Appaent viscosities of heavy media and the Driessen Cone. *Trans. Inst. Min. Met.*, 1957-58, **67**, 6, 229-249.

5. Lilge E.O. and Plitt L.R. The cone force equation and hydrocyclone design. *Proc. Interamerican Conf. on Metls Tech.*, May 1968, San Antonia, ASME, 108-118.

6. Schubert H. Sorting according to density in centrifugal force fields. *Bergakademie*, Leipzig, 1965, **17**, 5, 293-297.

7. Khaidakin V.I. Effect of the slurry content of the suspension on coal cleaning conditions in a cascade hydrocyclone. *Coke and Chem. (USSR)*, 1975, **7**, 12-13.

8. Sokaski M. and Geer M.R. Cleaning unsized fine coal in a dense-medium cyclone pilot plant. *USBM RE 6274*, 1963.

9. Fahlstrom P.H. Studies of the hydrocyclone as a classifier. *Proc. 6th Int. Min. Proc. Cong.*, Cannes, 1963, 87-108.

10. Cohen E. and Isherwood R.J. Principles of dense media separation in hydrocyclones. *Proc. Int. Min. Proc. Long.*, London, 1960, 573-591.

11. Upadrashta K.R. and Venkateswarlu D. Study of hydrocylone as heavy media separator. *J. Powder and Bulk Solids Tech.*, 1982, **6**, 3, 22-32.

12. Davies D.S., Driessen H.H. and Oliver R.H. Advances in hydrocyclone heavy media separation for fine ores. *Proc. 6th Int. Miner. Proc. Cong.* Cannes (Pergamon), 1965, 303-321.

13. Collins D.N., Turnbull T., Wright R. and Ngan W. Separation efficiency in dense medium cyclones. *Trans. Inst. Min. Met.*, March 1983, **92**, C38-C51.

14. Napier-Munn T.J. The mechanism of separation in dense medium cyclones. *2nd Int. Conf. on Hydrocyclones*, Sept. 1984, Bath, 253-280, (BHRA).

15. Napier-Munn T.J. Modelling and simulating dense medium separation processes - a progress report. *Minerals Eng.*, 1991, **4**, 3/4, 329-346.

16. Davis J.J. and Napier-Munn T.J. The influence of medium viscosity on the performance of dense medium cyclones in coal preparation. *3rd Int. Conf. on Hydrocyclones*, Sept. 1987, Oxford, 155-165 (BHRA).

17. Wood C.J., Davis J.J. and Lyman G.J. Towards a medium behaviour based performance model for coal-washing dense medium cyclones. *Dense Medium Ops. Conf.*, Brisbane, July 1987, 247-255 (Aus. Inst. Min. Met.).

18. Scott I.A. A dense medium cyclone model based on the pivot phenomenon. PhD thesis, 1988, *University of Queensland*.

19. Scott I.A. and Napier-Munn T.J. A dense medium cyclone model based on the pivot phenomenon. Paper submitted to *Inst. Min. Met.*, London, 1992.

20. Napier-Munn T.J. "Dense medium cyclones in diamond recovery". MSc thesis, University of the Witwatersrand, 1977.

21. Clarkson C.J. and Wood C.J. "A model of dense medium cyclone performance". 5th Austr. Coal Prep. Conf., Newcastle, Paper B3, Ed. P. Lean (Aust. Coal Prep. Soc.), May 1991, 65-79.

22. Napier-Munn T.J. "Pressure drop in dense medium cyclones". *Int. J. Miner. Proc.*, **16**, 1986, 209-230.

23. Napier-Munn T.J. "The mechanism of separation in dense medium cyclones". PhD thesis, University of London, 1984.

24. Holland-Batt A.B. "A bulk model for separation in hydrocyclones". *Trans. Inst. Min. Met.*, **91**, March 1982, C21-C25.

25. Napier-Munn T.J. "Density inversion in dense medium cyclones". *Trans. Inst. Min. Met.*, **94**, June 1985, C96-C98.

26. Vanangamudi M., Mitra J. and Rao T.C. "Analysis of Performance of a 76 mm Vorsyl separator". *Minerals Eng.*, **5**, 1, 1992, 93-101.

27. Davis, J.J. "A study of coal washing dense medium cyclones". PhD thesis, University of Queensland, 1987.

28. Lynch A.J. "Mineral crushing and grinding circuits". *Elsevier*, 1977.

29. Napier-Munn T.J., Reeves T.J. and Hansen J.O. "The monitoring of medium rheology in dense medium cyclone plants". *Proc. Aus. Inst. Min. Met.*, **294**, 3, May 1989, 85-93.

30. Tuteja R. "Mathematical modelling of dense medium cyclones". M.Eng.Sc. thesis, University of Queensland, 1991 (under examination).

31. Nageswararao K. "Hydrocyclone modelling and scale-up". PhD thesis, University of Queensland, 1978.

32. Lynch A.J. and Morrell S. "The understanding of comminution and classification and its practical application in plant design and operation", in *Comminution - theory and practice*, ed K. Kowatra, 405-426, SME (SME Annual Conference, Phoenix, Feb.), 1992.

APPENDIX 1

Ferrosilicon Classification and Medium Product Data for 200 mm Cyclone (Scott [18])

Test No.	Medium Densities (kg m^{-3})			Medium viscosity (Nsm^{-2} x 10^3)**	FeSi volume fraction	1 - C (%)	Classification Parameters	
	Feed	Overflow	Underflow				α	d$_{50c}$ (μm)
1.2	1456	1404	1597	22.9	0.039	27.4	3.07	66.3
1.2R*	1460	1402	1614	22.9	0.036	25.9	2.60	65.6
2.2	1538	1526	1594	46.0	0.052	30.9	3.66	116.6
4.2	1528	1336	2180	14.0	0.071	20.1	2.76	38.0
5.2	1732	1572	2206	20.7	0.102	23.0	3.54	58.1
5.2R*	1728	1574	2192	19.8	0.102	22.7	3.05	59.5
6.2	1770	1631	2170	23.9	0.104	23.7	2.39	69.8
8.2	2737	2626	3012	50.2	0.274	25.8	2.57	136.3
9.2	2411	2229	2953	22.9	0.226	22.4	2.76	87.3
10.2	2631	2442	3114	27.2	0.272	24.6	1.23	139.5

** Debex viscometer at bobbin speed of 400 rpm

* Repeat tests

THE EFFECT OF RHEOLOGY ON THE PERFORMANCE OF HYDROCYCLONES

RICHARD R. HORSLEY
Deputy Vice-Chancellor and Professor of Mechanical Engineering
Curtin University of Technology, Perth, Australia
QUOC-KHANH TRAN
Ph.D. student, Department of Mechanical Engineering
Curtin University of Technology, Perth, Australia
JOHN A. REIZES
Associate Professor
University of New South Wales, Sydney, Australia

ABSTRACT

The experimental work has been carried out on range of spigot diameters and two hydrocyclone sizes. The slurries used were mixtures of ball clay and silica which have been used as models for industrial slurries. Variations in the rheological properties were effected by varying the concentration of each of the components and with the use of viscosity modifiers. Viscosity modifiers were employed to significantly reduce the yield stress in the slurry, which led to an increase in the separation efficiency and improved the sharpness of cut significantly. A correlation for the prediction of hydrocyclone performance was developed. An agreement exists between the prediction and the experimental data. These results are very important for industrial processes involving separation of concentrated mixtures of particulate flows as they provide information for the design and operation of mineral processing plants in the mining industry.

INTRODUCTION

Hydrocyclones are extensively used in widespread applications for many industries [2,15]. They are used in closed circuit grinding operations as classifiers and have found many other uses such as desliming, de-gritting and thickening. The current design and operation of hydrocyclones are suitable for dilute slurries [2,10,15]. Slurries with high solids concentration and wide particle size distributions exhibit yield pseudoplastic characteristics in mineral processing plants [6], and makes hydrocyclone design and operation very difficult.

This work has concentrated on the effect of rheological properties and spigot diameters on the separation efficiency of a 4" and a 6" hydrocyclone. This paper also presents a mathematical model to predict the performance of hydrocyclones.

PREDICTION OF CUT SIZE FOR YIELD STRESS SLURRIES

A single number which is used to characterise the performance of hydrocyclones is the cut size d_{50}. The cut size is defined as the size which has an equal probability of being separated with either the underflow or the overflow. There are two approaches to the development of a hydrocyclone model of the cut size. The empirical model approach [2,14,16] and the fundamental mathematical modelling approach [1,4,5,7,13]. Most of the models were based on data obtained from systems using dilute pulp. Very little data [16] has been obtained for conditions of high concentrations of feed pulp.

At high solids concentration and/or flocculated slurries, the performance of the hydrocyclone becomes affected by the nature of the solids as much as the operating conditions and this complicates matters considerably [6]. Empirical models of varying complexity and applicability exist, but are inadequate. A hydrocyclone model needs to be developed to suit these conditions.

Rietema [14] proposed the residence time theory and the following relationship was developed

$$\frac{d_{50}^2 (\rho_s - \rho_L) L}{18 \mu U_z R_1} \int_{r_a}^{R} \frac{V_t^2}{r} dr = \frac{D_i}{2} \quad \text{-----} \quad (1)$$

in which ρ_s is the solid density, ρ_L is the liquid density, μ is the liquid viscosity, U_z is the axial velocity of particle, R is the hydrocyclone radius, r_a is the radius of air core, L is the total length of hydrocyclone, D_i is the inlet diameter, V_t is the tangential velocity of slurry and $R_1 = R - r_a$.

Kelsall's measurements [9] of the radial velocity showed that this velocity was very small for the hydrocyclone, so that Rietema [14] was able to assume that this radial velocity can be neglected. It follows that the main contributor of the total pressure drop across a hydrocyclone is the pressure drop due to the tangential velocity. However Xu et al [19] showed that the radial pressure drop differs considerably from that obtained from tangential velocity considerations and that the radial velocity is given by

$$V_r = \frac{k}{r^n} \quad \text{-----} \quad (2)$$

in which k and n are constants.

Taking into account the radial pressure drop, the total pressure drop across a hydrocyclone can be estimated as

$$\Delta P = \rho_L \int_{r_a}^{R} \frac{v_t^2}{r} dr + \rho_L \frac{k^2}{2} r_a^{-2n} \quad \text{-----} \quad (3)$$

Suppose that the additional pressure drop, ΔP_{ad}, is given by

$$\Delta P_{ad} = \rho_L \frac{k^2}{2} r_a^{-2n}, \quad \text{-----} \quad (4)$$

the pressure required to obtain a velocity k/r_a^n at the air core. This, of course, is not accurate but should represent an "average" radial velocity in the hydrocyclone.

The substitution of (3) into (1) gives

$$d_{50} = C_1 \left(\left(\frac{\rho_L}{\rho_S - \rho_L} \right) \left(\frac{Q}{\Delta P - \rho_L r_a^{-2n}} \right) \mu \right)^{1/2} \quad \text{-----} \quad (5)$$

in which $C_1 = \sqrt{\dfrac{C_{y50}}{L}}$ is a constant, Q is the feed volumetric flow rate, C_{y50} is the characteristic cyclone number and Xu et al [19] found that the value of k in equation (2) was different for different cross-sections along the hydrocyclone length and varied from 1.34 to 2.01, which gave the average value of 1.58. To simplify this model k is assumed to be $\sqrt{2}$.

The study of the effect of chemical additives in yield stress slurries on fall velocity of steel objects by Tran and Horsley [18] suggested the following relationship between the settling velocity U_s, and the concentration by volume C_v and yield stress of the slurry τ_y

$$\frac{U_S}{U_{ter}} = e^{-C_2 C_v \tau_y} \quad \text{-----} \quad (6)$$

in which U_{ter} is the terminal settling velocity of particle, and C_2 is a constant which is dependent on the particle size and shape.

Very concentrated slurries are discussed in this paper and experimental results show that there is very little concentration change between the inlet and either the underflow or overflow. It follows that the liquid viscosity would not be "seen" by the particle and it is replaced by the apparent viscosity of the slurry η_a. Therefore the following equation first proposed by Tran [17] becomes

$$d_{50} = C_1 e^{C_2 \tau_y C_v} \left(\left(\frac{\rho_{SL}}{\rho_S - \rho_{SL}} \right) \left(\frac{Q}{\Delta P - \rho_{SL} r_a^{-2n}} \right) \eta_a \right)^{1/2} \quad \text{-----} \quad (7)$$

in which ρ_{SL} is the slurry density and the radius of air core r_a [15] is determined by

$$r_a = \left(\frac{D - D_i}{2} \right) \left(\frac{\rho_{SL} v_i^2}{3 \Delta P} \right)^{1/3} \quad \text{-----} \quad (8)$$

in which D is the hydrocyclone diameter.

Following Rietema [14], Tran [17] suggested the mean shear rate of the particle in the hydrocyclone is

$$\dot{\gamma} = \frac{2V_r}{D_i} = \frac{V_i}{L_t} \quad \text{-----} \quad (9)$$

in which L_t is the total helical path length of the particle. Equation (9) together with the rheogram for the particular slurry are used to estimate η_a in equation (7).

The separation will not occur when either (i) the cut size is greater than the maximum size of feed solids or (ii) the term ($\Delta P - \rho_{SL} r_a^{-2n}$) ≤ 0.

EXPERIMENTAL PROCEDURE

Mixtures of ball clay and silica were used as models of slurries of commercial interest. The specific gravity of ball clay and silica were 2.67 and 2.65 respectively. The size distribution of particles was determined by a MALVERN MASTERSIZER as shown in figure 1. The yield stress of slurries was reduced by the addition of sodium hexametaphosphate.

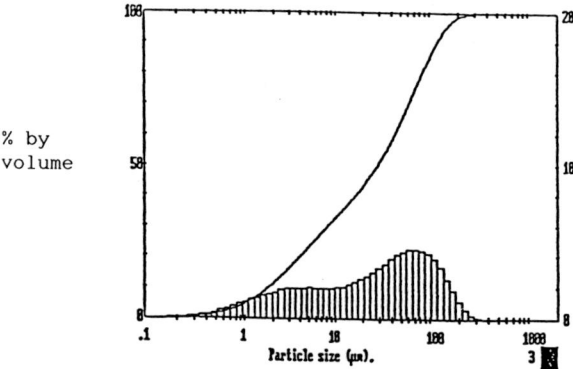

Figure 1: Particle size analysis.

A schematic of the experimental apparatus is shown in figure 2. It consists of a baffled tank and a discharge box, which collects overflow from the 6" and 4" hydrocyclone into the tank. Dimensions of hydrocyclones can be seen in table 1. A mixer was used to ensure a uniform suspension of the solids. Two conical receivers were used for the simultaneous sampling of the two products. An AMDEL digital density meter and a Mag-X flow meter were installed into the circuit to measure the density of feed slurry and the feed flow rate. A Mono pump was used to deliver the feed through the hydrocyclones.

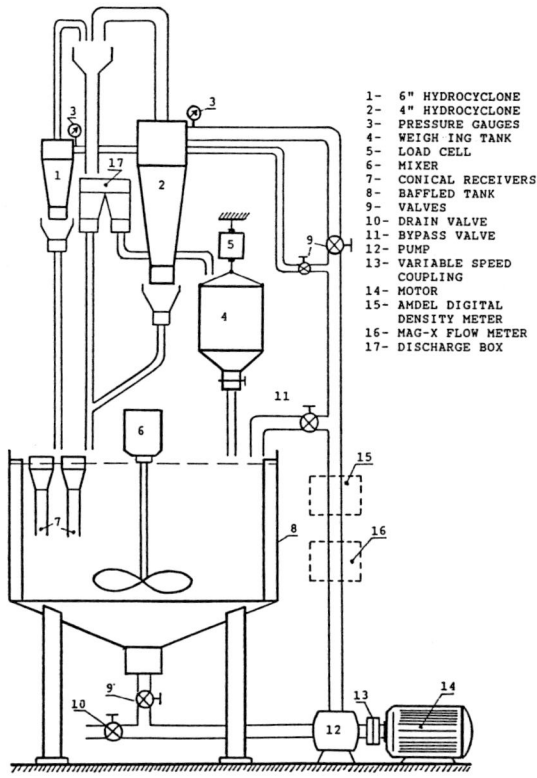

Figure 2: Schematic diagram of hydrocyclone rig.

TABLE 1
Hydrocyclone dimensions

HYDROCYCLONE TYPE	WARMAN CYCLONE 4" SERIES 'C'	WARMAN CYCLONE 6" SERIES 'C'
BORE (mm)	100	150
LENGTH (mm)	650	1000
VORTEX FINDER (mm)	25 DIA., 85 LENGTH	40 DIA., 115 LENGTH
SPIGOT DIA. (mm)	10, 15, 20	25, 30, 35, 40, 45, 50
CONE ANGLE (DEG.)	20	20

The hydrocyclone size was chosen and its dimensions were set as required. The correct proportions of ball clay, silica and water were mixed in the baffled tank with the mixer on. The system was allowed to run for a sufficient time to ensure thorough dispersion of solids before a sample of feed slurry was taken. The inlet pressure was set at the following values: 50 kPa, 100 kPa, 150 kPa and 200 kPa. At each feed pressure, the flow rate, density and temperature were recorded. The measurement of the mass flow rate of underflow and overflow was done via diversion systems. The overflow was diverted into the weighing tank where its mass was read from the indicator display connected to the load cell. While the underflow was diverted into a bucket located outside the tank where its mass was measured by a weighing scale. A stop watch was used to record the time taken to collect the slurries. The underflow, overflow and samples of inlet slurry before and after each test were collected in plastic bottles for rheological and particle size analysis. The rheology of feed slurry was measured in the MODIFIED VISCOMETER developed by Overend et al [12].

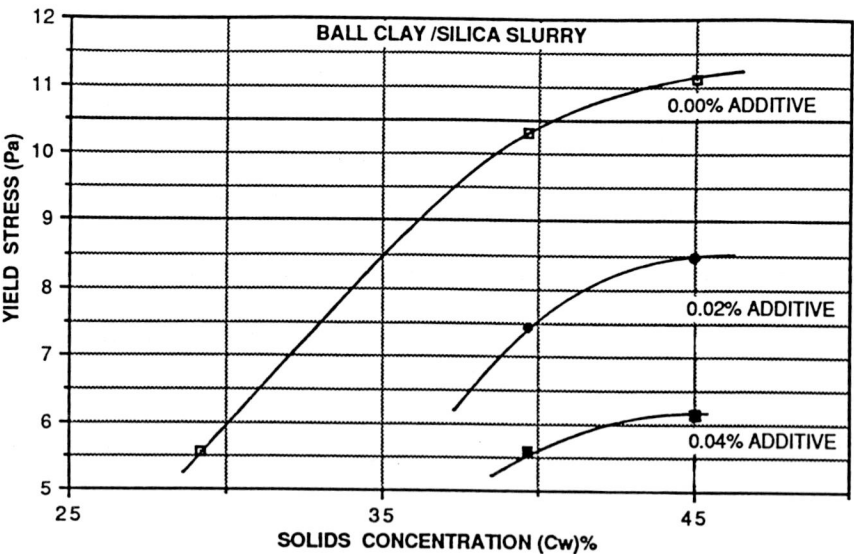

Figure 3: Graph of Yield Stress versus Solid Concentration.

RESULTS AND DISCUSSION

Figure 3 shows that the yield stress of ball clay/silica slurry depended on the solids concentration. As the solids concentration increased, the yield stress increased accordingly. With the addition of a viscosity modifier, $Na_2(PO_3)_n.Na_2O$, reduction of yield stress value occurred.

The most significant effect of the variation of spigot diameter (D_u) was the underflow-to-throughput ratio values (R_f). Figure 4 indicates that the R_f increased with the increase of D_u for a given inlet pressure. As the inlet pressure increased, the R_f value decreased. As the ratio $D_u/D_o>1$, the underflow volume flow rate became larger than the overflow volume flow rate, that is $R_f>0.5$. A higher R_f value implied there so as a greater amount of "dead flux", i.e. more particles are "trapped" in the coarse product. This is to be avoided in wet closed grinding circuits as further size reduction, which is not necessary, requires a high rate of energy consumption in the ball mills.

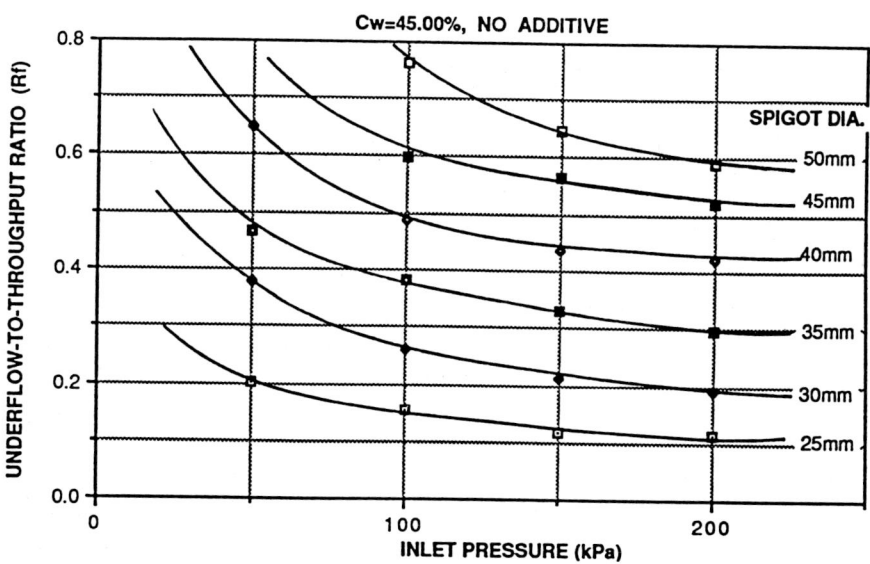

Figure 4: Graph of Underflow-to-throughput ratio versus inlet pressure (6" Hydrocyclone).

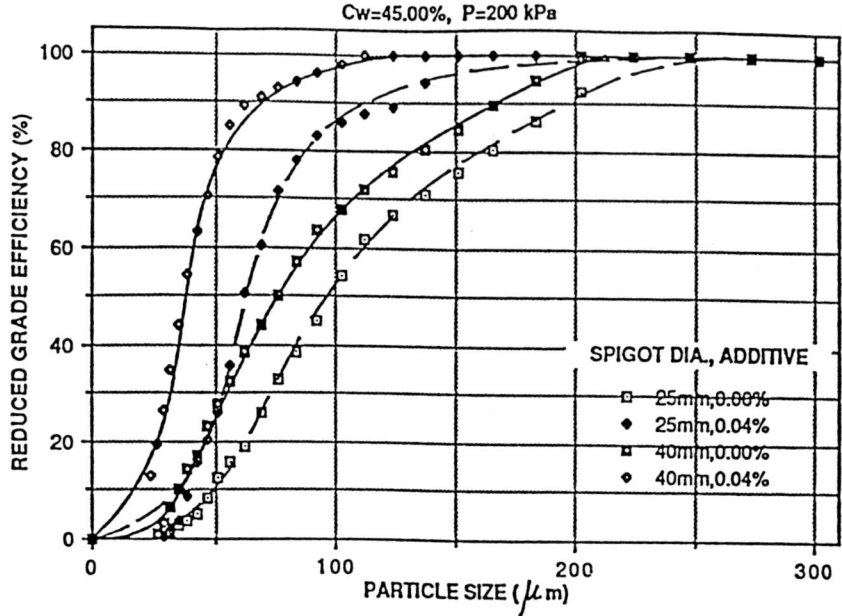

Figure 5: Reduced grade efficiency curves of two different spigots with and without additive (6" Hydrocyclone).

The variation of spigot diameter also had a profound effect on the reduced cut size d_{50C}. Figure 5 shows the comparison of reduced grade efficiency curves. As the spigot diameter increased, the reduced cut size decreased, which indicated that better separation efficiency was expected for larger D_u, and the sharpness of cut slightly increased. The difference of sharpness among the curves was not significant. This is explained due to a higher R_f for a larger D_u. However, with the addition of a viscosity modifier, the sharpness of cut improved significantly for a given spigot diameter and solids concentration. This can be seen in figure 6. Figure 6 also shows that the smaller size hydrocyclone gave a considerably lower reduced cut size. The viscosity modifier was employed to significantly reduce the slurry viscosity, which increased the settling velocity of the particles. This caused them to segregate more rapidly with fewer misplaced particles, thus improving efficiency. However the use of viscosity modifier did not change the R_f value significantly, as can be seen in figure 7.

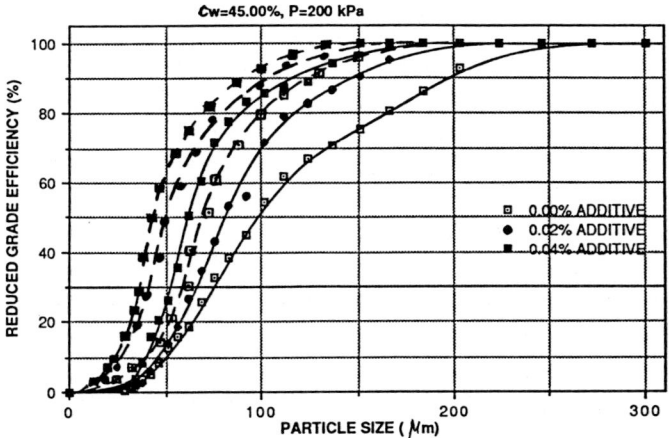

Figure 6: Reduced grade efficiency curves for three different Yield stress, $D_u/D_o = 0.6$ (--- 4" Cyclone, —— 6" Cyclone).

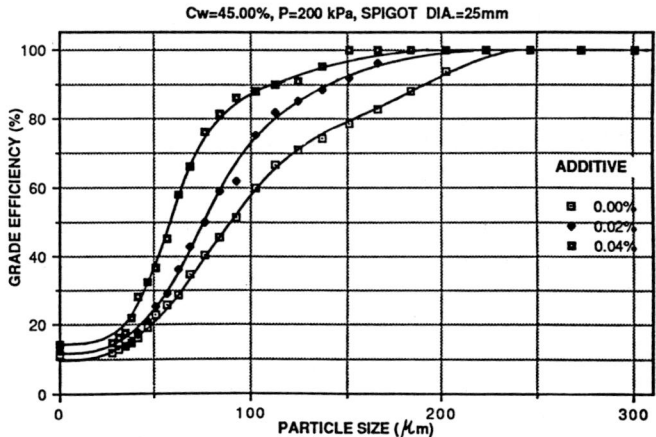

Figure 7: Grade efficiency curves for three different Yield Stresses (6" Hydrocyclone).

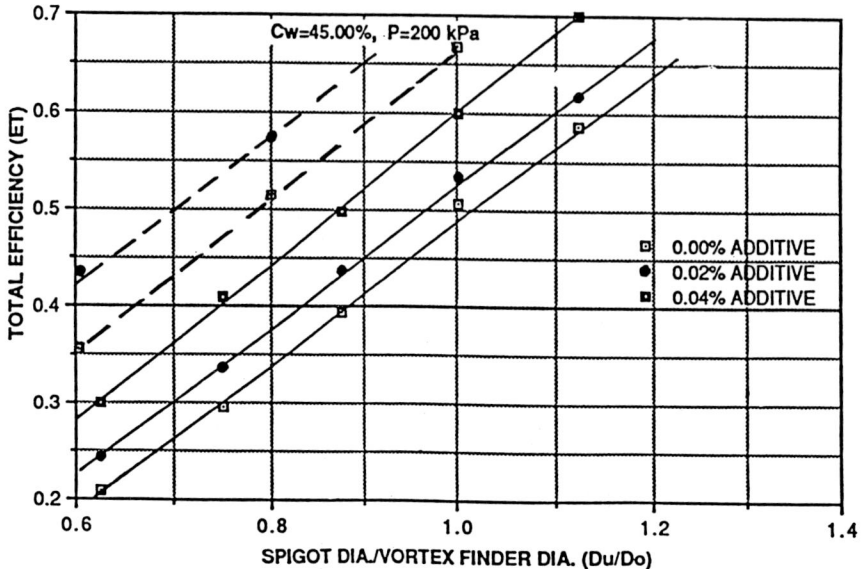

Figure 8: Graph of Total Efficiency versus D_u / D_o
(--- 4" Hydrocyclone, —— 6" Hydrocyclone).

Figure 8 shows the relationship between the total efficiency E_T and D_u/D_o. E_T is an important parameter because it indicates the actual amount of solids recovered in the underflow as a ratio of total mass input. An increase of the spigot diameter led to an increase in the total efficiency. The smaller size hydrocyclone gave a higher value of total efficiency. It appears that more solids is fed back to the ball mills from the underflow and therefore reduce the throughput to the next stage of the processing plant.

Agreement has been obtained between the experimental data and those predicted by Tran's [17] correlation in equation (6) for the tested slurries of 30% to 50% solids concentration by weight. This can be seen in figure 9. The constants in the correlation are tabulated in table 2. The values of n and C_1 are dependent on hydrocyclone dimensions. It was found that n was proportional to the spigot diameter and C_1 was inversely proportional to the spigot diameter. The ratio of ball clay to silica was kept constant so it was assumed that all particles had identical size and shape at cumulative percentage of 50. Therefore C_2 is a constant of 0.145.

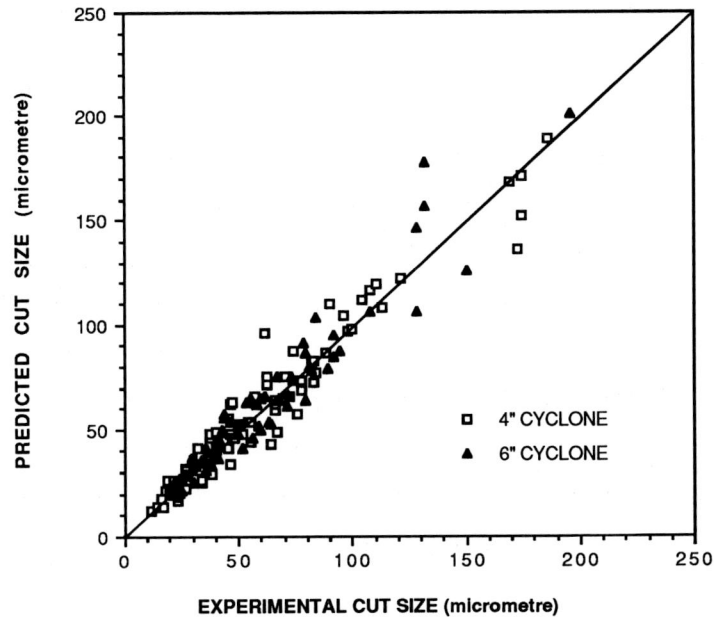

Figure 9: Experimental and Predicted values of d_{50}.

TABLE 2
Constants in equation (6)

HYDROCYCLONE	SPIGOT DIA. (mm)	C_1	$\bar{n}$
6"	25	71325	0.420
	30	54312	0.440
	35	41151	0.489
	40	29733	0.523
4"	10	258590	0.296
	15	147267	0.348
	20	76115	0.463

CONCLUSION

The correlation for predicting the performance of hydrocyclones reasonably holds for a wide range of feed conditions of yield pseudo-plastic slurries. The values of the constants in the correlation depend on the hydrocyclone size, hydrocyclone configuration and feed solids flow rate.

The separation efficiency is clearly dependent on rheological properties of the feed slurry and the spigot diameter. The addition of viscosity modifier led to an increase in the separation efficiency and improved the sharpness of cut significantly.

REFERENCES

[1] M.I.G. Bloor and D.B. Ingham (1973). *On the efficiency of the industrial cyclone.* Trans. Instn Chem. Engrs, vol.51, pp 173-176.
[2] D. Bradley (1965). *The hydrocyclone.* Pergamon, London.
[3] D. Bradley and D.J. Pulling (1959). *Flow patterns in the hydraulic cyclone and their interpretation in terms of performance.* Trans. Instn Chem. Engrs, vol.37, pp 34-45.
[4] M.R. Davidson (1988). *Similarity solutions for flow in hydrocyclones.* Chem. Eng. Sci., vol.43, no.7, pp 1499-1505.
[5] A.B. Holland-Batt (1982). *A bulk model for separation in hydrocyclones.* Trans. Inst. Min. Metall. Sect.C, 91, p C21.
[6] R.R. Horsley and D.W. Allen (1987). *The effect of yield stress on hydrocyclone performance in the mining industry.* 3rd Int. Conf. on Hydrocyclones, 30 Sept - 2 Oct., Oxford, England, pp 269-275.
[7] K.T. Hsieh and R.K. Rajamani (1991). *Mathematical model of the hydrocyclone based on physics of fluid flow.* AIChE J., vol.37, no.5, pp 735-746.
[8] S.K. Kawatra, T.C. Eisele, D. Zhang, M.T. Resesky and L.L. Sutter (1988). *Temperature effect on grinding circuit performance.* Soc. Min. Engrs, preprint no.88-1, Colorado,US.
[9] D.F. Kelsall (1953). *A further study of the hydraulic cyclone.* Chem. Eng. Sci., vol.2, pp 254-272.
[10] A.J. Lynch (1977). *Mineral crushing and grinding circuits: Their simulation, Optimization and Control.* Elsevier, Oxford
[11] T.J. Napier-Munn (1980). *Influence of medium viscosity on the density separation of minerals in cyclones.* Int. Conf. on Hydrocyclones. 1-3 Oct.,Cambridge, UK, paper 6, pp 63-82.
[12] I.J. Overend, R.R. Horsley, R.L. Jones and R.K. Vinycomb (1984). *A new method for the measurement of rheological properties of settling slurries.* Advances in Rheology, vol.2, pp 571-574.
[13] K. Rajamani and K.T. Hsieh (1988). *Hydrocyclone model: A fluid mechanics approach.* Soc. Min. Engrs, Preprint 88-163, Colorado, US.
[14] K. Rietema (1961). *The mechanism of the separation of finely dispersed solids in cyclones.* Cyclones in Industry edited by Rietema and Verver, Elsevier, Amsterdam, ch.4, pp 46-63.
[15] L. Svarovsky (1984). *Hydrocyclones.* Holt, Rinehart and Winston, London.
[16] L. Svarovsky and B.S. Marasinghe (1984). *Performance of hydrocyclones at high feed solids concentrations.* Int. Conf. on

Hydrocyclones, 1-3 Oct.,Cambridge,UK,paper 10,pp 127-142.
[17] Q.K. Tran (1989). *The effect of slurry yield stress upon the separation performance of the hydrocyclone.* M.Eng. Thesis, Curtin University of Technology, Perth, Australia.
[18] Q.K. Tran and R.R. Horsley (1990). *The effect of chemical additives on the drag and settling velocities of yield stress slurries.* 3rd European Conf. on Rheology, Sept.1990, Edinburgh, UK.
[19] J. Xu, Q. Luo and C. Deng (1987). *A study of the radial velocity in a hydrocyclone.* Mining Science and Technology, Trans. Tech. Publications. China Coal Industry Publishing House, pp 1277-1283.

Nomenclature

C_1, C_2, k	- constants
C_w	- solids concentration by weight
C_v	- solids concentration by volume
D	- hydrocyclone diameter
D_i	- inlet diameter
D_o	- vortex finder diameter
D_u	- spigot diameter
d_{50}	- cut size
d_{50c}	- reduced cut size
E_T	- total efficiency
P	- inlet pressure
ΔP	- pressure drop
Q	- feed volumetric flow rate
r_a	- radius of air core
R	- hydrocyclone radius
R_f	- underflow-to-throughput ratio
U_s	- settling velocity of particle
U_{ter}	- terminal settling velocity of particle
U_z	- axial velocity of particle
V_i	- inlet velocity of feed slurry
V_r	- radial velocity of particle
V_{ra}	- radial velocity at the air core
V_t	- tangential velocity of particle
$\dot{\gamma}$	- shear rate
η_a	- apparent viscosity
μ	- liquid viscosity
ρ_L	- liquid density
ρ_S	- solid density
ρ_{SL}	- slurry density
τ_y	- yield stress

DEVELOPMENT OF A CYCLONIC DE-GASSING SEPARATOR FOR USE IN A ROADSIDE FUEL DISPENSER

P D G MASSINGBERD-MUNDY
K B SNOOKS & J G GULLIVER
Department of Mechanical Engineering,
University of Southampton,
Southampton, Hampshire, SO9 5NH, UK

ABSTRACT

The work presented in this paper was prompted by impending changes in the legislation relating to the air separation requirements of roadside fuel dispensers - more commonly referred to as "Petrol Pumps". The paper is concerned with the development of a cyclone separator as an alternative to the gravimetric separation techniques currently used in the majority of fuel dispensers.

A small cyclone unit, 200mm long and 37mm principal diameter was manufactured and tested. It was designed to cope with fuel flowrates up to 70 litres/minute and air to fuel ratios ranging from 100% fuel to 100% air. It was carefully matched to a positive displacement pump typical of the fuel dispenser industry. Various configurations of vortex finder were explored and the pressure balance between the two outlets was investigated. Other options were explored to improve efficiency and reduce the pressure drop.

GLOSSARY

OIML	-	Organisation Internationale de Mètrology Lègale
PD	-	Positive Displacement
Va	-	Volume of air bled into the suction line (corrected to atmospheric pressure)
Vn	-	Volume of fuel delivered through the PD meter (including any unseparated air)

INTRODUCTION

There are three basic hydraulic components in a fuel dispenser; the pump, the air separator and the meter. The air separator is designed to remove any free gas, air or vapour from the fuel stream before it is metered and delivered to the customer. The presence of a separator is a legal requirement under UK and EEC regulations for all dispensing systems where the pressure can fall below atmospheric. Such a condition arises in the majority of European dispensing systems where each dispenser draws fuel from an underground tank.

Until recently the performance requirements of the separator were not clearly defined. However, with the economic integration of Europe and the work of OIML, separator standards have been defined which must be satisfied by European suppliers. Dispenser manufacturers are now required to demonstrate the effectiveness of their separator by confirming the accuracy of the meter while bleeding any quantity of air into the suction lines below the dispenser. Failure of the separator results in the customer "buying" air which is illegal.

Current technology uses gravimetric separators which are relatively bulky due to the slow velocity of the bubbles under gravity. Additional technology can be used to shut down the dispenser in the event of separator overload. The hydrocyclone offers much faster and more compact separation and the potential for eliminating the need for a gas detector.

This paper discusses the new legislative requirements and the associated test procedures before describing the design and development of a suitable cyclonic separator.

BASIC OBJECTIVE

The objective of the project was to develop a cyclonic separator capable of meeting the new standards without resorting to the use of an additional device such as a gas detector.

The new legal requirement is that the separator should be capable of separating any quantity of air present in the fuel stream. This is defined by stating that the meter in the dispenser should remain accurate to within a certain tolerance whatever the ratio of air volume to the total volume (fuel and air) at inlet to the dispenser. All volumes are corrected to atmospheric pressure and the tolerances are ±0.5% and ±1% for petrol and diesel respectively.

THE TEST RIG

A schematic diagram of the test facility used during the project is shown in Figure 1. It was based on the testing method specified by the OIML requirements.

Test Method to Measure Separator Performance

To demonstrate compliance with the new requirement, the accuracy of the positive displacement (PD) meter within the dispenser is measured while air is bled into the suction line upstream of the pump through an air flow meter. The fuel flow rate is taken directly from the PD meter within the dispenser. The accuracy of the PD meter is measured by calibrating it against a known volume. The calibration is achieved by using a standard forecourt nozzle and display to fill a narrow necked calibration vessel. The discrepancy between the volume measured by the PD meter and that measured by the vessel is used to calculate the percentage error according to the following relationship:

$$\%\text{error} = \frac{\text{Vol in the vessel} - \text{Vol registered by the meter}}{\text{Vol in the vessel}} \times 100$$

To eliminate any errors caused by non-linearity of the PD meter, it is first calibrated when no air is present across the whole range of fuel flow rates. As such the 'volume registered by the meter' in the expression above can be corrected for flow rate. The percentage error calculated is therefore the additional error due to air. If it exceeds the tolerance quoted above the separator has failed.

Throughout the tests presented in this report the pump speed remained constant and the nozzle was held fully open to ensure maximum flow rate and minimum back pressure on the cyclone - therefore presenting the worst case. Finally the inlet pressure to the pump was held constant at -0.34 bar gauge to simulate the vertical lift of the fuel on a typical forecourt installation.

Presentation of the results

The majority of the results in this paper are plotted as percentage error in the PD meter against the ratio or percentage of air in the total flow into the pump. The total flow is estimated from the sum of the volume of air (Va) and the volume of fuel delivered through the PD meter (Vn). This is the accepted method of presentation in the industry and is refered to as additional meter error plotted against Va/(Va+Vn). Alternatively the results can be summarised by stating that the separator will work satisfactorily up to a certain percentage of air in the total flow. Satisfactory performance being; that the additional meter error remains within the tolerances stated.

Discussion

This test method is relatively simple when compared with much of the testing performed on hydrocyclones: it is a cheap and practical method of verifying the performance of a separator in situ. The limitation is that once the separator has failed, the PD meter starts to measure fuel and air and hence overread. As such the estimate of the total volume at inlet to the pump is erroneous because it is based on the sum of the volume of air bled and the volume of 'fuel' delivered through the PD meter. Allowances can be made for the effect by measuring the pressure at the meter which, in conjunction with the percentage error measured against the calibration vessel, can be used to estimate the volume of air passing through the PD meter. This is rarely done in practice since it is sufficient to declare that the separator is failing.

All results presented in this paper were obtained using this test method without correcting the fuel flow rate reading (Vn) for the air passing through the PD meter. If correction were carried out on the results then the curves shown in Figures 3,5,7 and 9 (which include Vn) would show slightly better results with each point moving slightly to the right.

Details of the rig are presented in Figure 1. Note that the second stage separator (discussed in the following section) was omitted, the air/fuel mixture from the overflow being returned directly to the dump tank.

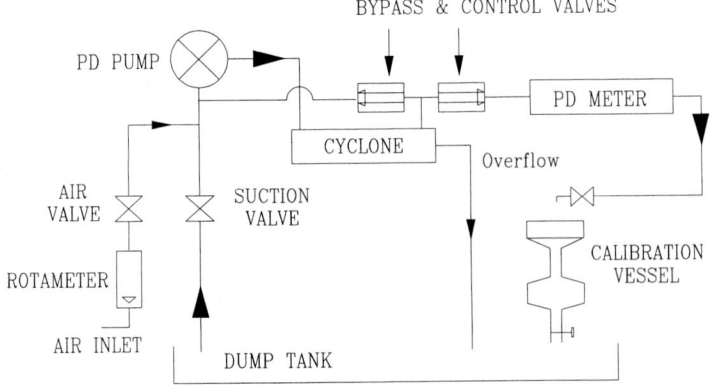

Figure 1. Schematic diagram of the test facility.

EXISTING TECHNOLOGY

Traditionally, the air separator in a fuel dispenser has been gravimetric, consisting of a small chamber where the velocity of the fuel is reduced so that the majority of free gas has time to separate out. Gas free fuel is extracted from the bottom of the chamber while a mixture of fuel and gas/vapour is extracted from the top and passed to a second stage separator. In the second stage, where flow velocities are significantly slower than in the 1st stage, the gas is allowed to settle out virtually completely. Fuel from the second stage is directed through a control valve to the suction side of the pump while fuel free gas/vapour is vented to atmosphere. There are various refinements to the system which do not warrant a full explanation for the purposes of this paper. It is sufficient to report that it is possible to detect the presence of gas in the pipe between the two separators so that, in the event of the separator failing, the dispenser can be turned off. This technique can be used to meet the new separator requirements.

ALTERNATIVE TECHNOLOGY

The overall objective of the Teaching Company Project was to consider alternative technologies to replace those currently used for dispensing fuel. The hydrocyclone clearly deserved attention as an alternative to the existing gravimetric separator for the following reasons:

Clearly the cyclone has the advantage of higher acceleration fields and therefore faster and more efficient separation making the unit more compact than an equivalent gravimetric separator.

There was also a potential cost saving because the improved performance might alter the requirement for the gas detector used to turn off the dispenser in the event of the separator failing. It was hoped that the cyclone would eliminate the need for a gas detector altogether.

The disadvantages of the cyclone separator were identified to be the increased pressure drop which would reduce the delivery rate and its inability to maintain satisfactory performance over the range of pressures and delivery flow rates (0 to 50 litres/min) experienced in a fuel dispenser. The fundamental limitation of the latter problem was overcome by including the cyclone in the bypass loop required by the PD pump (see Figure 1). As such the flow and therefore the acceleration field in the cyclone is maintained even when the delivery nozzle is closed. However, the pressure balance crucial to successful separation remained a problem.

DEVELOPMENT OF THE HYDROCYCLONE

Conical Cyclone

The basic geometry of cyclone developed during the project is shown in Figure 2. It included a single rectangular input (288mm²) into a 360° volute, a conical centre section running from a diameter of 37mm to 25mm with an overall length of 204mm and an outlet volute into an rectangular section (833mm²). The dimensions are summarised in Table 1.

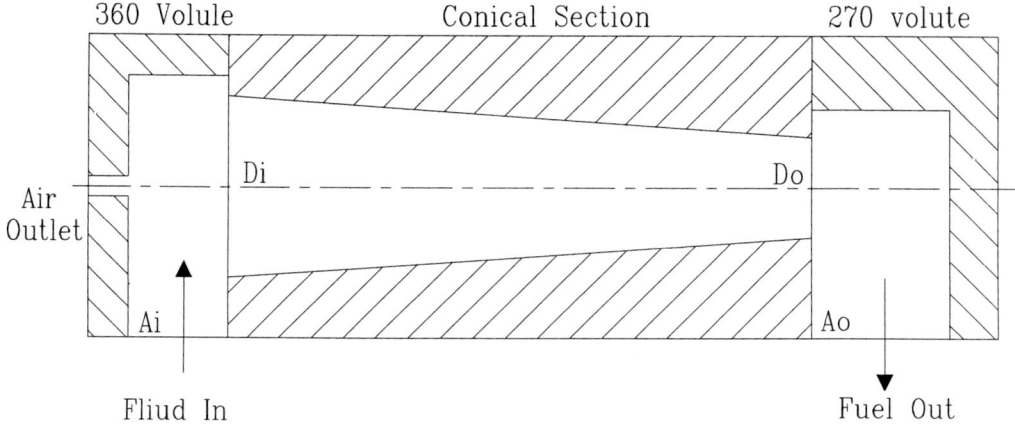

Figure 2. Basic geometry of the cyclone separator.

TABLE 1
Summary of Dimensions

Dimension	Symbol	Value	Area or x/Di ratio
Cone diameter	Di	37mm	1
Feed area	Ai	28x14mm	2:1
Inlet port		360° volute x 24mm deep	-
Cone diameter	Do	25mm	0.68
Outlet Area	Ao	24.5x34mm	3:4
Outlet port		270° volute x 34mm deep	-
Overall length	L	204mm	5.51
Cone angle	α	4.84° included angle	-

Cylindrical cyclone

The cyclone described above is the outcome of extensive development. Early experiments were conducted on a similar sized unit with a cylindrical central section. The cylindrical unit was used to compare reverse flow with through flow and was used in the development of the porting between the pump and the cyclone described below.

Through flow versus reverse flow in the cyclone

Gas free fuel is extracted from the outside of the cyclone via the outlet volute. The gas is extracted from the core of the cyclone through a tube called the vortex finder. This can be positioned at the outlet end of the cyclone resulting in through flow of the core or at the inlet resulting in reverse flow of the core relative to the main fuel flow.

The cylindrical cyclone was tested with the vortex finder in the two positions. The results are presented in Figure 3 and show clearly that the through flow cyclone is the more efficient of the two. However, for a variety of reasons, the development moved towards a conical cyclone incorporating reverse flow.

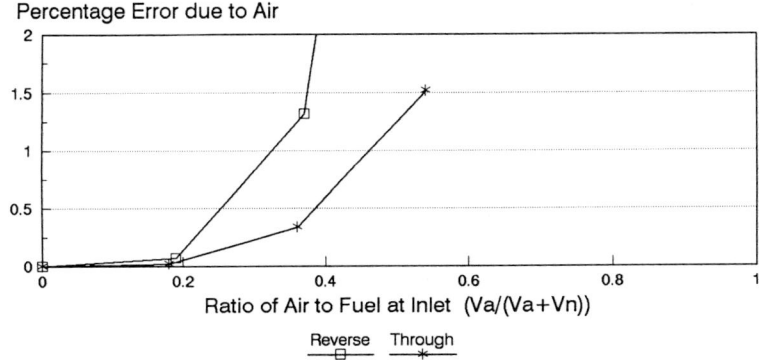

Test Fuel : Kerosene

Figure 3. Results from cylindrical cyclone comparing reverse flow with through flow.

Tests were also performed to assess the optimum dimensions for the vortex finder. These tests revealed that a vortex finder is superfluous in a reverse flow hydrocyclone and actually degrades the performance and increases the pressure drop due to the reduced cross-sectional area into the conical/cylindrical section.

Porting Development

Matching the cyclone to the PD pump as found in fuel dispensers proved to be very challenging. The first results, using a sliding vane pump were disappointing because the separator failed above about 12% air. Initially the poor performance was attributed to incorrect balance of the two output streams but it was later verified that the outlet pressure balance and outlet flow ratios were giving the results expected. Several developments were tried, including the introduction of flow straighteners, lengths of straight pipe and a simple coalescer, to ameliorate the performance but none showed any improvement. Successful results had been obtained with another sliding vane pump which had an uninterrupted path between the outlet of the pump and the inlet to the cyclone. This convinced us that the design of the pump-separator porting is crucial to the performance of the cyclone: changes in cross sectional area and changes in flow direction are to be avoided as much as possible.

Based on this experience, the cylindrical cyclone was attached to an internal gear pump with an outlet parallel to the pump axis. This configuration yielded promising results and led to the development on another internal gear pump with purpose built porting as sketched in Figure 4. The results from this configuration are shown in Figure 5 and were clearly the best results at the time.

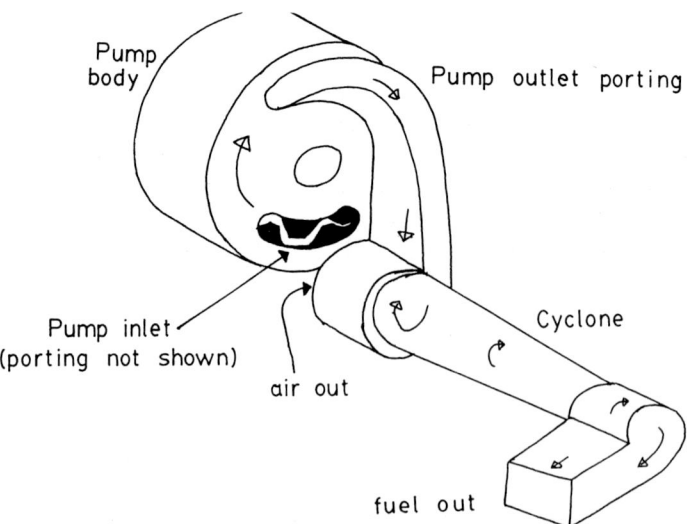

Figure 4. Sketch of the porting between the pump and the cyclone.

One particular feature of the final arrangement is that separation starts upstream of the cyclone volute due to the gentle bend in the porting. This bend was the opposite way in the previous design so any separation in the bend had a negative effect on the cyclone. In the final arrangement shown in Figure 4 any separation in the porting is advantageous.

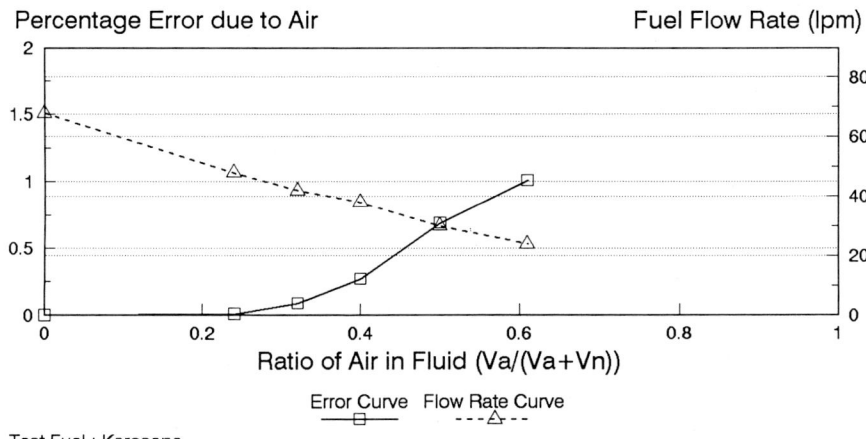

Test Fuel : Kerosene

Figure 5. Results from the cylindrical cyclone with optimised porting.

Optimising a variable system

As stated in the section above on the test method, the variables held constant for all the tests were; pump speed, a fully open nozzle and a constant inlet pressure to the pump. These constants are not well reflected at the cyclone where the flow rate, pressures and gas content vary in accordance with the configuration of the whole dispensing system.

For example, Figure 6 shows how the pump outlet pressure falls with increasing quantities of air introduced into the suction line. The fuel flow rate also falls from a maximum of about 70 litres/minute as air is introduced.

To maintain good separation the flow split between the two outlets of the cyclone must be suitably balanced for all conditions. There are essentially two controlling factors; the first is the control valve in the fuel outlet while the second is the size of the fixed orifice in the end of the vortex finder.

A control valve was included in the fuel outlet (underflow) for two reasons; firstly to act as a non-return valve and secondly to maintain the pressure in the cyclone in the event of excessive quantities of air. By maintaining the pressure the gas laden fuel is forced through the overflow to, what would be, the second stage separator. The valve operated on the differential pressure across it and hence was related to the fuel flow rate. The possibility of the control valve acting against atmospheric pressure was explored but results from a prototype valve showed no improvement in separator performance. This was probably due to the constant pressure valve which starts to close much earlier than the differential pressure valve so restricting the fuel flow to such an extent that the cyclonic action in the separator was insufficient to separate the phases.

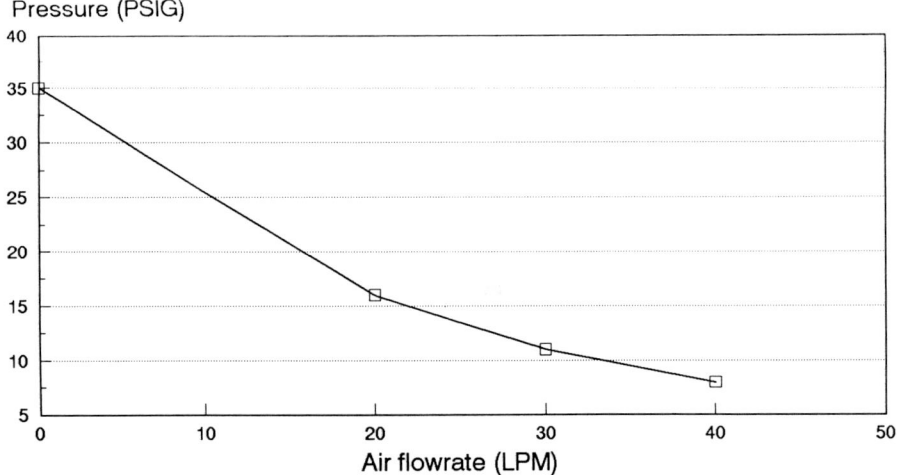

Test Fuel : Kerosene

Figure 6. Variation of pump outlet pressure with various air/flow rates.

The size of the orifice in the overflow was varied to explore its effect on the performance of the separator. The results shown in Figure 7 indicate that the performance increases with larger hole sizes. This appears to be logical since the extra flow through the orifice will convey more gas. However, the results were replotted against air flow rate (not air/total volume ratio) which reveals that the apparent improvement is due principally to the reduced delivery flow rate resulting from more fuel being recirculated through the overflow. This is shown in Figure 8.

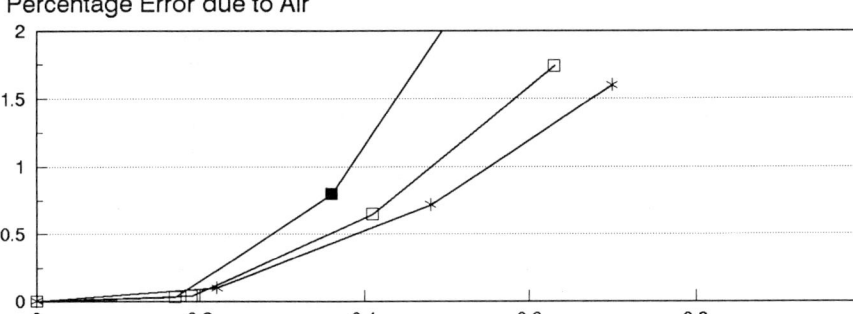

Figure 7. Results from the conical cyclone with various overflow orifice sizes plotted against air volume to total volume ratio.

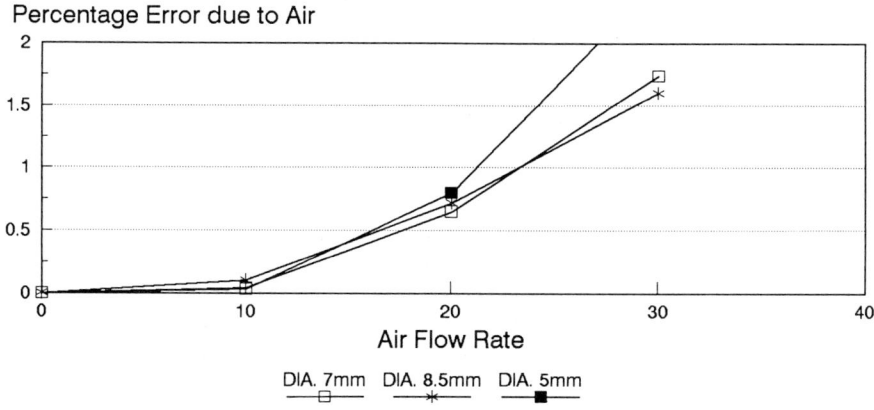

Figure 8. Results from the conical cyclone with various overflow orifice sizes plotted against actual airflow rate.

COMPARISON WITH EXISTING TECHNOLOGY

Since separation performance is viscosity dependant (because bubbles migrate more slowly in more viscous fluids) the following results are presented for the worst case - automotive diesel. Tests on a typical gravimetric separator in a UK dispenser show that unit fails (percentage error > 1%) at about 10% air in the total volume. In comparison the cyclones discussed in this paper fails above 30%. This is clearly a major improvement and is shown graphically in Figure 9 but the cyclone does not meet the standards for all air/fuel ratios as required and a gas detector would still be required.

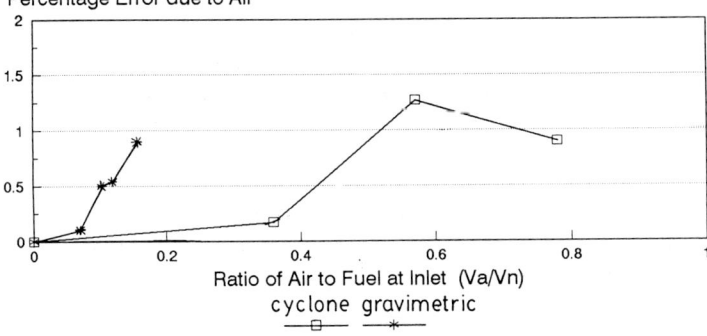

Test Fuel : Kerosene

Figure 9. Results from the conical cyclone compared with the existing gravimetric separator.

The cost of increased performance of a cyclone separator is an increase in the pressure drop. Pressure drops across a gravimetric separator are typically less than 0.14 bar gauge whereas the pressure drop measured across the cyclone was typically 0.34 bar gauge at full flow rate.

DISCUSSION OF ENVIRONMENTAL CONSIDERATIONS

The emphasis of the work presented in this paper was to satisfy the new separator standards without incorporating a gas detector. However, during the course of the project, as the new standards became more clearly defined, the emphasis in the dispenser industry moved towards protecting the environment.

A separator is essentially included to protect the customer from buying air under fault conditions such as priming, dry tanking and leaking suction lines. Only the latter condition results in continuous presence of gas. It is also the condition most damaging to the environment in that the leak could allow fuel to seep into the ground during inoperative periods and the gas separated during operation is simply pumped into the atmosphere. An additional device which prevents delivery under such a fault condition, is therefore considered to be a better option.

CONCLUSION

Cyclone technology has been shown to offer better separation performance than existing gravimetric techniques. However, the hydrocyclone developed does not achieve the performance necessary to meet the new standards for diesel at all air/fuel ratios.

The failure of the new cyclone to meet the new standards is a result of the range of flow rates, pressures and velocities that arise in a typical dispenser and the wide range of air/fuel ratios that must be considered. While further development of the control valve and the cyclone geometry might offer further improvements to the performance it is the authors belief that the new standards are unlikely to be met in full by a cyclonic separator without additional technology.

The penalty for increased separation performance is typically an increase in the pressure drop across the separator.

ACKNOWLEDGEMENTS

The project described in this paper was undertaken by the authors as part of a Teaching Company Project set up between Gilbarco Limited and the Department of Mechanical Engineering at the University of Southampton to explore alternative technology for roadside fuel dispensers. The project was originally set up with Avery Hardoll Limited who remained actively involved throughout the project. The research undertaken could not have been done without the generous support of Professor M T Thew and Dr I C Smyth of the University of Southampton.

PRACTICAL APPLICATIONS

REVOLUTIONARY METAL/CERAMIC AND POLYMER COMPOSITE TECHNOLOGIES
TO FORM NET SHAPE EROSION/CORROSION RESISTANT
HYDROCYCLONE APEXES AND LINERS

Jerry Weinstein, Matt Schreiner and Richard Webb
Alanx Products L.P.
101 Lake Drive
Newark, DE 19702

ABSTRACT

A unique erosion and corrosion resistant composite material has been developed by Alanx Products L.P. for use in hydrocyclones as apexes or spigots, cone and headliners, and vortex finders. Results of wear performance will be presented for lined hycrocyclones in the mining, mineral processing, and power generation industries.

This paper will present an overview of the revolutionary ALANX® ceramic/metal composite technology, and its application in the lining of hydrocyclones. Results will include both laboratory determination of compositions and properties and actual performance of components in hydrocyclones in installations on four continents.

Wear data for these ceramic/metal composites and for traditional metals, polymers and ceramics will be given a variety of laboratory slurry tests which simulate working conditions of hydrocyclones, in aqueous neutral and corrosive environments. Field results of cyclone liners used in hard-rock mining have shown more than 10 times the life of competitive ceramic, polymer and hard-metal liner materials. Field data from North and South American as well as Australian mining applications will be highlighted.

INTRODUCTION

While many ceramics have excellent wear resistance compared to metals, the acceptance of ceramic components in severe wear applications has been slow. This is due to ceramics' lack of toughness and propensity to fail catastrophically (i.e., low reliability) high cost, and limitations in producing large sizes and complex shapes. A new ceramic matrix composite technology* has been applied by Alanx Products L.P. to overcome

---

* Technology patented by Lanxide Corporation and tradenamed
 The DIMOX® directed metal oxidation process.

some of these shortcomings while maintaining the desired wear resistance of ceramics. This paper describes (1) the directed metal oxidation composite technology and its ability to make net shape parts of large and complex geometries size, (2) results of laboratory wear tests, (3) field test data hydrocyclones at mining and mineral processing end users.

The Ceramic/Metal Composite Process

The process to form ALANX® ceramic/metal composites is based upon Lanxide Corporation's invention of a new way to form ceramic composites. This approach produces wear resistant components through the directed oxidation of molten aluminum metal to form a ceramic matrix bond between filler grains of silicon carbide. (1, 2) Figure 1 illustrates the process. The silicon carbide reinforcing particles are formed into the final component shape by conventional ceramic fabricating techniques such as slip casting, pressing, or injection molding. The resulting preformed shape is placed in contact with molten aluminum metal in a heated oxidizing atmosphere. Under controlled conditions, the aluminum oxidizes forming a reaction product matrix which "grows" through the preform, infiltrating the silicon carbide reinforcement without displacing it. Barrier material placed on the surface of the preform stops the matrix from growing past the desired shape, thus defining the component geometry. In this way, a fully dense near-net-shape composite with a relatively smooth surface finish is obtained without the shrinkage typically associated with the densification of ceramics. The size capability is only limited by the reaction kiln, which presently is 60" diameter and 40" height. The resulting ceramic/metal composite consists of a three dimensionally interconnected ceramic matrix grown throughout the silicon carbide reinforcement with some residual metal left in the matrix.

A microstructure of a typical ALANX® silicon carbide reinforced aluminum oxide metal composite is shown in Figure 2. The mechanical characteristics of ALANX® ceramic/metal composites are dominated by the interconnected ceramic matrix and silicon carbide reinforcement, thus the composites exhibit extreme hardness and excellent wear resistance. The brittleness usually associated with ceramics is offset by the improved fracture toughness imparted by the ductile metal component (<10% of the composite) that is distributed throughout the matrix.

The directed metal oxidation process has been scaled up by Alanx Products L.P. for the production of large and complex parts, as seen in Figure 3.

Laboratory Slurry Erosion Tests

The slurry erosion resistances of ALANX® SiC particulate reinforced ceramic/metal composites were determined using a slurry pot test device to simulate erosion conditions found in hydrocyclones. Good correlation has been seen with these tests and performance in hydrocyclone liners in hard rock mining applications.

Test Procedure

Test samples were rotated through a highly loaded (40 wt. %) sand slurry. Test sample surfaces were impacted by the abrasive particles in the slurry. Erosion rates were calculated from measured weight losses. Two test specimen geometries were used; pins and fluted disks.

In the pin test, four 0.5 in. diameter by 2.5 in. long cylindrical pins (13.7 mm dia. x 63.5 mm) were attached to a steel flywheel as shown in Figure 4. The pins and flywheel were immersed in the sand aqueous slurry (in a water cooled pot) and rotated at 32 ft/sec (10m/sec) velocity for 20 hours. The sand slurry particle size was 0.01 to 0.02 in. (300 to 600 micron). A 96% sintered alumina pin was included in all test sets as a standard for normalizing the wear rates of other test pin materials under various test conditions.

Corrosion-erosion pin testing was done by using various acids and salt additives to the sand-water slurry mix as described above. For the acid slurries, the ph levels were maintained by automatically titrating the acid into the slurries as needed based on continuous measurements from a pH meter in the sand slurry.

Advantages of the pin test arrangement are the simplicity of the test specimen geometry and the ability to test three different material pin samples simultaneously. However, the erosion imparted to the pin is due almost entirely to high angle (>45°) sand particle impingement. (3)

The fluted-disk test was used to complement the pin tests. The fluted-disk specimen is a 4.7 in. diameter by 0.8 in. high (120 mm dia. x 20 mm high) disk having six flutes, each with a 0.3 in. (7.5 mm) radius fabricated and/or machined on its outside diameter. The arrangement used for the fluted-disk test specimen is shown in Figure 5. The disk was attached directly to the shaft of the motor drive and rotated at 59 ft/sec (18m/sec) velocity. Weight loss measurements were made after one, two, and four hours. Two particle sizes of sand were used: a "fine" sand with particle sizes from 0.01 to 0.02 in. (300 to 600μm), and a "coarse" sand with particle sizes from 0.04 to 0.07 (1000 to 1700μm).

The advantage of this disc test is that the three major modes of erosive wear can be identified on specific surfaces: directional impact with high angle impingement (leading edge of the flute), random impact (disk periphery) and sliding bed with low angle impingement (top and bottom surfaces). Huggett and Walker, who introduced this specimen geometry, report that it gives a very close correlation with wear rates measured in pump field tests for various materials. (4)

Slurry Test Results

Using these slurry tests, the effect of ALANX® composite microstructural variables on slurry erosion rates was determined and directed the development of a composite with maximum slurry erosion resistance (5).

In general, it was found that the erosion rate decreases with increasing SiC filler particulate size and increasing SiC filler packing density(s). (5)

In addition, it was found that when the rotating-pin slurry erosion tests were extended to durations longer than 20 hours, the erosion rates decreased to even lower levels. Observation of the wear surfaces of ALANX® composite pin and fluted disc samples after testing indicated that a portion of the matrix phases were preferentially removed, leaving the highly erosion resistant SiC filler grains sticking up "proud" on the surface. This is seen in the SEM micrograph of a composite wear surface in Figure 6.

This kind of wear pattern is an example of the microstructural effect known as "shadowing." Shadowing has been recognized as a significant factor in the erosion of other heterogeneous materials such as ceramic refractories.(6) Shadowing is a geometrical effect where the more erosion-resistant aggregate particles shield the less erosion-resistant matrix from erosive impacts. This kind of wear pattern results in an optimal material for slurry abrasion resistance. In this composite material, there is excellent bonding between the remaining matrix and the SiC particles. This prevents the removal of the SiC particulates with ongoing erosion. In addition, the high packing density of the SiC particulates prevents further access to the recessed matrix by the abrasive erodent particles of the test slurry.

The worn surfaces on the pin samples from the rotating-pin test are almost exclusively the result of high angle wear. Observation of wear surfaces on the fluted disc samples from the rotating-disk test shows the visually evident effect of a wide range of erodent impingement angles. In Figure 6, an SEM micrograph of the leading edge of one of the worn flutes is shown. This leading edge surface (to the left side in the micrograph) was subjected to high angle and high kinetic energy particle impacts. Note the severe wear seen by the eroded and rounded SiC filler grains standing out in high relief against the matrix. In contrast, the surface seen in the right part of the micrograph is the top surface of the disk which has been subjected to the sliding bed erosive (low angle) mode. This surface and the periphery of the disk which had been subject to random low angle impacts are similar in appearance. There is little wear evident on these surfaces as shown by the low relief due to only slight preferential removal of the matrix. This scanning electron micrograph illustrates the large difference in erosion rates due to erodent impingement angle. The much lower erosion rates at low angles are due to the lower effective kinetic energies of low angle impacts. Therefore, these SiC particulate reinforced ceramic composites would be expected to perform better at low angles than at high angles, but the "shadowing" wear resistance effect contributes at all angles. Operating optimally, the inner surfaces of hydrocyclones see mostly low angle wear.

Based on this work of optimizing SiC particle size and particle packing density, SiC reinforced ceramic/metal matrix composites were developed by Alanx Products L.P. with exceptional erosive wear resistance over the full range of impact angles. Alanx has designated these composites ALANX® 2K+ and CG896 and has scaled up production of them. Typical mechanical properties of ALANX® wear resistant composite are shown in Table 1. The ALANX® FGS composite contains a finer grain of SiC filler and has been formulated for shaft sleeve applications in centrifugal slurry pumps. The ALANX® CG273 composite has been developed for extremely corrosive/erosive applications.

Comparisons of ALANX® Ceramic/Metal Composites With Other Wear Resistant Materials

The laboratory slurry tests were used to compare the erosion resistance of ALANX® composites to commercially available wear resistant materials. The relative wear resistance of several materials, as determined in a rotating-pin slurry test, are shown in Table 2. A wear resistance factor, calculated as the volume loss of a sintered 96% Al_2O_3 standard divided by that of the test material, is used as a figure of merit for erosion resistance. As shown in Table 2, the ALANX® ceramic/metal composites and a newly developed ceramic/polymer composite exhibit superior erosion resistance compared to competitive materials tested, including nitride-bonded silicon carbide, Fe-27Cr (High Chrome Iron), Ni-Hard® IV and 96% sintered alumina. All competitive materials tested are those commonly used for abrasion resistant hydrocyclone liner parts.

Relative corrosion-erosion resistance of several materials was also measured on the same rotating pin slurry test, as shown in Figure 7. The pin testing was done in sulfuric and phosphoric acid environments (pH=1) as well as in a 3% salt solution. As shown in Figure 7, the Alanx® CG273 composite exhibited superior corrosion-erosion resistance compared to nitride-bonded silicon carbide, Fe-27Cr (High Chrome Iron), CD4MCu alloy, and 96% sintered alumina.

Comparisons of ALANX® CG896 to nitride-bonded silicon carbide and high chrome iron (Fe-27Cr) alloy in the rotating-disk test confirm the superior erosion resistance of ALANX® composites. As shown in Table 3, the rotating-disk test indicates that ALANX® CG896 is 2.5 to 3.5 times more erosion resistant than other materials tested. Based on the discussions above, if these tests were extended to longer durations, the erosion resistance of ALANX® CG896 relative to nitride-bonded SiC and Fe-27Cr would be even higher. It is expected that the ALANX® 2K+ composite would out perform all other materials previously tested as seen in the pin test data (Table 2), although 2K+ was not disc tested at the time of this writing.

Finally, the relative slurry abrasion resistances of ALANX® composites and several other wear resistant materials were determined using the ASTM test G-75-82 for measuring the abrasivity of slurries. (7) Tests were conducted using a Miller machine as shown in Figure 8. A 50% by weight slurry of standard AFS 50-70 mesh sand in water served as the abrasive media. Four sequential four-hour tests were run on each material. The results are shown in Figure 9. ALANX® CG896 demonstrated superior performance in this sliding abrasion test compared to competitive materials. The preferential removal of matrix early in the test sequence produces a pronounced initial dependence of slurry abrasion rate on cumulative test duration as discussed previously.

Hydrocyclone Field Tests

Hydrocyclone liner inserts made of ALANX® 2K+ ceramic/metal composite have proven their operational cost effectiveness in the field. With well over forty documented successes from actual field installations around the world, many major hard rock mining and mineral processing operations are now specifying this Alanx composite material to replace the more traditionally used liners of rubber, hard metals and other ceramics. It has been determined through these tests that the 2K+ composite provides not only longer component service, but also more efficient metallurgical separation in a closed-circuit grinding operation due to the slower, more linear wear behavior of the composite material.

At a moderate sized Cu-Ag mining operation (8,500 tons per day) in the pacific northwest, ALANX® 2K+ cyclone apexes are now specified for service in Krebs D-20 hydrocyclones. The cyclones are handling -5/8" ball mill discharge of 92% silica ore with a slurry consistency of 65% solids. The composite apexes are replacing rubber apexes based on a documented life improvement of 16 times (from 7 days useful life to 115 days). Economic calculations verified component cost-effectiveness and more consistent separation was observed as a result of a non-oval wear pattern on the 2K+ cyclone apex throat, a previous problem with rubber liners.

A gold producing operation in western Australia has recently completed an extensive evaluation of 2K+ cyclone apexes in a Linatex 225mm hydrocyclone. Data for this trial is presented in Table 4. In this trial, a group of seven nitride bonded silicon carbide ("Refrax")* ceramic apexes were run with two 2K+ apexes in a bank of nine cyclones. The cyclones were handling ball mill discharge of -¼ size of silica slurry of 65% solids. The two 2K+ apexes provided an average life improvement of six times (average hours per millimeter of wear) over that of the lower quality ceramic parts thus proving their cost effectiveness over the previously specified "Refrax" apexes. An additional benefit of the 2K+ apexes verified in this trial was the consistency of wear behavior.

As the data indicates, the two 2K+ apexes experienced nearly identical levels of wear following 715 hours of operation while the "Refrax" apexes exhibited very erratic, inconsistent wear behavior. This consistent wear behavior for the 2K+ composite can be attributed to the homogeneous microstructure of the material throughout its cross-sectional thickness. ALANX® 2K+ cyclone liners are now specified in this operation.

At 4,000 tons per day mining operation in the Carlin Trend gold deposit of northeast Nevada, ALANX® 2K+ apexes have successfully proven their cost effectiveness over a high grade of reaction-bonded silicon carbide ceramic ("KT")**. This well monitored trial was conducted in a bank of six Krebs D-15 cyclones operating in circuit with primary ball mills and handling a 60% solids slurry of -3/8" silica ore. Careful measurements of apex diameter were taken on a monthly basis of three 2K+ units and three units of the "KT" ceramic operating in series in the same cyclone bank. While the wear results indicated nearly identical and linear wear rates for the two materials, a post test economic analysis clearly identified the 2K+ material as the most cost effective performer. This analysis considered both wear life and material cost.(8)

While these results highlight only three specific field successes, there are numerous others. Specifically, results for ALANX® 2K+ cyclone liners in copper, gold, and nickel mining operations around the world have been equally positive.

A 1,400 ton per day Silver-Gold mining operation in the Pacific Northwest has completed an evaluation of ALANX® 2K+ hydrocyclone lower cones and apexes in their Krebs D26 cyclones. These cyclones handle minus 1/2" ball mill discharge of 60% solids aqueous slurry. The ALANX® 2K+

*   "Refrax" is a registered trademark of Carborundum Company.

**   "KT" refers to "KT-Hexalloy" which is a registered trademark of Carborundum Corporation.

material was evaluated versus two types of reaction bonded silicon carbide ("KT" and "CH" are the Krebs OEM designation for these ceramics) and proved to perform equally to both. Post test analysis of both performance and cost data proved the ALANX® 2K+ to be more cost-effective by a 35% margin. This analysis only considered component life and cost, since no additional benefits related to labor savings, downtime, and production efficiency could be identified since all parts tested exhibited similar lifetimes.

At a 40,000 ton per day copper operation in Chile, South America, ALANX® 2K+ cyclone apexes were evaluated on Krebs D26 hydrocyclones handling minus 1/2" ball mill discharge of 60% solids aqueous slurry. This modern milling operation was intent on implementing a planned maintenance scheduling approach in their mill which required their cyclones to achieve 100 days of uninterrupted operation. Prior to their evaluation of the 2K+ composite, they were only capable of achieving an average of 45 days of operation from "Refrax" apexes. Three 2K+ apexes installed in a bank of six cyclones achieved the 100 day life target and actually showed very little dimensional change (wear) after 100 days of continuous operation. While the apexes proved their capability to achieve the life target with the D26 sized cyclones, this operation is now replacing all of their standard cyclones with a new flat bottom cyclone design. Additional testing is currently underway to verify the 100 day life capability of the 2K+ material as an apex in this different type of cyclone design.

SUMMARY

Wear resistant SiC particle reinforced ceramic/metal composites are being produced to net shape using a new ceramic/metal composite technology. The SiC particle size and loading have been varied to optimize the erosion resistance of these composites. The ALANX® composites showed superior laboratory slurry erosion and field pump test results in comparisons with those of commercially available wear resistant materials. Field data indicates that these ALANX® composites out perform many currently used materials in a variety of hard rock mining and mineral processing industrial applications, but especially in field trials of hydrocyclone and pump parts in gold and copper mining applications. ALANX® ceramic/metal composites are supplied exclusively by Alanx Products L.P. of Newark, Delaware and licensed representatives and distributors worldwide.

REFERENCES

1. M. S. Newkirk, A. W. Urquhart, H. R. Zwicker, and E. Breval, "Formation of Lanxide™ Ceramic Materials," J. Mat. Res. 1 (1) 81-89 (1986).

2. M. S. Newkirk, H. D. Lesher, D. R. White, C. R. Kennedy, A. W. Urquhart, and T. D. Claar, "Preparation of Lanxide™ Ceramic Matrix Composites: Matrix Formation by the Directed Oxidation of Molten Metals," Ceram. Eng. Sci. Proc., 8 (7-8) 879-85 (1987).

3. H. Hojo, K. Tsuda and T. Yabu, "Erosion Damage of Polymeric Material by Slurry," Wear, 112, 17-28 (1986).

4. P. G. Huggett and C. I. Walker, "Development of a Wear Test to Simulate Slurry Erosion," Proc. 11th Int. Conf. on the Hydraulic Transport of Solids in Pipes, Stratford, England, October 1988.

5. J. G. Weinstein and B. Rossing, "Application of a New Ceramic/Metal Composite Technology to Form Net Shape Wear Resistant Components," S.K. Das, et.al Eds., High Performance Composites for the 1990's, pp. 339-60, 1991.

6. S. M. Widerhorn, "Erosion of Ceramics," Proc. of the Conf. on Corrosion-Erosion-Wear of Materials in Emerging Fossil Energy Systems, Berkeley, CA, January 27-29, 1982, National Association of Corrosion Engineers, 444-479, (1982).

7. J. E. Miller, "The SAR Number for Slurry Abrasion Resistance," in Slurry Erosion: Uses, Applications, and Test Methods, ASTM STP 946, J. E. Miller and F. E. Schmidt, Eds., Am. Soc. Testing Materials, Philadelphia, 155-66 (1987).

8. Kyran Casteel, Ed., "Alanx Success at Brenda," World Mining Equipment, (1990).

LIST OF FIGURES

1. The directed metal oxidation process for ceramic/metal composites.

2. Typical microstructure of a SiC particulate reinforced ceramic/metal composite consisting of a three dimensionally interconnected Al_2O_3 matrix (also containing some metal) grown around the SiC reinforcement.

3. Wear resistant components made of ALANX® ceramic/metal composites.

4. Rotating-pin slurry erosion test.

5. Rotating-disc slurry erosion test.

6. SEM photograph of wear surfaces of rotating disk test sample of ALANX® CG896 after 4 hours in 300-600 μm SiO_2 slurry.

7. Relative slurry corrosion-erosion resistance of several wear resistant materials in corrosive environments (rotating-pin test).

8. Sliding block (Miller) slurry abrasion test rig.

9. Results of sliding block (Miller) slurry abrasion test.

LIST OF TABLES

1. Typical properties of ALANX® ceramic/metal composites.

2. Relative slurry erosion resistance of several wear resistant materials (rotating-pin test).

3. Wear resistance of ALANX® CG896, Fe-27Cr, and nitride-bonded silicon carbide (rotating-disk slurry erosion test).

4. Hydrocyclone field test data for ALANX® 2K+ liners run at an Australian Gold mine.

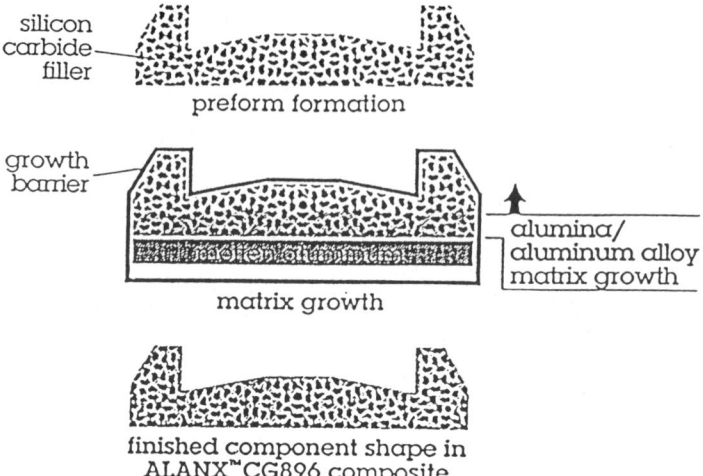

FIGURE 1

The directed metal oxidation process for ceramic/metal composites.

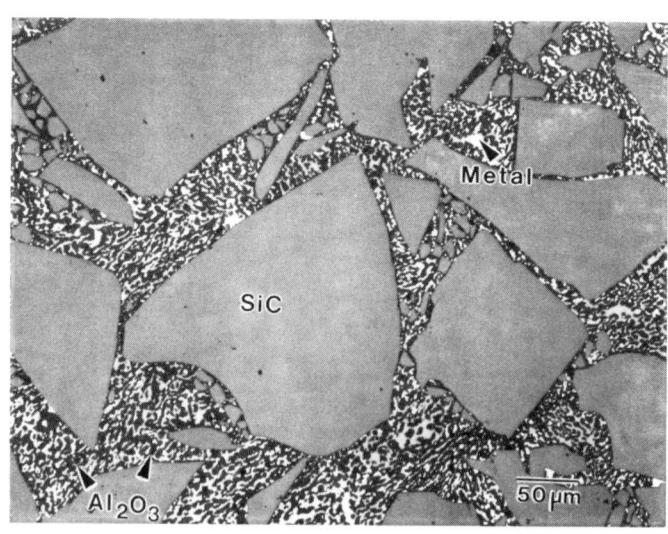

FIGURE 2

Typical microstructure of a SiC particulate reinforced ceramic/metal composite consisting of a three dimensionally interconnected Al_2O_3 matrix (also containing some metal) grown around the SiC reinforcement.

FIGURE 3

Wear resistant components made of ALANX^R ceramic/metal composites.

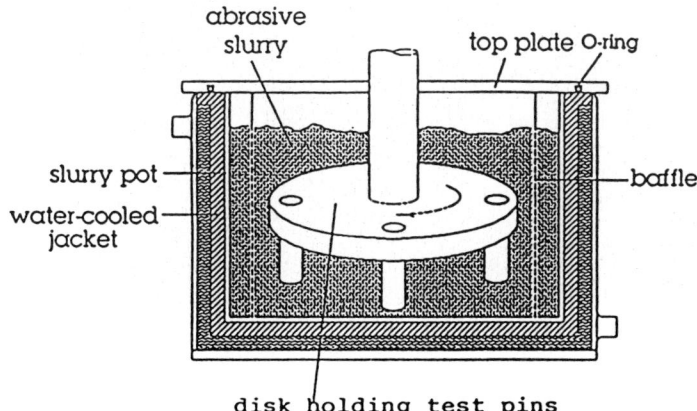

FIGURE 4

Rotating-pin slurry erosion test.

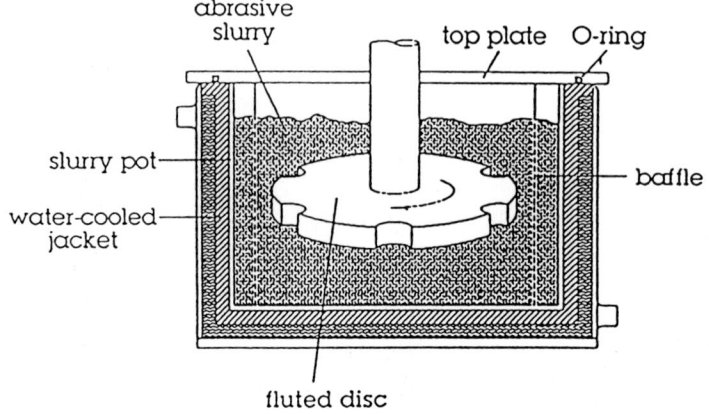

FIGURE 5

Rotating-disc slurry erosion test.

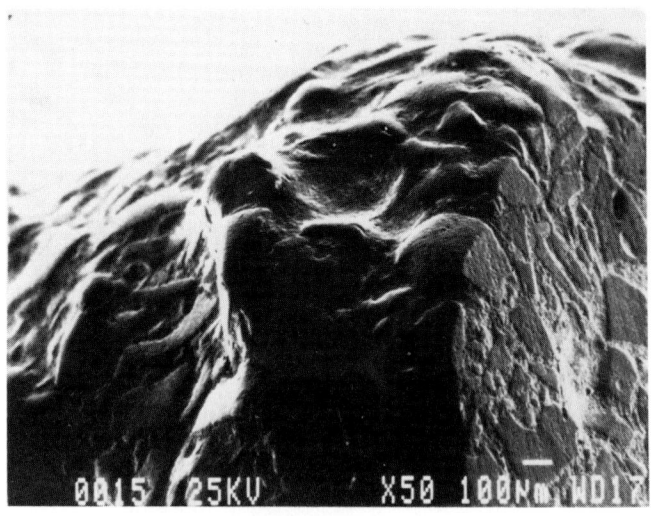

FIGURE 6

SEM photograph of wear surfaces of rotating disk test sample after 4 hours in 300-600 μm SiO_2 slurry.

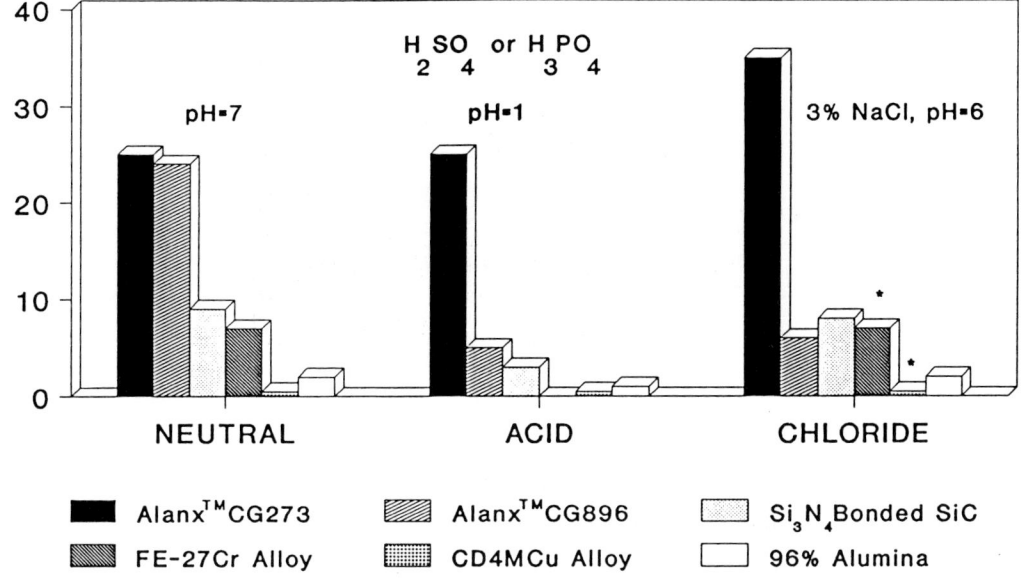

FIGURE 7

Relative slurry corrosion-erosion resistance of several wear resistant materials in corrosive environments (rotating-pin test).

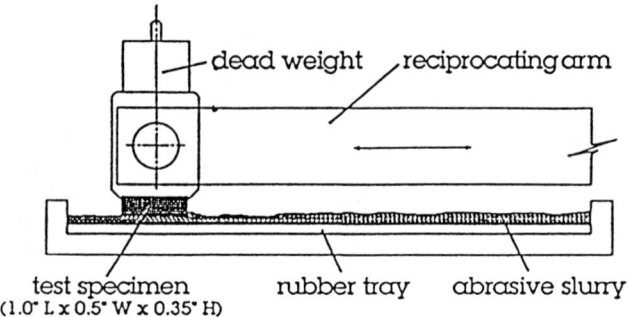

FIGURE 8

Sliding block (Miller) slurry abrasion test rig.

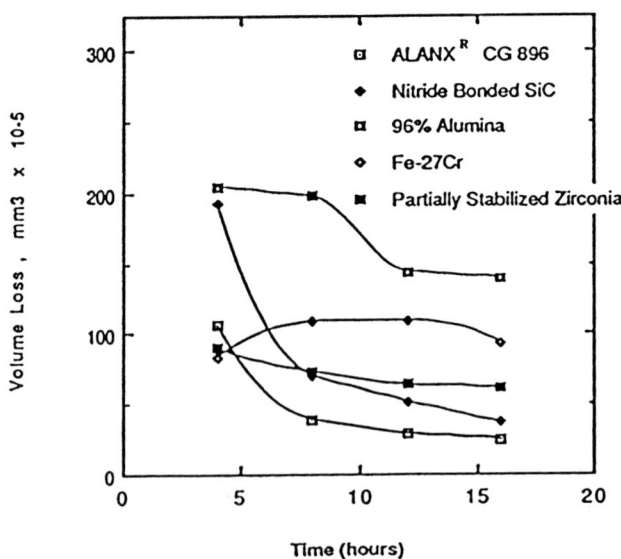

FIGURE 9

Results of sliding block (Miller) slurry abrasion test.

	ALANX 2K+	ALANX CG896	ALANX CG273**	ALANX FGS
Wear life factor*	45	28	24	12
pH range	5-10	5-10	1-14	5-10
Density g/ml (lb/ft^3)	3.26 (203)	3.28 (205)	3.42 (213)	3.32 (207)
Hardness (R_A)	80	80	83	80
Flexural strength (MPa)	90	140	100	220
Fracture toughness K (MPa·M$^{1/2}$)	5.5	6.0	4.4	5.4
Shear modulus (GPA)	131	131	144	144
Young's modulus (GPa)	313	313	357	366
Poisson ratio	0.19	0.19	0.23	0.27
Thermal expansion (10^{-6}/K)	5.40	5.43	5.20	5.89
Thermal conductivity (W/m·K)	150	147	125	140

* The wear life factor compares the service life of competitive materials. The larger the number, the longer the wear life.

** CG273 has excellent corrosion resistance.

TABLE 1

Typical properties of ALANX^R ceramic/metal composites.

Material	Wear Life Factor
ALANX (R) 2K+	45.0
ALANX (R) CG273	24.0
ALANX (R) CG896	24.0
ALANX (R) FGS	12.0
Nitride-bonded silicon carbide	9.0
Partially Stabilized Zirconia	9.0
High chrome iron (Fe-27Cr)	6.5
NI-Hard IV	5.3
Stellite 6B	4.9
Alumina (96%)	1.0
Gray cast iron	0.9

* The higher the number, the higher the wear resistance

Test conditions
  Specimen: pin (0.54″ O.D.x2.5″ length)
  Velocity: 10m/sec
  Slurry concentration: 40 wt. % solids
  Slurry particle size: 300 μm – 600 μm silica
  Slurry pH: 7-8 (neutral)
  Test duration: 20 hours

$$\text{Average wear resistance factor} = \frac{\text{Vol. loss standard of alumina}}{\text{Vol. loss test material}}$$

Alanx ceramic/polymer has been recently developed and is now being tested in pump and cyclone applications in the field.

TABLE 2

Relative slurry erosion resistance of several wear resistant materials (rotating-pin test).

	Volume Loss, mm^3	
Material	300–600 μm SiO_2	1000–1700 μm SiO_2
ALANX® CG896	410	1036
Fe-27Cr	1006	3891
Nitride-Bonded SiC	1386	2860

Test Conditions:

    Specimen: Fluted Disk (120 dia. x 20 mm)
    Velocity: 18 m/sec
    Slurry Solids Concentrations: 40 wt. %
    Test Duration: 4 hours

TABLE 3

Wear resistance of ALANX® CG896 ceramic/metal composite, Fe-27Cr, and
nitride-bonded silicon carbide (rotating-disk slurry erosion test).

Cyclone Position	Apex Material	Test Duration (Hours)	Increase in Apex Diameter (mm)	Hours/mm
1	Refrax*	274	4.2	65
2	Refrax	254	4.3	59
3	Refrax	331	5.1	65
4	ALANX 2K+	715	3.0	238
5	Refrax	141	13.7	10
6	Refrax	239	13.0	18
7	Refrax	161	10.1	16
8	ALANX 2K+	715	3.6	199
9	Refrax	255	11.5	22

Average hours/mm for Refrax 36
Average hours/mm for ALANX 2K+ 219

* "Refrax" is a registered tradename for a nitride-bonded SiC product manufactured by the Carborundum Company

TABLE 4

Hydrocyclone field test data for ALANX 2K+ liners run at an Australian gold mine

LARCODEMS Separator - Development of Three-product Unit

C L Shah, BSc, CEng, FMES
British Coal, Technical Services and Research Executive
Ashby Road, Stanhope Bretby, Burton on Trent, DE15 OQD, UK

SYNOPSIS

Coal Preparation Engineers at British Coal Corporation's Technical Services and Research Executive (TSRE) developed the world's largest centrifugal dense medium separator, the LARCODEMS, capable of treating up to 250t/h of 100-0.5mm raw coal. The first production plant was commissioned in 1986 and a number of plants are now in operation. The successful application of the two-product vessel encouraged its further development into a three-product separator. A single unit with just one medium input at low relative density can now produce clean coal, middlings and reject products.

This paper describes the two-product LARCODEMS and the development of the three-product unit. Some commercial and operational advantages of the development are highlighted.

INTRODUCTION

The LARCODEMS Separator[1,2] was developed to enable 100-0.5mm raw coal to be treated at up to 250t/h at efficiency levels significantly better than Baum jigs but at reduced cost. The first production installation, Figure 1, was commissioned in August 1986 and there are now five plants in production. It has been demonstrated that the LARCODEMS plants are:-

a)   small and simple
b)   efficient and reliable
c)   require less power and maintenance
d)   inexpensive to construct and operate

Figure 2, shows the schematic design and the flow patterns in the separator. The unit consists of a cylindrical chamber mounted at $30°$ to the horizontal. The separating medium is introduced, either by pump or by gravity feeding, under pressure into the chamber via an involute inlet at the bottom end. The liquid thus entering causes the rotation of the liquid in the separator and a vortex is generated around an aircore along the axis of the chamber. The medium nearest to the wall is carried via the involute outlet at the top end of the chamber to be discharged via a vortextractor. The vortextractor is a device successfully used in Vorsyl separators for discharging solids with a minimum volume of medium.

The raw coal is introduced into the separator with the help of a small amount of flushing medium. On entering the chamber, the material of the lowest relative density is carried by the vortex and discharged as clean coal, but the material of the highest relative density is sent to the separator wall by the centrifugal force and ejected via the vortextractor as reject. The material that is only of a slightly higher density than the medium tends to be carried lower down the vortex before it moves to the separator wall and then to the vortextractor.

The first production installation of the LARCODEMS plant at Point of Ayr Colliery demonstrated that the separator had fulfilled all design and development expectations[3,4,5]. The first unit was cast in nihard IV iron and lasted three years. The unit is now cast in high chrome iron, which is expected to increase the wear life to about 5 years. The LARCODEMS unit has no moving parts and therefore, requires no regular maintenance.

Because of the compact design of the LARCODEMS plants, it has been possible to develop modular plants[6,7] which can be factory constructed in small transportable modules and erected on site in a short period. This construction method can reduce design and building costs. The first modular plant was constructed in just 21 weeks, giving some 40% capital cost saving. The following LARCODEMS plants are in operation as this paper is written:

Point of Ayr	Commissioned August 1986
Silverdale	Commissioned April 1989
Goldthorpe	Commissioned April 1991
Rossington (two units)	Commissioned September 1991
Asfordby	Commissioned November 1991

THREE-PRODUCT LARCODEMS SEPARATOR CONCEPT

It soon became apparent that there are many applications where low relative density (RD), less that 1.5, separations are necessary. In such cases it would be necessary to recover a middlings product from the primary reject by retreating it in a secondary separator operating at a higher RD. This adds to the complexity of the plant and increases the cost of preparation. The LARCODEMS separator design is such that it is possible to introduce the reject directly into a specially designed second separator, which will give a separation at a higher RD, producing a middlings and a reject product. Further development work was undertaken by Coal Preparation Branch at TSRE to prove the theory of the three-product LARCODEMS and demonstrate the principle.

In all centrifugal dense medium separators using magnetite, there is a natural segregation of the medium solids. These natural characteristics were utilised to develop the three-product separators. The medium coming out of the clean coal outlet is of lower RD and that coming out of the reject outlet is of higher RD than the feed medium RD. Hence, the medium entering the second separator would be of higher RD than the feed medium. Therefore, the separation RD in the second separator will be higher than that in the primary separator. This principle is used to enable three-product separation to be made in the three-product LARCODEMS separator, Figure 3.

DESIGN AND OPERATION

In the three-product LARCODEMS, the reject vortextractor of the two-product LARCODEMS is replaced by the secondary chamber. The flow patterns as well as the separation of solids in the first chamber are unchanged. The reject medium from the first chamber enters the second chamber tangentially, thus causing rotation of the medium in the chamber and a vortex is generated around an aircore along the axis of the chamber. The medium around the aircore discharges through the upper axial outlet and the medium around the chamber wall discharges through the bottom axial conical outlet.

In operation the reject material from the first chamber enters the second chamber with the medium. The material of the higher relative density, shale, is kept near the separator wall by the centrifugal force and discharged via the bottom outlet. Whereas the lighter material, the middlings, are pulled to the vortex and discharged via the top outlet.

The design of the second chamber had to take into account the reduced pressure available at the inlet and the need for the coarse material to successfully pass into and out of as well as within the separator. As in the development of the two-product LARCODEMS, it was possible to eliminate the vortexfinder by adopting an involute entry.

PILOT UNIT

The three-product LARCODEMS was initially tested using a 300mm diameter model, Figure 4, capable of treating 25-0.5mm raw coal at upto 10t/h. A summary of results obtained from three tests can be seen in Table 1.

The results show that there is a sufficiently large (between 0.354 and 0.411) difference in separation RDs between primary and secondary separations. It can be seen that throughout the tests the reject ash content was high indicating that coal is not lost to reject even when operating at very low medium RD. A much higher separation cut-point can be reached than in a two-product separator.

500mm DIAMETER UNIT

A 500mm diameter unit, Figure 5, capable of treating 60t/h of 50-0.5mm raw coal, was installed at Orgreave Coal Preparation Plant for testing over a period of 4 weeks. The unit was installed to replace a gravity fed cyclone separator. It was therefore necessary to use a much greater volume of raw feed flushing medium to move the deslimed raw feed to the LARCODEMS inlet. This volume was initially controlled but because of oversize material present in the medium the butterfly valve kept blocking and therefore was later removed. This over-flushing caused problems in the primary separation as explained later.

Table 2 gives the results of the three-product LARCODEMS operating under different operating conditions. The two variables were, feed medium RD and amount of flushing medium/water used.

The results show that the separator performed very well even under adverse conditions. The excessive amount of flushing medium used because of the main plant operational problems, affected the performance in the primary chamber but not in the secondary chamber.

When excessive flushing medium is used, it causes a disturbance in the vortex and the raw feed travels much further down the vortex before the separation occurs. Hence, there is a much greater chance of high ash/high RD material short circuiting into the clean coal product. The result is a higher ash clean coal product. The samples 9 and 10 demonstrate this effect. The clean coal and middlings ash contents were changed from 6.7% and 36.7% respectively to 5.5% and 46.9% when the use of flushing medium was minimised.

The results also show that if low RD water is used in place of the high RD medium to flush the raw feed, the result is a low RD separation in the primary and a higher RD separation in the secondary. This can be seen in samples 2 and 3.

PERFORMANCE

Table 3, gives a summary of 5 performance test results, treating 38-0.5mm raw coal, under various conditions.

The results confirm the findings of the spot samples. In spite of the installation constraints, highlighted before, the results clearly demonstrate the flexibility of the separator. It is possible to lower the primary separation RD without incurring coal losses to the reject. The results also indicate that the three-product LARCODEMS can enable much higher RD separations to be made than in a two-product unit because the feed medium used is of relatively lower RD and therefore, less prone to problems related to high medium viscosity. The unit, if the middlings are combined with clean coal, will give additional clean coal capacity for the LARCODEMS separator.

It is important to note that the separator relies on the segregation effect of the medium and therefore, as with all other dense medium separators, the separation efficiency and the separation RDs will be affected by the proportion of slimes present in the medium. The increase in amount of slimes present reduces the separation density difference between the primary and the secondary separation in the three-product LARCODEMS.

POTENTIAL APPLICATIONS

a) Simple Flexible Coal Preparation Plants

   Where the market requirements for the clean coal product is variable, the three-product LARCODEMS will enable maximum recovery of the combustible material at a relatively low cost.

b) Sulphur Removal

   Most coals in the UK have higher sulphur content in the relative density ranges between 1.4 and 1.9 (middlings). It has been possible to carry out low RD separations to recover low sulphur coal and then recover the middlings by separation of the primary reject at a higher RD. The recovered middlings can be crushed, to liberate some of the pyrite sulphur, and retreated to discard the liberated sulphur. The three-product LARCODEMS will provide a cheaper means to recover the middlings.

c) Improved Feed to Coal Liquefaction Plant

   At the request of the BCC, Coal Liquefaction Project at Point of Ayr a number of coals were treated in the three-product LARCODEMS to prepare feed for their experimental work on coal liquefaction. The results were as given in Table 4.

It is not yet possible to quantify the advantages to the liquefaction process, but it has been claimed[8] that in commercial liquefaction economic benefits would be expected from processing only the clean product from the separation.

CONCLUSIONS

The LARCODEMS development has shown that large diameter hydrocyclones can enable low cost efficient coal preparation plant to be built giving added advantages of low operating and maintenance costs.

Further development of the LARCODEMS to a three-product separator will increase the flexibility of coal preparation plants. It enables a secondary high RD separation to be carried out, maximising the recovery of combustible material, at a minimal cost.

ACKNOWLEDGEMENT

The views expressed in this paper are those of the author and are not necessarily those of British Coal.

REFERENCES

1. SHAH, C.L. et al,
   'A 250t/h Centrifugal Dense Medium Separator for minus 100mm Raw Coal.' 10th International Coal Preparation Congress 1986, Edmonton, Canada

2. SHAH, C.L.
   'A New Centrifugal Dense Medium Separator for Treating 250t/h of Coal Sized up to 100mm 3rd Internations Conference on Hydrocyclones, 1987, Oxford, England p91-10 BHRA The Fluid Engineering Centre Published by: Elsevier Applied Science Publishers

3. CAMMACK, P.
   'The LARCODEMS - a New Dense Medium Separator for 100mm-0.5mm Raw Coal" AUFBEREITUNGS - TECHNIK 8/1987 p427-434

4. SHAH, C.L.
   'The LARCODEMS Separator for 100-0.5mm Raw Coal' MINE & QUARRY JUNE 1988 p48-52

5. HOWELLS, A et al,
   'The Rossington Coal Preparation Plant, MINE & QUARRY, APRIL 1992 p14-22

6. FIRTH, J.
'The "MODULAR LARCODEMS" - A further development in the cost effective use of the revolutionary Dense Medium Separator for 4" to 0.02" Raw Coal'. COAL PREP 91, LEXINGTON, KENTUCKY APRIL 30 - MAY 2 1991 p4-14 MacLean Hunter Presentations, Inc

7. TYLER, C.
'LARCODEMS UPDATE' - COAL PREP 92, 9th International Coal Preparation Exhibition and Conference, CINCINNATI, OHIO, MAY 5-7, 1992 p66-70 Maclean Hunter Presentations, Inc

8. MOORE, S.A. et al,
'Partial Maceral Separation in Dense Medium Coal Preparation Equipment and its Effects on Direct Liquefaction' 1991 International Conference on Coal Science Proceedings, p715-718 16-20 September, University of Newcastle upon Tyne, UK. Edited by IEA Research. Publisher : Butterworth-Heinemann Ltd, 1991

Test No	1	2	3
Medium RD	1.35	1.45	1.60
Cleaned Returning Medium RD	1.22	1.30	1.39
Middling Returning Medium RD	1.30	1.40	1.57
Reject Returning Medium RD	1.70	1.85	1.95
Raw Feed Ash %	40.5	45.4	38.2
Clean Coal Yield %	48.4	42.5	59.7
Clean Coal Ash %	6.0	6.1	9.7
Middlings Yield %	10.8	11.9	5.6
Middlings Ash %	33.8	35.9	45.9
Reject Yield %	40.8	45.6	34.7
Reject Ash %	83.0	84.4	85.9
Primary Separation RD	1.396	1.434	1.638
Secondary Separation RD	1.750	1.845	2.019
Separation RD Difference	0.354	0.411	0.381

Table 1 Summary of Test Results 300mm Diameter Separator

Sample No	Medium RD	Clean Coal		Middlings		Reject		Flushing Medium/Water
		% Sinks at 1.6 RD	Ash %	% Sinks at 1.6 RD	Ash%	% Float at 1.6 RD	Ash%	
1.	1.564	1.2	5.1	62.1	41.4	0.3	85.4	Controlled Medium
2.	1.548	0.9	4.8	63.9	40.4	0.3	83.9	Controlled Medium
3.	1.548	0.5	4.4	71.5	44.4	0.1	85.7	Controlled Water
4.	1.507	0.2	4.1	58.1	39.3	0.5	86.9	Controlled Medium
5.	1.486	1.1	4.9	81.1	50.9	0.2	88.5	Extra Water
6.	1.365	1.9	6.6	23.4	20.0	0.4	87.7	Excessive Medium
7.	1.339	2.4	5.3	10.6	20.9	0.4	84.8	Excessive Medium
8.	1.301	2.4	4.5	5.3	9.0	0.3	88.2	Excessive Medium
9.	1.550	4.1	6.7	52.1	36.7	0.6	86.8	Excessive Medium
10.	1.550	1.3	5.5	72.9	46.9	0.1	87.5	Controlled Medium

Table 2. Spot Sample Results Under Different Operating Conditions

Test No	1	2	3	4	5
Medium RD	1.548	1.486	1.365	1.339	1.301
Flushing Liquid	Medium	Water	Medium*	Medium*	Medium*
Raw Feed Ash %	36.8	45.0	42.1	38.7	44.1
Clean Coal Yield %	63.8	55.3	58.0	51.6	28.3
Clean Coal Ash %	5.4	4.5	4.4	7.8	8.2
Middlings Yield %	1.4	2.6	3.8	5.2	18.3
Middlings Ash %	41.9	47.8	28.7	18.4	10.9
Reject Yield %	34.8	42.1	38.2	43.3.	53.4
Reject Ash %	85.0	87.2	87.0	84.9	88.0
Primary Separation RD	1.613	1.542	1.465	1.435	1.300
Secondary Separation RD	1.780	1.894	1.773	1.675	1.682
Separation RD Difference	0.167	0.352	0.308	0.240	0.382

Table 3 Summary of Performance Results 500mm Diameter Separator

* Excessive amount of flushing medium

Sample	Medium RD	Clean Coal		Middlings		Reject	
		Wt %	Ash %	Wt %	Ash %	Wt %	Ash %
Point of Ayr	1.45	74.3	4.0	2.6	35.1	23.1	80.9
Point of Ayr	1.40	70.6	2.7	7.8	20.3	21.6	77.8
Point of Ayr	1.35	67.3	2.1	16.4	8.5	16.3	67.7
Point of Ayr	1.30	32.5	1.9	37.4	4.5	30.1	78.9
Calverton	1.35	42.9	2.0	5.6	25.4	51.5	83.5
Cotgrave	1.35	20.1	2.4	28.9	9.8	51.0	86.0
Treeton	1.35	43.9	2.5	18.9	9.0	37.2	83.3
Baddesley	1.35	46.6	2.9	21.1	9.4	32.3	78.3
Westerholt	1.35	38.1	2.1	24.5	9.1	37.4	79.9
Bilsthorpe	1.35	44.7	2.5	18.7	7.9	36.6	84.8
Littleton	1.35	29.8	2.2	38.4	7.1	31.8	87.5

Table 4 Low RD Separation for Coal Liquefaction

Figure 1. LARCODEMS Separator at Point of Ayr

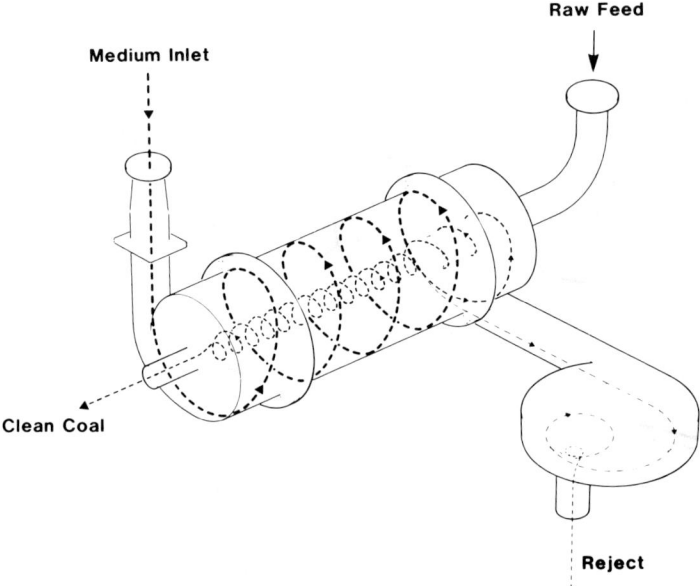

Figure 2. Two-product LARCODEMS Flow Patterns

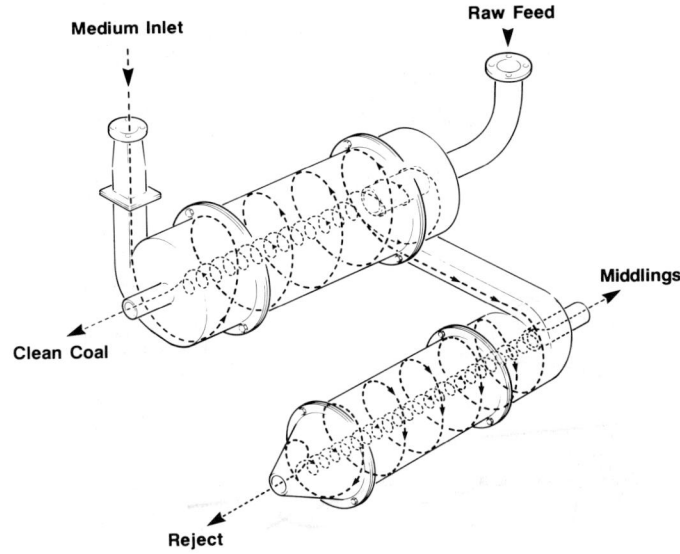

Figure 3.   Three-product LARCODEMS Flow Patterns

Figure 4.   300mm Diameter Pilot Unit

Figure 5. 500mm Diameter Three-product Unit at Orgreave Coal Preparation Plant

LIQUID HYDROCYCLONE SEPARATION SYSTEMS

M.F. Schubert, Conoco Inc., Vortoil* Separations Division

F. Skilbeck, Vortoil Separation Systems Limited

H.J. Walker, Vortoil Separation Systems Limited

ABSTRACT

Since introduction in 1983, the Vortoil* Hydrocyclone has become widely accepted in the oil and gas industry, with treating capacity of nearly 8 MM barrels of water per day sold to date. This success is largely due to the ability of this technology to consistently meet environmental discharge standards while offering significant installed weight and space savings and reduced operating costs.

However in order to maximise the benefits of hydrocyclones in cleaning oily water correct system design is critical. Lessons gained from actual field experience have been applied to full size system design, significantly improving oil removal efficiencies.

Further research has enhanced the separation efficiency of this technology in anticipation of more stringent environmental standards as well as adapting the hydrocyclone for primary separation of produced oil and water.

The applications of the hydrocyclone technology to be discussed in this paper are:

1) Hydrocyclone system design,

2) Hydrocyclone designs for enhanced separation,

3) Low pressure operation,

4) Hydrocyclone systems for primary production separation.

INTRODUCTION

The deoiling hydrocyclone provides a cost effective method to handle increasing water production, alleviating bottlenecks in many production systems. The ability to de-bottleneck water treating systems has in many instances allowed operators the flexibility to increase water floods and production rates, thus increasing oil production and improving oil recovery.

By improving separation efficiency, operators have an economic solution to new environmental standards. Efficiency improvements come not only from enhanced hydrocyclone separation but also from better system design and complementary separation equipment.

The ability to operate hydrocyclones at lower pressures, coupled with the associated space and weight savings, offers greater application of the technology for older production with the potential of extending field life. Using correctly designed centrifugal pumps to elevate feed pressure is a cost effective option when system pressure is inadequate.

The application of hydrocyclones in primary separation has now been tested in an attempt to further increase production processing capacities over conventional equipment and potentially increase oil reserves. Typically, expanding process capacity of production platforms is cost prohibitive due to limitations on platform space. Hydrocyclones for bulk oil water separations can be used to dramatically increase process capacity with minimal topsides modification.

HYDROCYCLONE SYSTEM DESIGN

In oilfield operations produced water is often generated from various stages of the oil/water separation system. In conventional water treatment systems (e.g. induced gas flotation etc) it is common practice to combine these flows into one stream feeding the treatment system (**FIGURE 1**).

However, the most suitable location for a hydrocyclone system is upstream of the level control valve, where the oil droplet size is at its greatest and hence hydrocyclone oil removal efficiency is maximised (1).

In addition the feed pressure at this location is the highest available and as hydrocyclone capacity is directly related to feed pressure the system capacity and individual unit turndown are also maximised. As Vortoil* Separation Systems' hydrocyclones have always been housed within a pressure vessel it is possible to supply units to any pressure rating, and hence this is the optimal location for the hydrocyclones. To date hydrocyclones have been tested and supplied upto ANSI class 1500lb pressure rating with operating pressures ranging from over 100 bar down to below 1 bar.

A further process improvement was gained with the use of a flotation device downstream of the hydrocyclones. This is attributed to the hydrocyclones perceived ability to coalesce the very small oil droplets that are not removed in the hydrocyclone. When the pressure is reduced downstream of the hydrocyclone the dissolved gas liberated attaches to these coalesced oil droplets, floating them to the surface.

This effect can be utilised by installing a low retention time skim tank (downstream enhancement vessel or DEV) downstream of the hydrocyclones (see **FIGURE 2**) and can substantially improve overall oil removal efficiencies (Table 1) particulary when treating difficult emulsions. It is this complete system design which allows hydrocyclone system oil removal efficiencies to meet or exceed environmental standards, worldwide, in the oil and gas production industries.

In addition this process enhancement has also enabled hydrocyclone systems to be successfully utilised in multifeed systems where upstream pressure losses are minimal i.e. the differences in operating pressures of the upstream separators are similar.

ENHANCED HYDROCYCLONE PERFORMANCE

The Vortoil liquid hydrocyclone geometry is based on research work conducted at Southampton University in England in the late 1970's under the direction of Martin Thew. Over the last years, Vortoil Separation Systems has undertaken an extensive research program to examine the operational parameters and major geometric components in order to improve the oil removal efficiency of the hydrocyclone system. The test program included the following parameters:

System Conditions	Geometric Parameters
Droplet Size	Inlet Area
Temperature	Nominal Diameter
Fluid Density	Length
Pressure	Geometric shapes
	Diameter Ratios

Due to the amount of data collected, details of the research work will not be presented here. Instead, the resulting geometry chosen from this work will be compared to a conventional 35 mm design.

The results of the test work has enabled Vortoil Separation Systems to introduce an optimized geometry which has a model designation of "K" liner. Upon selecting the appropriate geometric configuration, a direct comparison was made between the K liner and a conventional 35 mm design. The units were compared operating side by side while varying pressure, concentration and droplet size. A schematic of the test facility is shown in **FIGURE 3**.

Comparative results shown in **FIGURE 4** indicate the percentage increase in efficiency of the K Liner over the 35 mm units. In this particular test the two units were operated using a median feed droplet size of 9 and 15 microns. The results show that over the operating pressure range tested, the K liner shows a 10% to 25% increase in efficiency at the 15 micron drop size. At the smaller 9 micron drop size which represents a more difficult separation, the K Liner is 22% to 55% more efficient.

This is more significant at the low differential pressures (below 20 psi) where the K liner efficiency is dramatically higher. It should be noted that these tests were conducted at the specific conditions listed and indicate only relative efficiency between the units under those conditions.

TABLE 2 lists field test results conducted using the K liner together with comparative data on a 35 mm unit.

LOW PRESSURE OPERATION

Low Pressure operation of hydrocyclones can be addressed in two ways, 1) by improving the separation efficiency of the hydrocyclone or 2) elevating the system pressure to provide adequate hydrocyclone performance and operational flexibility.

Hydrocyclone Operation

The test results described above show that the K liner is efficient at pressures as low as 15 psig.

Generally the absolute efficiency at the low pressure will be determined by the temperature and droplet size of the produced fluids. It should also be considered that, if under certain conditions the hydrocyclone outlet does not meet the discharge standard, the combined system efficiency using a downstream skimmer usually will.

Pumped Systems

Extensive testing has been conducted to determine methods to pump produced fluids without creating significant droplet shearing. As mentioned before, any significant droplet shear will affect the separation of the fluids. Previous work conducted on low shear pumps reported that the use of progressive cavity pumps offered a low shear method of elevating system pressure.(2) Although this has proven a good technical solution, the cost and operational problems of progressive cavity pumps has led to further investigation for a simple, low cost alternative.

The most cost effective method was deemed to be centrifugal pumps which are widely used in the industry. A cooperative test program was established with a major pump manufacturer to determine the optimal design of centrifugal pump for low shear operation.

It was found that by limiting the pump head and correctly selecting the involute type, diameter and RPM, low shear pumping could be obtained. The pump was tested by measuring the droplet distribution on the inlet and outlet of the pump using crude oil under controlled concentration and droplet conditions. The pump was tested at various speeds and pressures to find the optimum operating conditions.

FIGURE 5 shows typical results where the change in droplet size across the pump is measured at various flowrates. This depicts the change in volumetric concentration of droplets at or below the ranges indicated as measured at the inlet and outlet of the pump. A ratio of 1 would indicate no change in droplet size while a ratio of 0.9 indicates a 10% reduction in droplet size at the outlet.

The results show that although there is a reduction in droplet size for the 25 micron range, there is only a small change in 5 micron droplets. The smaller droplets are most difficult to separate and would therefore be an indication of the hydrocyclone performance. It can be concluded that the minimal shear that takes place would have only a small impact on the hydrocyclone efficiency.

Further evidence that properly designed centrifugal pump systems can produce low shear has been seen in field installations. Most recently, a 240,000 bpd onshore produced water system was commissioned using centrifugal pumps followed by Vortoil* hydrocyclones. The hydrocyclone system is currently handling inlet concentrations averaging 300 to 500 ppm with the hydrocyclone outlet averaging less than 20 ppm. The low outlet concentration is a clear indication that minimal shear is taking place in the pumps.

Process Control

The test results showed that to maximise oil removal efficiencies it is best to operate the pumps and hydrocyclones at a constant flowrate.

In order to accommodate variations in the feed flowrate a process design for pumped systems as shown in **FIGURE 6** has been developed. In this system, the pump and hydrocyclone are designed to operate at the maximum system flowrate. Flow control is provided by a control valve downstream of the hydrocyclones whilst a recycle line between the hydrocyclones and control valve recycles part of the flow to upstream of the pump. A pressure control valve regulates the fluid recycle to ensure constant flow through the pump and hydrocyclones.

An added advantage of this design is that the recycled fluid is repeatedly passed through the hydrocyclone where further separation can take place improving overall system efficiency.

PRIMARY SEPARATION USING HYDROCYCLONES

Through continuing research and field testing the hydrocyclone design has been adapted to include it's use in bulk water removal in high water cut production, termed "Preseparation" and also in removal of residual water from crude down to BS&W (Base Sediment and Water) specifications termed "Dehydration" (3). Both processes enable the expansion of process trains to handle increased production rates or rising water cuts in a light weight, compact system. Ultimately, these process designs enable the recovery of oil reserves that could otherwise not be realised cost effectively using conventional three phase separation techniques. Outlined below are basic descriptions and typical applications for these systems.

Vortoil* Preseparator

The Preseparator Hydrocyclones are designed to remove bulk water from a water continuous mixture (usually 50% plus water), the intent being to substantially reduce the liquid loading on conventional treatment systems whilst giving water either to discharge specifications or suitable for treatment in conventional deoiling hydrocyclones.

A typical application would be to upgrade an existing system design consisting of two, 3-phase separators originally sized to handle a maximum fluid throughput of 100,000 BFPD with a maximum water handling capacity of 40,000 and 10,000 BWPD as shown in **FIGURE 7**. Reservoir development plans have shown that increasing total fluids throughput could improve oil production rates and hence recoverable reserves. In order to do this, however, total production rates will have to be increased to 140,000 BFPD at 70% water cut. At these production rates, the first stage separator is unable to effect good oil/water separation due to a greatly reduced residence time. This results in poor water quality and excessive water carry over to the second stage separator.

To overcome this, the first stage separator could be converted to a 2 phase unit, feeding all the liquids directly into a Preseparation Hydrocyclone system as illustrated in **FIGURE 8.** The Preseparation Hydrocyclone system handles the bulk oil/water separation enabling both the produced water treatment deoiling system and the downstream crude dehydration system to function according to specification. The underflow is virtually oil free (typically less than 1000ppm oil) and is suitable for further treatment by a conventional deoiling hydrocyclone system. The overflow contains virtually all the crude oil with a small quantity of water. This stream also contains any free gas that has come out of solution, due to the pressure drop in the hydrocyclones. It is then fed directly into the second 3-phase separator for final separation of the product fluid.

By utilising this technology it should be possible to upgrade existing systems to handle increased throughputs in much the same way that Deoiling Hydrocyclones have been used to increase the capacity of existing produced water treatment systems. As with Deoiling Hydrocyclones the Preseparator Hydrocyclones are of a modular construction and are relatively easily retrofitted into the existing topsides without extensive platform modifications which would cause unnecessarily long shutdowns.

In addition to their wide acceptance in the oil industry, liquid/liquid hydrocyclones are now being evaluated for a variety of industrial applications. To date field testing has shown good potential in several areas. In years to come liquid/liquid hydrocyclones could reach the same level of acceptance in other industries as in oil field operations.

REFERENCES

1. Meldrum, N; Hydrocyclones: A Solution to Produced Water Treatment; paper OTC 5594, 19th Annual Offshore Technology Conference, Houston, Tx, April 27-30, 1987.

2. Flanigan, D.A., Skilbeck F., Stolhand J.E., Shimoda E.; "Use of Low-Shear Pumps in Conjunction With Hydrocyclones for Improved Performance in the Cleanup of Low-Pressure Produced Water"; paper 119743, 64th Annual SPE Conference, San Antonio, TX, October 8-11, 1989.

3. Schubert, M.F.;"Advancements in Liquid Hydrocyclone Separation Systems"; Paper OTC 6869, 24th Annual OTC, Houston, Tx May 4-7, 1992.

TABLE 1 - FIELD TEST RESULTS WITH & WITHOUT DOWNSTREAM DEGASSER

LOCATION	HYDROCYCLONE INLET (ppm)	HYDROCYCLONE OUTLET (ppm)	DEGASSER OUTLET (ppm)	COMMENTS
NORTH SEA	400	17		35 MM NO DEGASSER
NORTH SEA	600	25		60 MM NO DEGASSER
NORTH SEA	400	45	30	35 MM
USA	520	39	20	K-LINER PUMPED SYSTEM
USA	2310	160	6	35 MM PUMPED SYSTEM
DUBAI	327	88	39	35 MM
NORTH SEA	102	30	17	G-LINER PRODUCTION SYSTEM

TABLE 2 - FIELD TEST RESULTS K VS 35MM

HYDROCYCLONE TYPE	HYDROCYCLONE INLET (ppm)	HYDROCYCLONE OUTLET (ppm)	DEGASSER OUTLET (ppm)	COMMENTS
K	510	49	22	NO CHEMICAL TREATMENT
K	798	21	8	WITH CHEMICAL TREATMENT
35MM	697	65	42	NO CHEMICAL TREATMENT
35MM	731	22	10	WITH CHEMICAL TREATMENT

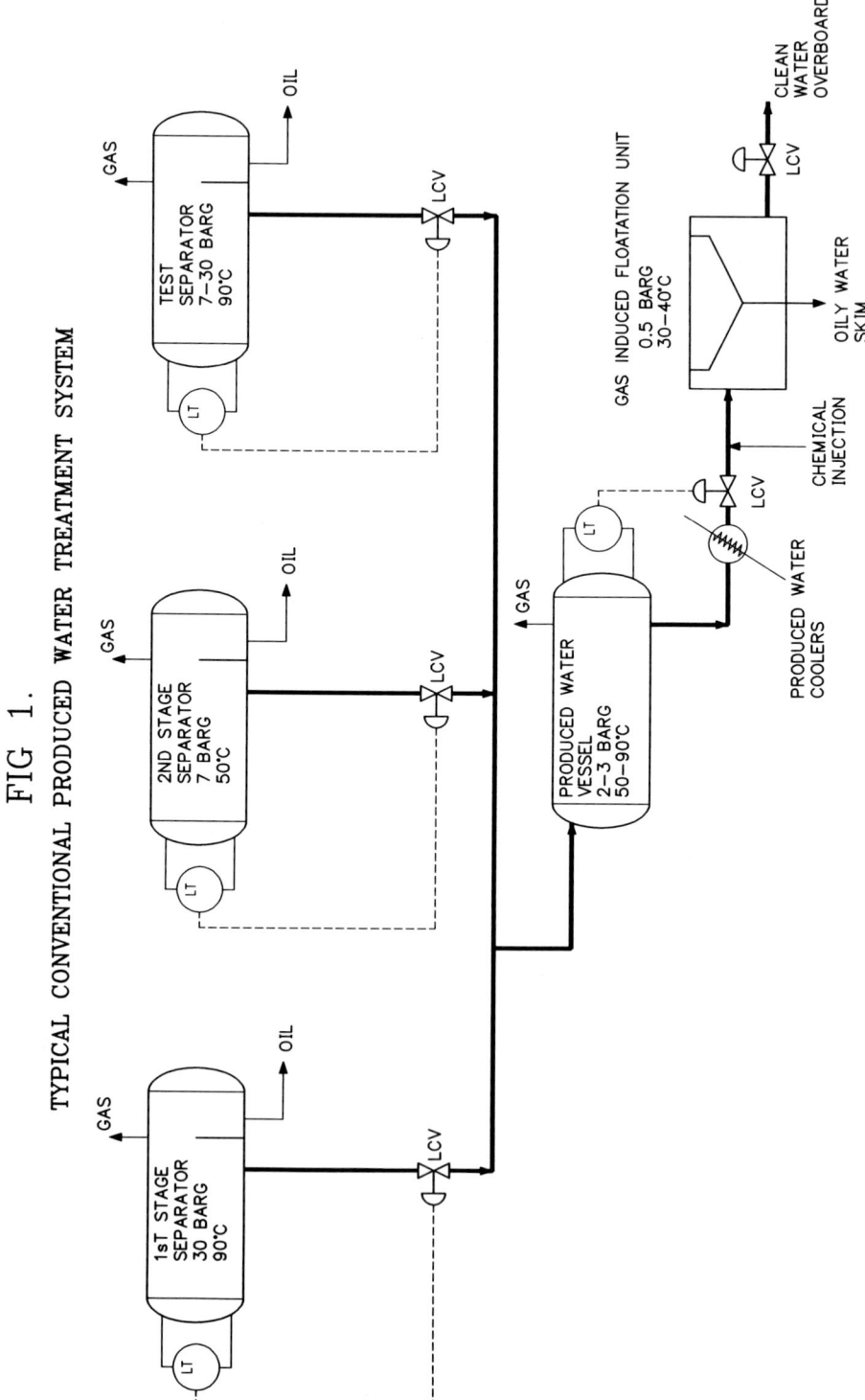

FIG 1. TYPICAL CONVENTIONAL PRODUCED WATER TREATMENT SYSTEM

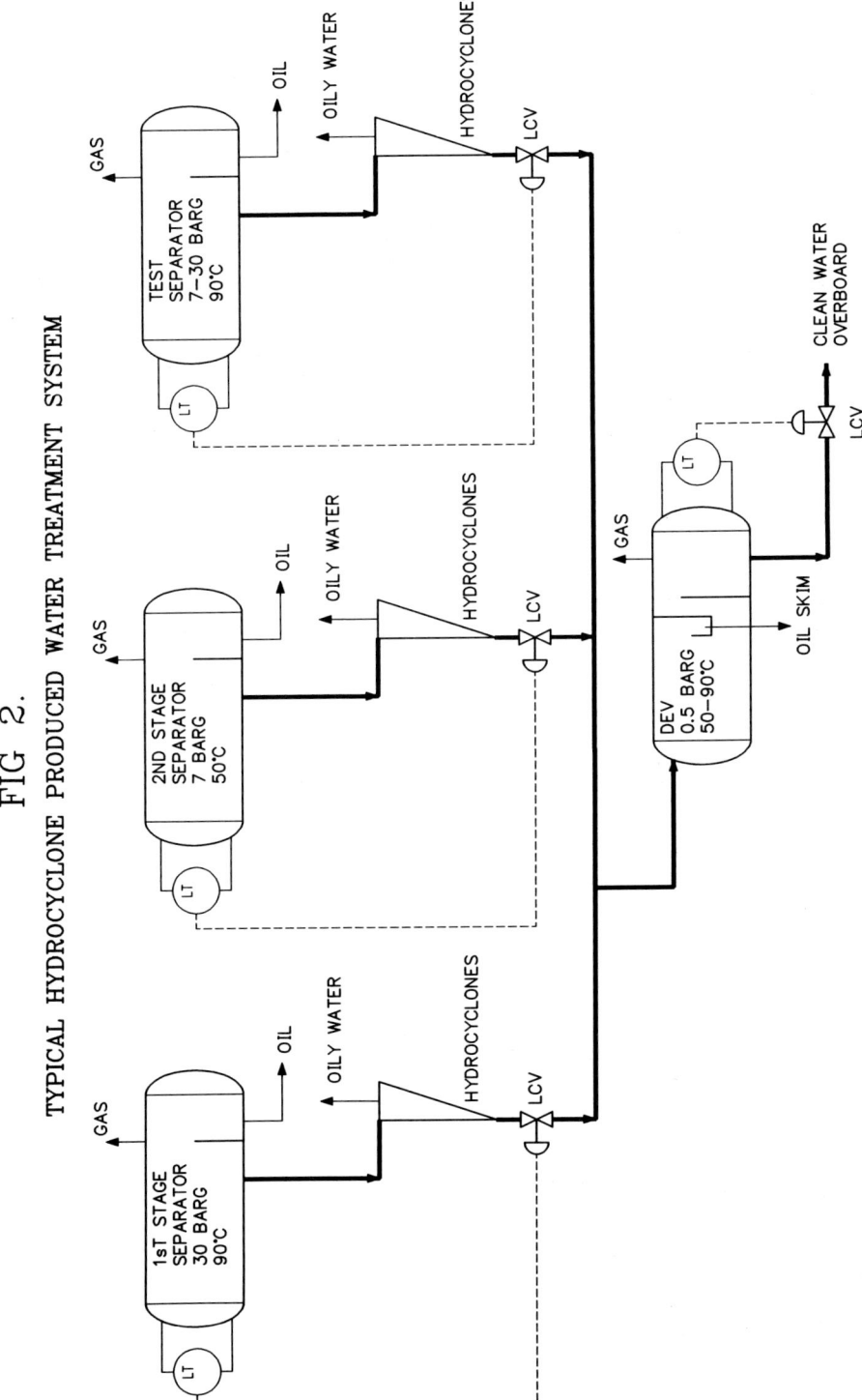

FIG 2.
TYPICAL HYDROCYCLONE PRODUCED WATER TREATMENT SYSTEM

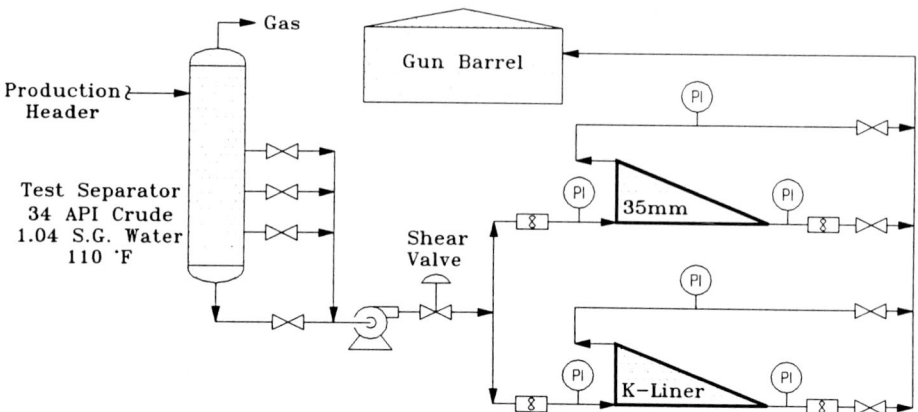

Figure 3, Hydrocyclone Test Facility

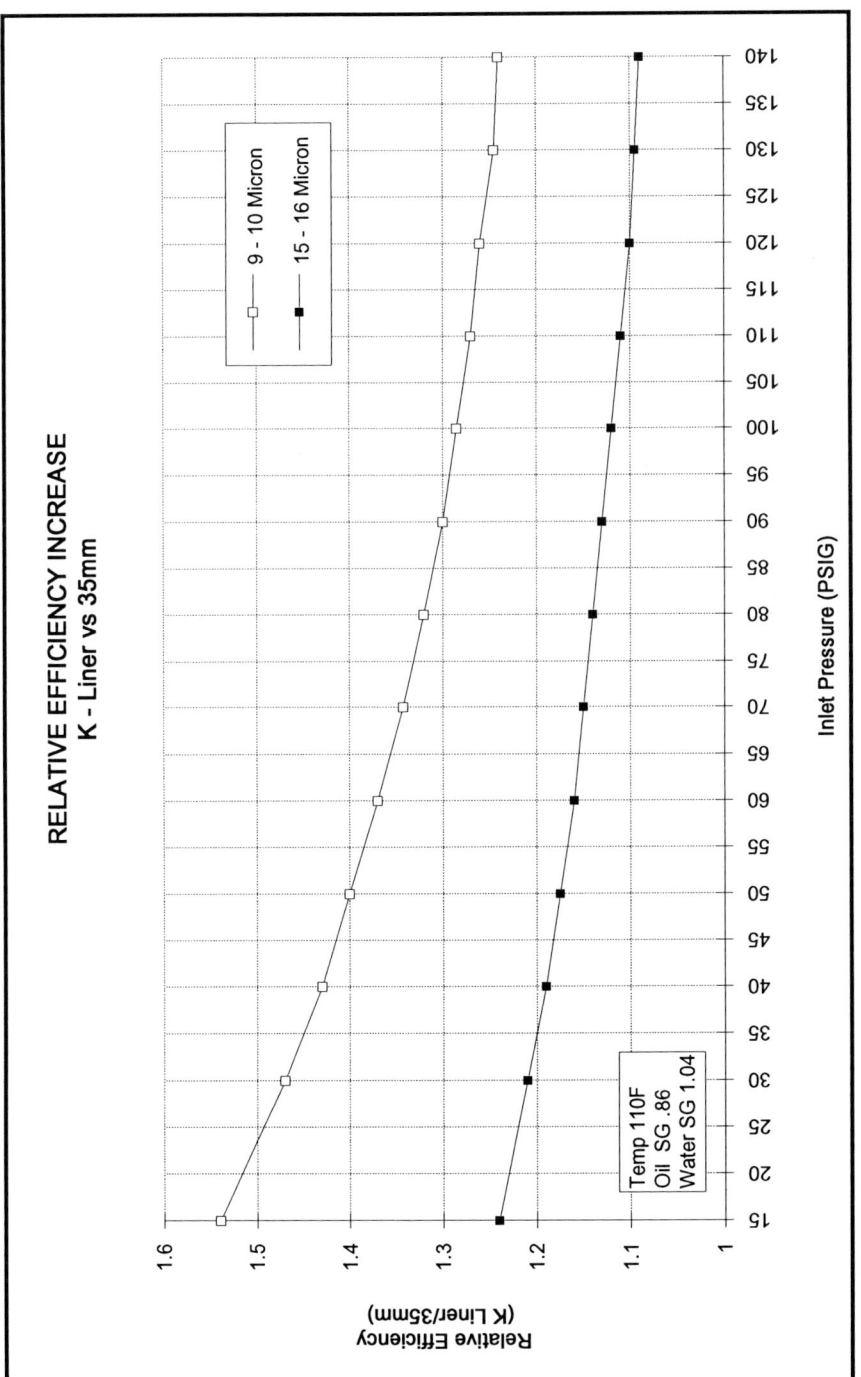

Fig. 4

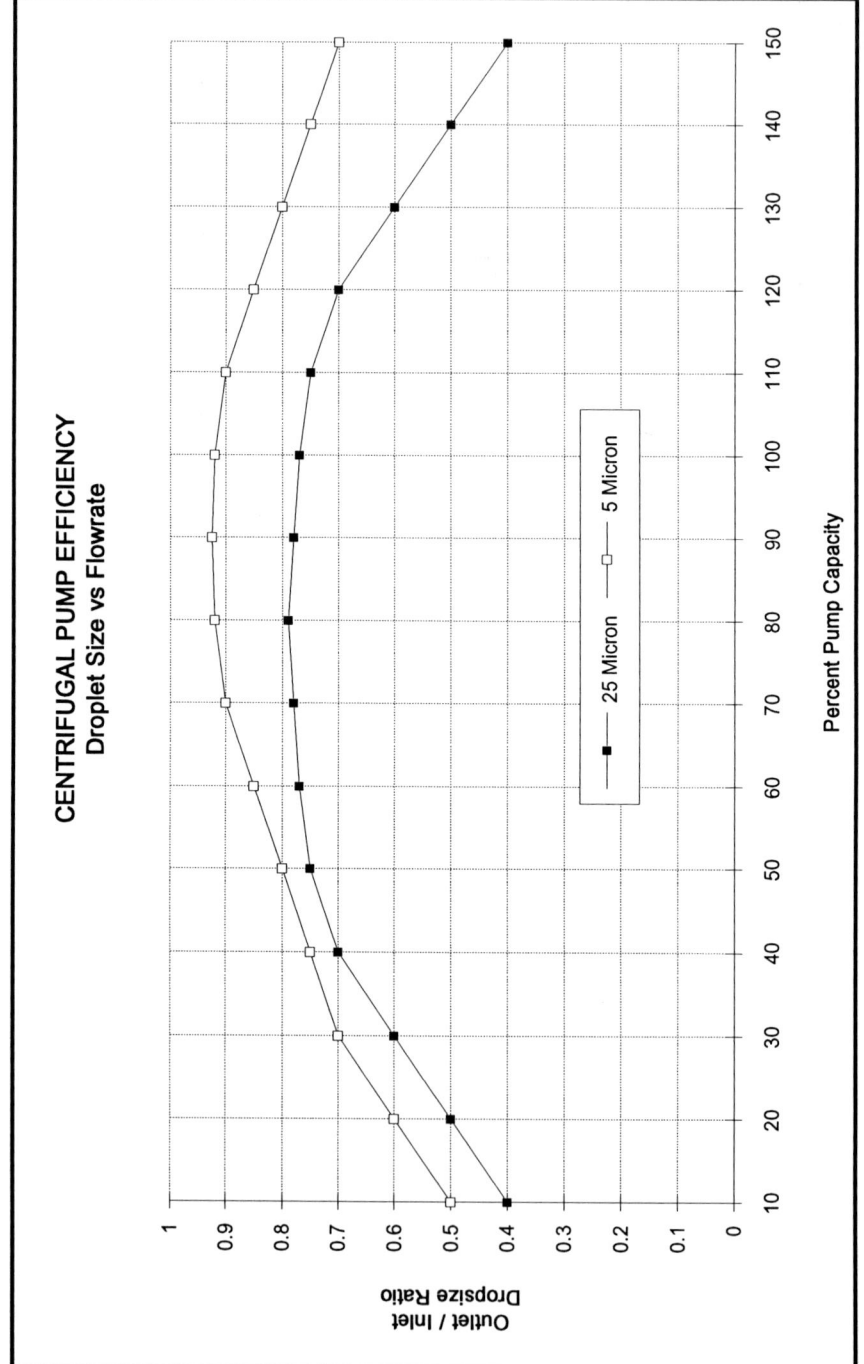

Fig. 5

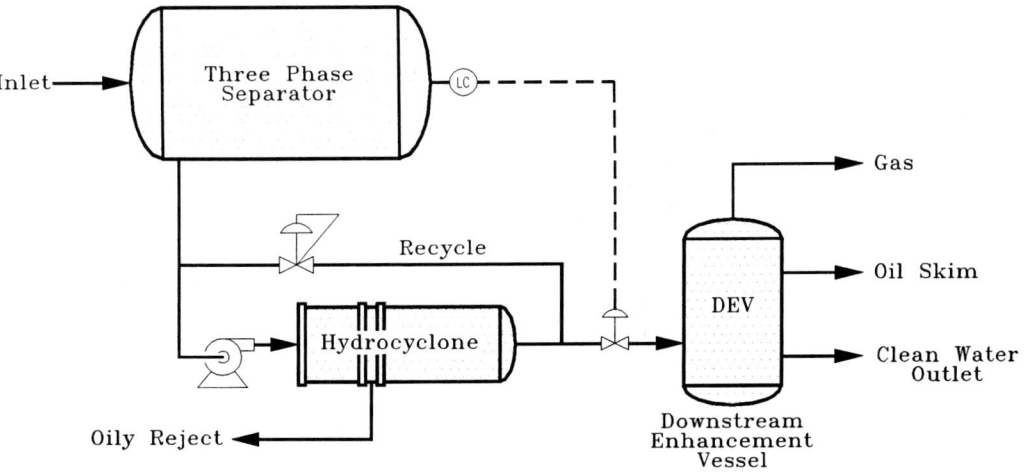

Figure 6, Hydrocyclone Pump Process

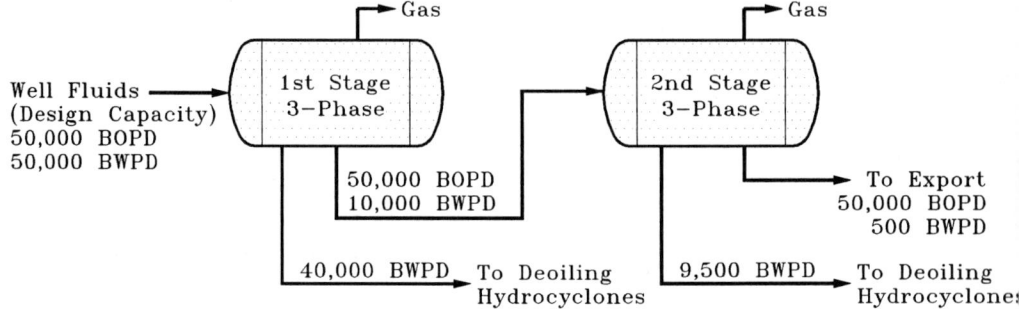

Figure 7, 3-Phase Conventional Separation Train

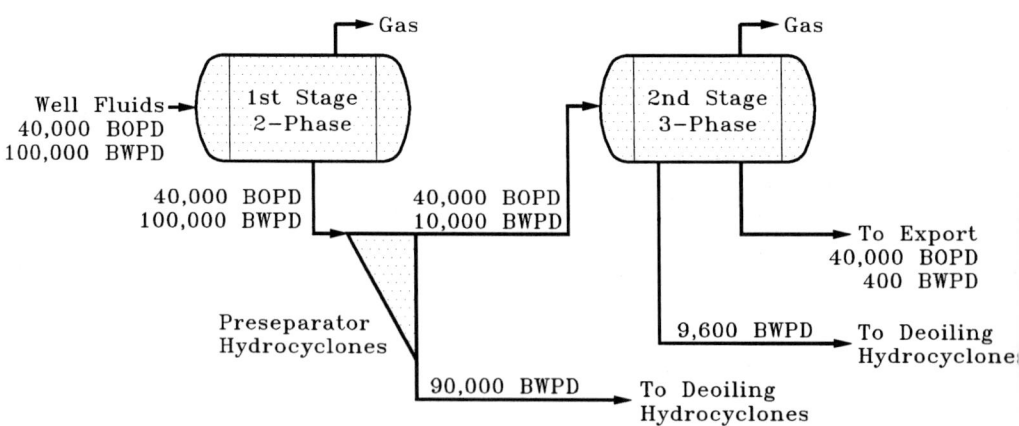

Figure 8, Preseparator Application

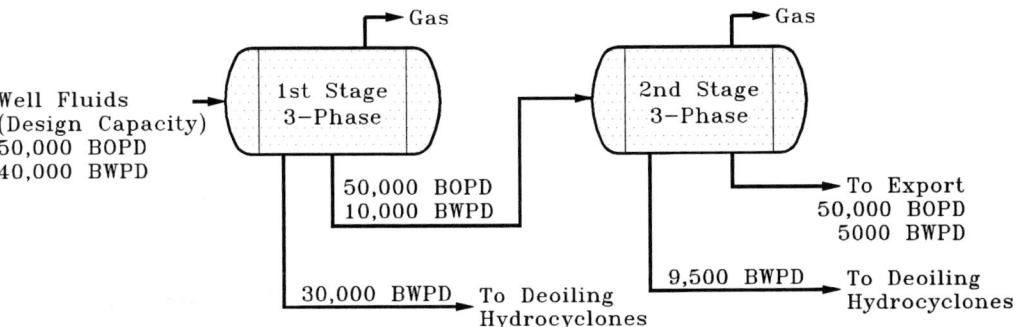

Figure 9, 3-Phase Conventional Separation Train

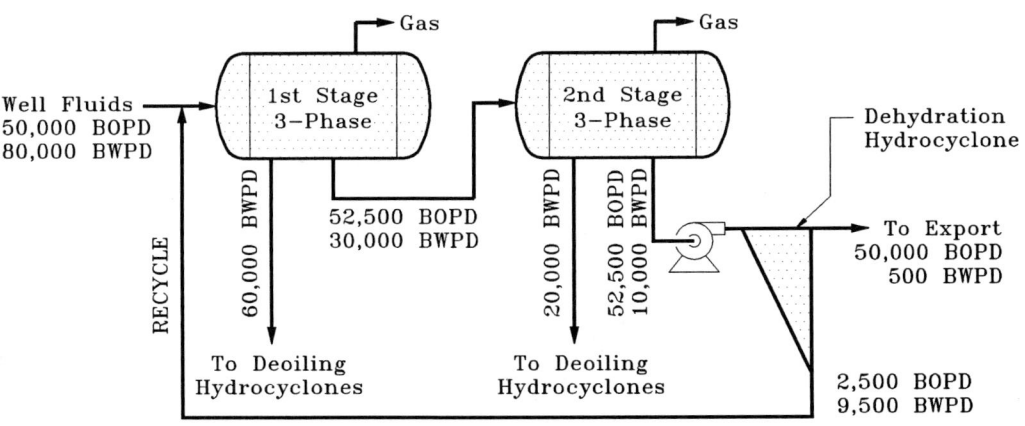

Figure 10, Dehydrator Application

TESTING OF THE VORTOIL DEOILING HYDROCYCLONE USING CANADIAN OFFSHORE CRUDE OIL

K.M. SIMMS[1], S.A. ZAIDI[1], K.A. HASHMI[2], M.T. THEW[3], and I.C. SMYTH[3]

1. Wastewater Technology Centre, Canada
2. Energy, Mines & Resources, Canada
3. University of Southampton, UK

ABSTRACT

A single CSPI Vortoil hydrocyclone was evaluated under laboratory conditions using simulated Canadian offshore produced water. Test water was comprised of crude oil dispersed in a sodium chloride solution. Oil removal capability was determined under various flow rate, feed oil concentration, oil drop size and temperature conditions. The effect of selected treatment chemicals, solids and free gas was also investigated. Test results indicated oil removal rates approaching 90% for dispersions of 25 μm mean drop size at room temperature, with modest improvements in performance at increased temperature. A number of difficulties were encountered in the laboratory test program and caution is advised in applying test results of this nature to the real offshore environment.

NOTATION LIST

d	particle diameter, μm
$\bar{d}$	volumetric mean particle diameter, μm
d_{50}	cut size or droplet diameter for which MP = 50%
Eu	Euler number or pressure coefficient = $\Delta P/0.5\rho v_i^2$
K	concentration of dispersed phase, % or ppm (volume), mg/L (mass/volume)
MP	oil drop migration probability to upstream outlet
N	shearing pump speed, rpm
p(d)	dispersion % undersize (or oversize) volumetric size distribution function
Q	flow rate, L/min
R	oil recovery = $K_{us}Q_{us}/K_iQ_i$
Re	Reynolds number = $v_i\rho D/\nu$
v	velocity, m/s
ρ	density, kg/m^3
ν	molecular viscosity, mPa.s
subscripts	
ds	downstream or water discharge
i	inlet or feed
us	upstream or reject

INTRODUCTION

Removal of oil from produced waters at offshore production facilities is generally required prior to marine discharge or injection. While hydrocyclone technology is being increasingly applied for removal of oil from produced waters on offshore oil production platforms, the technology has not been systematically tested for the Canadian offshore situation. At the time of this study, in 1990, several offshore development projects were being initiated in Canada and hard data on relevant produced water treatment options were required to assist decision makers in the oil industry and the regulatory agencies.

The Wastewater Technology Centre has been actively investigating various approaches for the treatment and monitoring of produced water since 1982. Under a study jointly funded by several oil companies and the federal government, the Wastewater Technology Centre contracted the University of Southampton to conduct an experimental evaluation of a static deoiling hydrocyclone under controlled laboratory conditions. The test work was designed to examine the effect of flow rate, oil concentration, droplet size, temperature, and interferences on separation efficiency, and to identify any anomalous behaviour associated with the Canadian crude oil used.

This paper details of the experimental work conducted and reviews the data generated and its relevance to offshore oil production installations in Canada.

BACKGROUND

There are a number of regions located off the Northern and Eastern coasts of Canada with sufficient oil potential to warrant development plans. Currently, development projects are being initiated at two East Coast regions, the Scotian Shelf and the Grand Banks. An initial light oil recovery operation, the Cohasset/Panuke Development Project on the Scotian Shelf, is scheduled to come on stream in summer 1992. With recoverable reserves of about 8×10^6 m^3, the project may generate up to 6400 m^3/d of produced water over a 6 to 7 year life [1]. Substantially larger reserves (approx. 80×10^6 m^3) of conventional oil are located in various reservoirs in the Grand Banks region. Here, initial production is expected in 1997 with substantial volumes of produced water requiring treatment (14,000 m^3/d for 18 years at the Hibernia Development Project[2]). Consequently, guidelines [3] are in place regarding allowable levels of oil in discharges of produced water to the marine environment. They recommend a maximum monthly average level of 40 mg/L oil and grease. Conoco Specialty Products Inc. Vortoil hydrocyclone units have been specified for produced water treatment at Cohasset/Panuke and are being considered for Hibernia. For the purposes of this study, it was decided to evaluate hydrocyclone technology for application at offshore oil recovery operations in the Grand Banks region. A typical crude oil from this region was therefore selected for use in this study.

EXPERIMENTAL

Study Scope

The test program was divided into two phases. Phase 1 included two parts: (i) an examination of the effect of flow rate on separation efficiency and (ii) generation of data on migration probability and separation efficiency at various inlet oil concentrations. In Phase 2, the effect of various additives on the hydrocyclone performance was measured. Table 1 highlights the operating conditions for each phase of the study. As hard data was not available on characteristics of the produced water from the East Coast of Canada, test conditions were selected in consultation with the Canadian oil industry.

TABLE 1
Operating Conditions for Each Phase of Testing

Operating Variables			Phase 1.1	Phase 1.2	Phase 2
Temperature	T	°C	25	25, 55	55
Inlet Concentration	K_i	ppm	500	50,150,500,1000, 5000	500
Mean droplet size	d_i	μm	20, 30	---	---
Shearing pump speed	N	rpm	2500, 2700	2000, 2700	2700
Inlet flow rate	Q_i	L/min	40,60,75,90,100, 110,120,130	90	90
Additives			none	none	gas, solids, chemicals

Additives included in Phase 2 of the study were free gas, solids and selected treatment chemicals as follows:

- nitrogen at 5 and 20% volume at inlet conditions
- silica flour (British Sand HPF2) with a mean diameter of 15 μm at 100 mg/L concentration
- water soluble treatment chemicals (Petrolite, UK) described in Table 2.

The test philosophy for the treatment chemical study was to first establish the influence of reverse demulsifier (RD) on the deoiling process and then work at the optimal RD dosage and monitor the effect of individual additions of other chemicals.

TABLE 2
Treatment Chemicals Tested

Chemical Type	Trade Name	Chemical Description	Dosage (ppm)
Reverse Demulsifier	Tretolite J-229	cationic polyamine quaternary compound in water	0, 1, 5
Corrosion Inhibitor	Kontol KW10	amide/imidazoline in naptha	2
Scale Inhibitor	SP-2500	phosphate polymer blend	5
Antifoaming Agent	VEZ-D-1935	water soluble silicones	2

Fluid Specifications

Simulated produced water samples were generated for the test work by addition of crude oil to a brine solution. The brine solution comprised of sodium chloride in tap water (density 1048 kg/m^3 @ 25 °C, pH 7.5). Crude oil from the Terra Nova reservoir located off the East Coast of Canada was supplied by Petro-Canada Resources. The properties of the crude oil are listed in Table 3.

TABLE 3
Properties of Test Oil

Density (kg/m^3) at 15°C	864
Gravity (°API)	32
Pour Point (°C)	< 0
Flash Point (°C)	16
Wax Content (wt%)	7.9
Viscosity (mm^2/s) @ 30°C	16.3

Test Unit

A 35 mm "high swirl" 3-section Vortoil unit with standard upstream reject (2 mm diameter), supplied by Conoco Specialty Products Inc., was used for the testing. The high swirl format was originally chosen because it was anticipated that the oil dispersions mechanically generated using Terra Nova crude would be very fine. The split ratio, which is the ratio of volumetric flow at the upstream outlet to the inlet feed flow, was set at 1% during the test work.

Test Rig Configuration

Test work was conducted using the University of Southampton once-through deoiling separator test facility. The system comprised of two main flow loops: the hydrocyclone test loop (Figure 1), containing various storage tanks, a heat exchanger, the main flow pump, the test unit and associated instrumentation, and the constant flow oil dispersion loop (Figure 2) containing the oil reservoir and oil injection system and the shearing pump for droplet size control.

Test Procedure

A sodium chloride solution was pumped into the water supply tank and circulated by a multi-stage centrifugal pump through the hydrocyclone test loop to ensure complete mixing. The feed flow rate and split ratio were established by setting the pump speed and control valves downstream of the hydrocyclone. When oil was injected into the hydrocyclone test loop, the effluent stream was diverted to a dump tank rather than being recirculated. After a minimum of 12 hours settling time to allow gravity separation of any entrained oil, the brine from the dump tank was pumped to the supply tank for reuse. Every two to three weeks, the brine was replaced with a fresh batch. For test runs above ambient temperature, the brine was recirculated through the heat exchanger which was switched out of the circuit when the test temperature had been reached.

A concentrated oil dispersion was generated in the constant flow loop (15 L/min) by shearing of the injected oil through a variable speed turbine pump. Some problems were encountered during the initial phases of the testing due to precipitation of wax from the crude and blockage of injection lines. The oil injection system was subsequently heat traced and operated at a temperature of 60°C to maintain flow stability and facilitate metering of the oily dispersion into the hydrocyclone test loop. The injection rate and pressure were controlled by adjusting the speed of the gear pump and the needle valve setting. For low oil flows (< 15 mL/min), a piston pump was used.

Solids and gas were injected at a point immediately ahead of the test hydrocyclone unit in the main flow

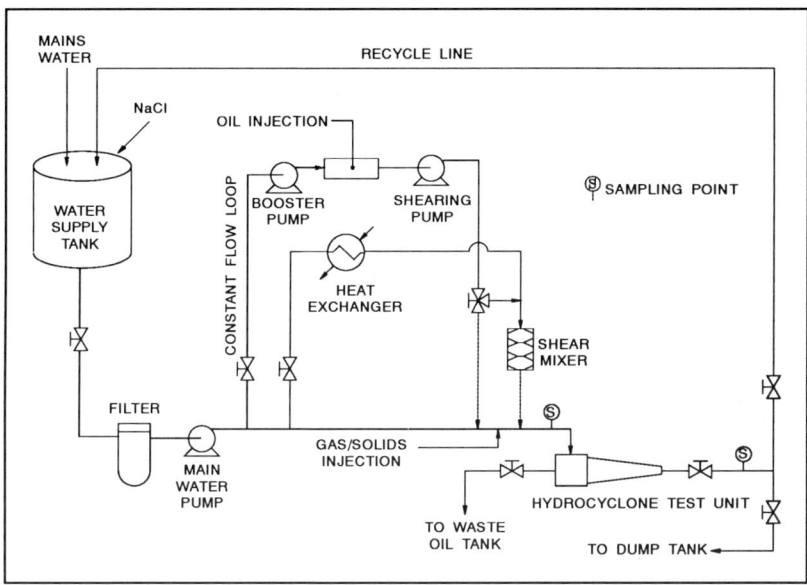

Figure 1. Schematic diagram of hydrocyclone test rig.

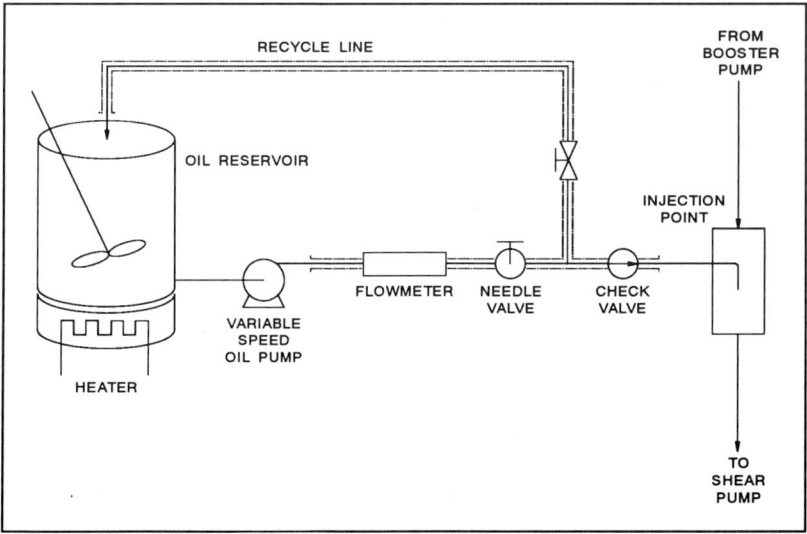

Figure 2. Schematic diagram of oil injection system

loop. The nitrogen was metered into the system as coarse (mm size) bubbles and test runs were undertaken with the upstream outlet at atmospheric pressure. Dilute solutions of treatment chemicals were injected through a hypodermic needle just downstream of the shearing pump in the oil dispersion loop. Residence time of the chemicals in the system before entering the hydrocyclone unit ranged from 4 to 5 seconds under turbulent flow conditions.

Sampling and Analysis
Sampling of the feed and downstream effluent was undertaken isokinetically with an initial portion of the sample being run to waste. To obtain drop size samples, part of the flow in the sample line was rapidly isolated and then drained out under gravity through a full bore valve to prevent sample shearing. Sample volumes of 5 mL were analyzed immediately (within one minute) for determination of droplet size distribution using a Coulter Counter TA-II unit.

Sample volumes of 50 to 100 ml were analyzed for oil and grease content by a partition infrared procedure employing CFC-113 as the extracting solvent and utilizing a Foxboro Miran fixed filter infrared spectrophotometer. Some additional samples were collected and analyzed when the brine in the hydrocyclone test loop was being recirculated to determine and compensate for background contaminant levels before the introduction of the dispersed oil.

Performance Criteria
The criteria for evaluating the hydrocyclone performance were separation efficiency and migration probability. Separation efficiency (E_{ds}) is defined in (1) as the percentage reduction of oil concentration in the water phase.

$$E_{ds} = 1 - (K_{ds}/K_i) \qquad (1)$$

Migration probability (MP) refers to the probability of a droplet of a particular diameter being separated out to the reject flow, see (2).

$$MP(d) = 1-(1-R) \times [dp_{ds}(d)/dp_i(d)] \qquad (2)$$

TEST RESULTS

Test Rig Debugging
Initial problems encountered with consistent metering of the waxy crude oil were alleviated by maintaining the oil dispersion loop at 60°C. Wax was also suspected as the cause of occasional plugging of the hydrocyclone outlet orifice during system start-up and may introduce similar problems in full scale operation if high temperature conditions are not maintained.

Generation of a fine and reproducible oil dispersion was found to be problematic with the Terra Nova crude oil. For the Phase 1.1 test work, two shearing pump speeds (2500 and 2700 rpm) were chosen in an attempt to obtain a 20 μm mean droplet size dispersion. In subsequent testing, shearing pump speeds of 2000 and 2700 rpm were employed to generate two different oil droplet dispersions. Figure 3 illustrates the mean oil drop sizes obtained at different shearing pump speeds on two test dates. Data previously obtained for a typical North Sea crude (Forties) demonstrates a smoother relationship between drop size and shearing pump speed with mean drop size of 15 μm attainable perhaps due to differences in viscosity and interfacial characteristics between the two crudes.

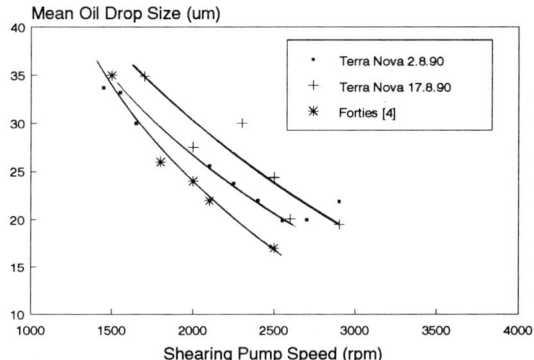

Figure 3. Feed drop size vs. shearing pump speed.

The crude oil was decanted after heating the original drum to 70°C and added to the reservoir of the oil injection system. Periodically the remaining oil would be replaced by a new batch. For each batch of crude oil, samples taken from the oil injection system as the testing progressed showed a tendency for the crude oil density to increase with time. This appeared to be largely a result of water uptake by the crude oil rather than a loss of volatiles as evidenced by an increasing discrepancy between measured feed oil concentrations and targeted values based on the oil flowmeter readings. A correction factor was employed to compensate for this drift. The water may have been introduced due to leakage of brine through the check valve which separates the oil injection system from the main test loop. This uptake of water by the crude oil reduced the effective density difference between the crude oil and water phases by up to % (from 192 to 158 kg/m^3 at 25°C and 199 to 165 kg/m^3 at 55°C), generally raised feed droplet sizes and increased the crude oil viscosities. At the injection temperature of 60°C, oil viscosities were commonly between 4 and 8 mPa.s.

Phase 1.1 - Effect of Flow Rate

A progressive increase in separation efficiency with flow rate was observed, as shown in Figure 4. From this graph, it is evident that peak performance was not reached at the maximum tested flow rate of 130 L/min. It is anticipated that for a standard 35 mm Vortoil hydrocyclone geometry (termed the F-liner), a similar efficiency level would be attained at around 175 L/min and the performance curve would level off to a constant E_{dk} value, indicating a high turn down ratio [5].

At the maximum tested flow rate of 130 L/min and with a nominal 25 μm mean drop size, the inlet oil concentration of 500 ppm was reduced to approximately 60 mg/L in the downstream effluent, representing an 88% separation efficiency. It is expected that this effluent concentration would fall to below 50 mg/L on reaching peak separation performance, possibly at a flow rate of around 150 L/min.

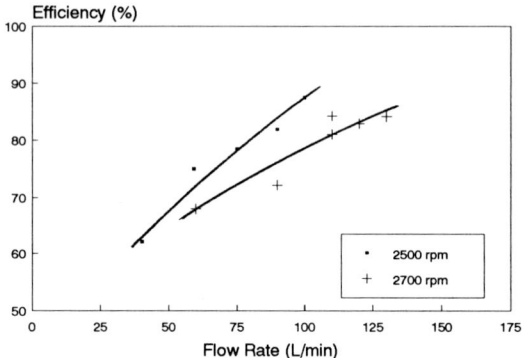

Figure 4. Separation efficiency vs. flow fate (500ppm, 25°C).

Pressure drop requirements for the test hydrocyclone are shown in Figure 5 and it is notable that these marginally increase at elevated temperatures. This is consistent with the small positive exponent found for the non-dimensionalised relationship between hydrocyclone pressure drop and throughput, commonly expressed in terms of Euler and Reynolds number (Eu $\propto$ Re^n). The maximum test flow of 130 L/min represented a pressure limitation of the test rig with an inlet to reject pressure drop requirement of approximately 1300 kPa. In comparison, a pressure drop of only 900 kPa occurs with the standard Vortoil geometry at approximately 175 L/min [5].

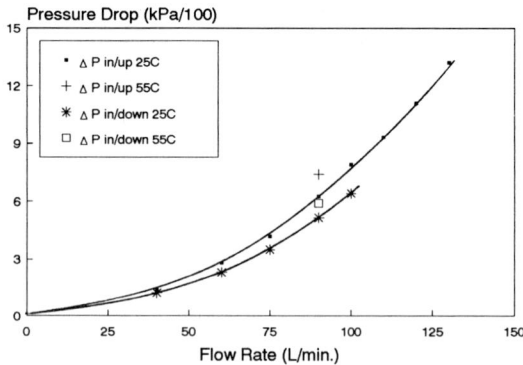

Figure 5. Pressure drop vs. flow rate.

Phase 1.2 - Effect of Oil Concentration

A range of inlet oil concentrations from 50 to 5000 ppm was tested at temperatures of 25 and 55°C. Testing was carried out at two shearing pump speeds, 2000 and 2700 rpm, which had produced nominal 30 and 20 μm dispersions respectively at 500 ppm inlet concentration and 25°C.

Drop size analysis of the inlet flow at the test conditions, presented in Figure 6, showed that oil concentration has a much greater effect on the degree of dispersion than either shearing pump speed or temperature, particularly above 1000 ppm. This may reflect a tendency for the oil to coalesce as droplet densities increase, especially as concentrations in the oil flow loop (15 L/min) are six times those in the hydrocyclone feed stream (90 L/min).

In terms of gross oil removal, a progressive improvement in performance with increasing feed concentration (K_i) is evident from Figures 7 and 8, with a tendency for the 25 and 55°C curves to converge at high oil levels. Some anomalous results were observed in these tests. For example, given that the increase in

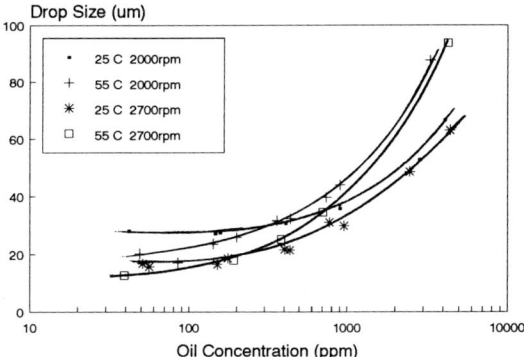

Figure 6. Feed drop size vs. oil concentration.

operating temperature from 25 to 55°C represents a 40% reduction in viscosity with all other factors being equal, improved separation performance would be expected at 55°C for both shearing conditions. This was clearly not the case, with little difference observed between the 25 and 55°C tests at 2000 rpm. Similarly, the finer oil drop dispersion produced at 2700 rpm would be expected to result in notably poorer separation than at 2000 rpm under both temperature conditions. This was observed for the 25°C tests, however, at 55°C, performance was virtually independent of the degree of shearing.

Test results are plotted, in Figures 9 and 10, in terms of residual oil levels in the downstream discharge flow. The data in Figure 9 shows that a discharge level of 40 mg/L can be achieved with feed concentrations up to 350 ppm with 2000 rpm dispersions of Terra Nova crude oil. Figure 10 (2700 rpm dispersions) more clearly illustrates the effect of operating temperature, in that to achieve a 40 mg/L discharge level, the maximum allowable feed concentration increases from 150 mg/L at 25°C to 300 mg/L at 55°C.

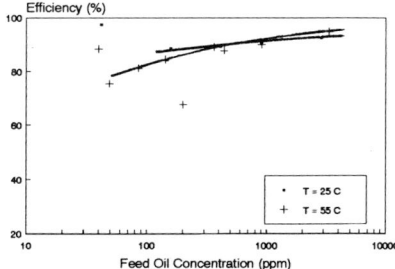

Figure 7. Efficiency vs. feed concentration at 2000 rpm.

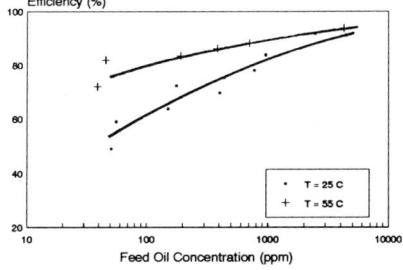

Figure 8. Efficiency vs. feed oil concentration at 2700 rpm.

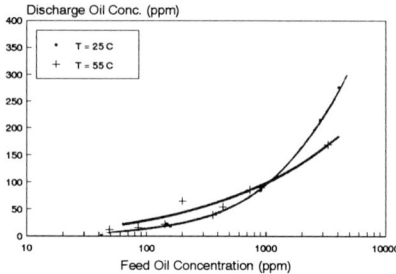

Figure 9. Residual oil levels at 2000 rpm.

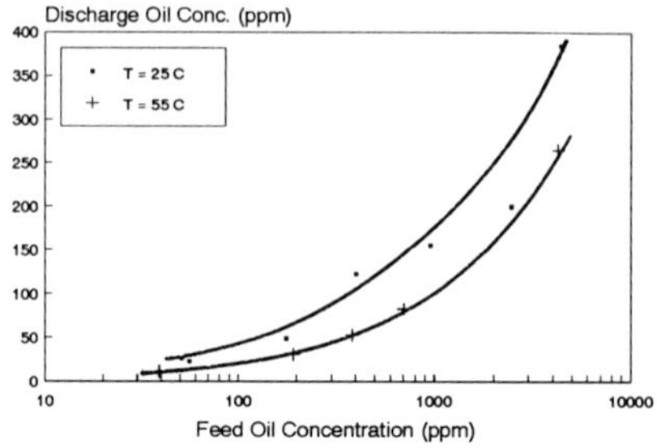

Figure 10. Residual oil levels at 2700 rpm.

Typical data on the feed and downstream oil drop dispersions achieved in this test work are plotted on a log scale in Figure 11. The figure shows data for oil concentrations of nominally 500 and 5000 ppm at 25°C.

Migration probability curves are given in Figures 12 and 13 at shearing pump speeds of 2000 and 2700 rpm respectively, for K_i values of nominally 500 and 5000 ppm at the two test temperatures. Assuming minimal droplet interaction, changes in shearing pump speed and oil concentration would not be expected to significantly impact on droplet migration probability. The lower separation potential for smaller drops (<25 μm) at 5000 ppm, shown in the migration probability curves, is therefore indicative of droplet break up in the hydrocyclone. Shear induced rupture of large oil drops would result in increased numbers of small drops in the system and therefore

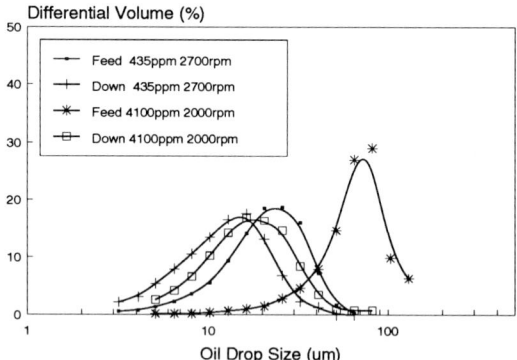

Figure 11. Oil drop size distributions at 25°C.

in the downstream effluent. A corresponding increase in apparent migration probability for the drop sizes subject to break up can also be anticipated, but was not clearly detected in this study.

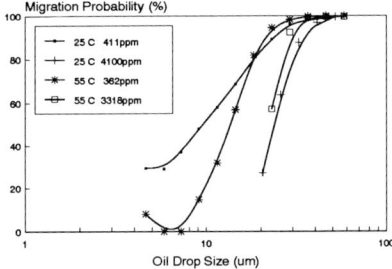 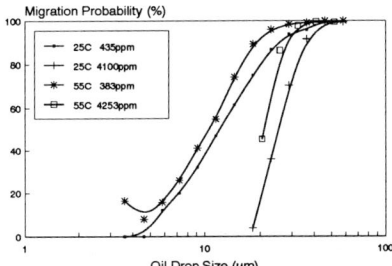

Figure 12. Oil drop migration probability at 2000 rpm.　　　Figure 13. Oil drop migration probability at 2700 rpm.

The phenomenon can be tracked by the drift in cut size (d_{50}) with oil concentration, shown in Figure 14. Only above about 1000 ppm feed concentration does the d_{50} value rise substantially, at which point the maximum drop size in the feed is 65 to 70 μm. This, then, can be interpreted as the maximum stable drop size for this particular oil/water system, set of operating conditions and hydrocyclone geometry and explains the relatively modest improvements in separation efficiency with oil concentration (Figure 7 and 8) despite large increases in drop size (Figure 6).

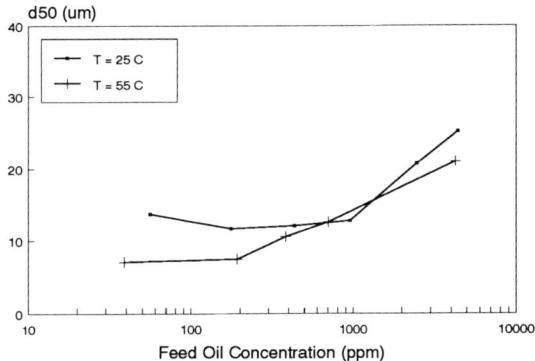

Figure 14. Drift in cutsize with feed oil concentration.

It is presumed that separation would peak at lower throughputs for these coarse dispersions. A hydrocyclone with a larger inlet and hence lower feed velocities for a given flow rate might also be preferred. However, it is difficult to predict where the optimum balance occurs between maximising the acceleration field and minimising the risk of shear induced droplet break up.

Phase 2 - Effect of Interferences

The separation efficiency of the hydrocyclone after addition of gas and solids is compared with control results in Table 4. The introduction of 20% nitrogen by volume to the feed flow produced a complete loss of deoiling capability in the hydrocyclone, as the upstream outlet was completely throttled by the gas. A second test was carried out at 5%, in which case the separation efficiency was still less then half that of the control condition. In both cases some gas was detected in the downstream as well as the upstream outlet. The disruptive action of the gas phase reflects the unsuitability of the test hydrocyclone geometry and operating conditions. For example, efficient deoiling hydrocyclone geometries with larger inlet and upstream outlet sizes have been tested for up to 7% Q_i air at inlet pressures with negligible effect on the separation performance [6]. Operating with greater back pressure would have reduced the potential for gas expansion within the hydrocyclone still further and therefore have allowed even higher levels of gas in the feed to be handled. Tests evaluating the effect of silica show a small, repeatable drop in efficiency. The significance of this change should be judged in relation to the separation efficiency value of 81.8% which represents the control level over the entire period of the interferences test phase.

TABLE 4
Effect of Gas and Solids on Hydrocyclone Performance

	Separation Efficiency (%)			
	Control	5% Gas	20% Gas	Solids
Test 1	81.8	---	0	80.3
Test 2	81.3	36.0	---	79.4

The effect of the selected oilfield treatment chemicals on hydrocyclone separation efficiency is shown in Table 5. The lower reverse demulsifier dosage proved most beneficial and this nominal 1 ppm rate was carried through to the second phase of testing with corrosion inhibitor, scale inhibitor and antifoaming agent at targeted dosages of 2, 5 and 2 ppm respectively. Generally, the 2 - 3% improvement in efficiency induced by the addition of reverse demulsifier was sustained for the other additives and associated with a 10 - 25% increase in feed drop sizes.

TABLE 5
Effect of Treatment Chemicals on Hydrocyclone Performance

	Control	RD(1.3ppm)	RD(5 ppm)	RD+CI	RD+SI	RD+AF
$E_{ds}(\%)$	81.3	83.7	82.0	84.1	84.5	83.2
$d_i(\mu m)$	23.0	26.9	26.5	25.2	28.4	29.0

CONCLUSIONS

The static hydrocyclone geometry tested showed oil removal rates approaching 90% for Terra Nova dispersions of 25 μm mean drop size at 25°C with a 500 ppm feed concentration. This reflected a better than 50% removal efficiency for droplets of 10 μm. Effluent oil levels below 40 mg/L were achieved for inlet oil concentrations up to 350 ppm at 90 L/min, taking the most favourable temperature and dispersion conditions tested.

Considering oil contents in the range of 50 to 5000 ppm, increasing temperature (to 55°C) resulted in a modest improvement in performance, although generally more significant improvements occurred due to the rapid increase in drop size associated with higher oil concentrations. This was checked, however, by the development of droplet break-up in the hydrocyclone.

The deoiling action of the hydrocyclone was only marginally reduced by the addition of fine solids (100 mg/L), but the presence of free gas proved to be disruptive. Separation was slightly enhanced by using a reverse demulsifier and remained so with the further addition of other production chemicals.

It is anticipated that the use of a standard Vortoil geometry (larger inlet than the test hydrocyclone) would generally allow comparable separation to be achieved for a lower pressure requirement. Improved performance could also be expected at high oil concentrations and if gas were present in the feed.

RELEVANCE OF TESTS TO FIELD PERFORMANCE

Given that only one hydrocyclone geometry and a limited range of operating conditions were tested, the test results obtained are not extensive enough to provide much more than a general guide and some points of reference as to how hydrocyclones could perform in the field. One observation that may be relevant to Vortoil installations on Canadian offshore oil production platforms, for example, is the potential for plugging of the upstream outlet at start up or if the process temperature drops too low.

A brief review of the literature can provide some insight into actual field performance of hydrocyclone units for the purposes of comparison with laboratory results presented in this paper. In recent trials on a North Sea platform [7], standard geometry 35 mm Vortoil units were tested on produced water having a mean oil droplet size of 20 μm. The density difference between the oil and the water was 0.19 g/cm^2. At inlet concentrations ranging from 100 to 350 ppm, the Vortoil units achieved an 85 to 90 % oil removal efficiency; results that are very comparable with our observations. Indeed, taking into account the lower density difference in the laboratory tests, it would appear that the high swirl unit demonstrated relatively better separation than the standard unit. In earlier tests on another North Sea platform [8], standard 35 mm Vortoil units were shown to achieve oil removal efficiencies approaching 99 % at inlet oil concentrations up to 3300 ppm. In this case, the laboratory results indicated relatively lower separation at high inlet oil concentrations. These data seem to corroborate our observations of droplet break-up in the high swirl Vortoil unit.

In comparing offshore tests and the work described in this paper, a number of limitations of the laboratory model of the deoiling process should be highlighted. First, the high swirl geometry hydrocyclone unit was not optimal for the oil droplet dispersions that were generated in this study. However, the short life of the artificially created oil droplets in the test rig (4 to 5 seconds) is likely to make them less stable and more prone to break up than oil droplets entrained in actual produced water. The presence of dissolved gas in the field may also act to enhance oil removal by flotation mechanisms. All these factors indicate a tendency for the laboratory results to underestimate the potential of the hydrocyclone in a field situation. In general, as the complex chemistry of real production fluids can never be fully reproduced in a laboratory, aspects of separation efficiency that are dependant on parameters such as drop stability can only be properly assessed in on-site tests.

ACKNOWLEDGEMENTS

This study was supported, in part, by the Federal Panel of Energy Research and Development. The technical review by Sandra Kok is acknowledged with appreciation.

REFERENCES

1. Personal Communication, Mr. Cal Ross, Lasmo Nova Scotia Inc., Halifax, Nova Scotia, Canada, March 1992.

2. Personal Communication, Mr. Rick McCubbin, Hibernia Management and Development Corporation Inc., St. John's, Newfoundland, Canada, March 1992.

3. Canada Oil and Gas Lands Administration, Ottawa, Ontario, Offshore Waste Treatment Guidelines, January 1990.

4. Nezhati, K. and Thew, M.T. Deoiling Hydrocyclone Optimisation for Two-Stage Geometry. Report ME/86/19, Dept. Mechanical Engg., University of Southampton, UK, October 1986.

5. Triponey, G. and Woillez, J. Recent Experience in Water Separation from Gas Condensate. Gas Processing Association Meeting, Nice, France, June 1990.

6. Colman, D.A., Thew, M.T. and Corney, D.R. Hydrocyclones for Oil/Water Separation. Paper 11, Proc. 1st Int. Conf. on Hydrocyclones, Cambridge, UK, Oct. 1980, Pub. BHRA (Cranfield, UK), 1980.

7. Flanigan, D.A., Skilbeck, F., Stolhand, J.E., and Shimoda, E. Use of Low-Shear Pumps in Conjunction with Hydrocyclones for Improved Performance in the Cleanup of Low-Pressure Produced Water, Paper SPE 19743, 64th Annual Technical Conference & Exhibition of the Society of Petroleum Engineers, San Antonio, Texas, October 1989.

8. Meldrum, N. A Solution to Produced Water Treatment, Paper OTC 5594, 19th Annual Offshore Technology Conference, Houston, Texas, April 1987.

A NEW METHOD OF STARCH
PRODUCTION FROM POTATOES

Stanislaw Bednarski
Technical University
Cracow - Poland
Institute of Chemical Eng.
  and Ph. Chemistry

SUMMARY

After a short introduction a pilot installation of multi-hydrocyclones for research of starch extraction process from pulp of mashed potatoes has been described.

Design solutions of new types hydrocyclones and multi-hydrocyclones as well as bihydrocyclones are presented. The scope of conducted research is discussed and some results obtained during several potatoes compaigns are given. On the base of these results a new method of starch production from potatoes is made know.

INTRODUCTION

Poland is the largest producer and exporter of potato starch in the world. Hence the interest in a cheap and simple method of its production, what in turn stimulated the introduction of more advanced technologies and equipment. Already in the second half of the 1950's as a result of conducted works by the author (1,2) and cooperating with him co-workers (8,4) the starch milk process of refining and thickening has been changed. In place of appliances utilizing the gravity force such as refining troughs, thickeners or washers, equipment based on the action of centrifugal force like hydrocyclones or multi-hydrocyclones as well as bihydrocyclones and next screen jet washers rotary screens and de laval separators (5) have been introduced.

The first in the world refining of starch sludge on a three-stage multi- hydrocyclone installation has started during the autumn compaign in 1957 at Lobon near Poznan in the largest at that time potato starch plant in Europe (6). At that time the author has carried out the first in the world trails of starch washing (extraction) from potato pulp on hydrocyclones of 30 mm diameter (7). At the same time also two new centrifugal appliances have been designed by the author, one for de-sanding of the starch milk named bihydrocyclone BHC (8) and second for defoaming of the starch milk called centrifugal froth breaker (9).

The process of starch washing out (extraction) from mashed potatoes and next the refining of starch milk is at present carried out in 17 to 19 hydrocyclone stages or on 7 to 9 arc screens stages and 8 to 10 hydrocyclone stages (10 - 13). Such installations require the application of the same number of pumps and the necessity of having a fully automated system.

In the beginning of the 1980's the author in coorporation with the central Laboratory of Potato Industry after two years of research work conducted in the industry on a pilot installation have worked out a considerably simpler technological system of mashed potatoes starch extraction and refining (6, fig. 2). That type of potato starch technological process system have reduced down to seven stages the washing-refining multi-hydrocyclones, to three stages the clarification multi-hydrocyclones and to two stages the de-sanding hydrocyclones.

In the case of applying modern potatoes washers the necessity of having de-sanding hydrocyclone does not exist any more. This way the total number of applied hydrocyclones stages has been reduced to about nine, i.e. almost by half compared with other solutions (15, 16).

Further research work has shown the possibility of lowering the number of multi-hydrocyclone stages and pump interstages in the process of starch production from potatoes. This was possible first of all as a result of application specially designed multi-stage hydrocyclones and multi- hydrocyclones and the introduction of a proper process flow sheet.

In the paper the latest design solutions of hydrocyclones and a three-stage multi-hydrocyclone are presented by using which the process flow sheet was significantly simplified what shall be discussed in the last part of the paper.

DESIGN OF HYDROCYCLONE AND MULTI-HYDROCYCLONE

The construction of the hydrocyclone and its working principle in case of application for densimetric separation is not a perfect one. The occuring in it undesirable directions of flow and the forming of a concentrated zone close to the active surface which acts similarly as self building up auxiliary filtration layer are the reasons of decreased separation sharpness.

This lack of sharpness occurs especially in cases when separated are suspensions containing solid particles of different densities and shapes, and mainly when the lighter particles have an enlarged surface whereas the heavier particles are of the grain shape. With such features of suspensions we have to deal in the process of potatoes starch extraction and refining. During the rotating motion occurring in the hydrocyclone boundary layer the solid fibrous particles are captivated and hindered by the grainy particles of starch which are resisting their flow towards the axis of the swirl from where they would have the possibility to go over into the overflow fraction. This phenomenon causes that through the underflow together with the thickened starch quite a significant amount of fibrous particles is passing through as well, thus lowering the separation efficiency of the hydrocyclone. The result is such one that it forces us to apply multi-stage hydrocyclone systems in which after multiple starch washing the starch will be free of fibres.

To eliminate or at least partly limit the above described phenomenon of fibres hindrance relevant works have started with the aim to change the shape of the internal or so called active surface of the hydrocyclone.

Part of the obtained on different suspension results have been presented in the paper given at the 2nd International Conference on Hydrocyclones in Bath. (17, 18) Further research works have led to other design solutions of the hydrocyclone inside surface what resulted in a better separation of suspensions especially those consisting of fibrous and grainy particles. (19)

In Fig. 1 shown is the lengthwise cross-section of a conical-cylindrical (a) hydrocyclone with a corrugated surface which is illustrated on the cross-section (b). On the whole or certain part of the cylindrical height and almost on the whole length of the conical section there are on the circumference properly shaped cavities and splines situated longitudinally in relation to the hydrocyclone axis. These cavities and splines in the conical part may run along the whole length or on part of it. The shape of the cavities in the cross-section can by any e.g. rectangular, prismatic, triangular, however advantageous are parabolic or arched according to Archimedes Spiral. However, the splines are better if the cross-section has a pointed or rounded off shape of saw blade teeth.

The cavities and splines may also assume shapes of an undulated surface made up best of regular curves close to a sinusoidal curve. The number of splines and cavities on the circumference can be any choice.

In Fig. 2 presented is another hydrocyclone construction showing the longitudinal cross-section, the left half of it is a three product unit without a conical part, however, the right half is having a short cone.

The internal surface of the cylindrical and short conical part is also undulated. The proper shaping of splices enables the toss of the fine grain towards the hydrocyclone axis. This leads to origination of small micro swirls which are loosening the thickened boundary layer structure freeing the tied up fibres.

As a result of the suspension swirling motion on the internal surface furnished with cavities and splices an intensification process of solid particles dislocation in relation to each other follows a better washing of granular particles in liquid medium also occurs.

The main advantage of a hydrocyclone with an undulated internal surface is its ability to give a sharp cut of solids according to their sizes and density. Moreover such a shaping of the internal hydrocyclone surface on which the thickened fraction of suspension flows across like on a wash-board, have made possible a significant reduction of the separation stages number used so far in multi-stage installations.

In the starch industry these hydrocyclones have secured a better washing out of the starch from mashed potatoes and also much simplified the process flow sheet.

Likewise the simplification of starch process flow sheet and production cost lowering were achieved by application of new designed multi-hydrocyclones with radially positioned hydrocyclones in a one (20) and three stage system (21) which have been partly discussed during the last conference (6).

In Fig. 3 a design solution of a multiple multi-hydrocyclone is shown, in this one appliance it was possible to separate two times the introduced suspension and simultaneously to carry out washing of the underflow fraction from the first stage of separation. This multi-hydrocyclone consists of a body which constitutes a vertical cylinder "1" closed by a lower "2" and upper "3" cover, in which one over the other are build in three sets of hydrocyclones. In succesion from below there are: third set of hydrocyclones III, feeding head "4" assembly of the 1st stage and the clarifying set of the 2nd stage. Eacho f the hydrocyclones sets I, II, III has two concentric walls made of ring elements placed one over the other. These walls in each set mark out respectively: inlet chambers 9I, 9II, 9III of hydrocyclones "12" fixed radially between these walls, collecting overflow chambers 10I, 10II and 10III and placed inside collecting underflow chambers 11I, 11II and 11III.

The feeding head "4" contains a ring chamber "5" equipped with a side feed inlet connector pipe. Chamber "5" is connected through holes in the upper wall with the inlet chamber 9I. Inside the feed head "4" there is a two side open transfer chamber "7" to which through the ring chamber "5" washing water connector pipe 8 is connected. Placed under the head "4" the third set of hydrocyclone III, has an inlet chamber 9III connected through the transfer chamber "7" with underflow chamber 11I.

The proper density of the underflow fraciton is achieved by supplying water through connector pipe "8" and its homogenizing take place in the static mixing elements "21" mounted in the space connecting the transfer chamber "7" with inlet chamber 9III.

The underflow fraction from the second stage of hydrocyclone II collected in the underflow chamber 11II flows outside through the central pipe "16" led along the axis of casing "1" through the set of the first stage transfer chamber "7" and third assembly of hydrocyclonesIII. Through the bottom cover "2" are led out further more: Connector pipe "14" from overflow chamber 10III and connector pipe "15" from the underflow chamber 11III.

The construction of the multi-hydrocyclone with the presented set of functional working chambers is put into one piece through axial pressure of individual elements. The central pipe "16" has in its upper part the head "17" which rest on the cover of the first stage set and at the down end has a helical element "18" clamping through cover "2" the threaded through sets III and I as well as the feed head "4". The elements of the 2nd stage are connected and fixed in the multi-hydrocyclone by a tie rod "19" which lower end is in turn fastened in the head "17", and the upper end by nut "20" acting on the cover of the second stage. The wall of casing "1" in the part over the feed head has a telescopic construction made of few rings sealed against each other.

The presented three-stage multi-hydrocyclone assembly design solution enables the connection in particular stages any number of the same or different diameter hydrocyclones. Through the placing on each other or removal of the individual ring-shaped elements containing radially placed hydrocyclones building up proper coils it is possible by using their multiple to change freely on successive stages the flow rate. The number of hydrocyclones in the coils may be constant or what is more beneficial may vary. In the case of 30 mm hydrocyclones the number which can be placed in one coil can be 16 to 32, on the average 24 pieces; and if 15 mm hydrocyclones should be installed then their number may vary from 30 to 60 on the average 45 pieces.

The maximum height of the assembly is about 2000 mm, the diameter of its lower part 680 mm, while the total flow rate with maximum number of coils and 30 mm diameter hydrocyclones in the first stage and 0.5 MPa pressure is about 360 m3/h.

DESCRIPTION OF POTATO STARCH PROCESS FLOW SHEET

Fig. 4 illustrates one of the most simple technological process flow sheets of potato starch production until now known in the world. On a similar system of appliances installed directly in the industry research works were carried out which have confirmed the high efficiency of such installation. This installation consists of the following main items:

- plant for raw material preparation
- de-sanding of starch sludge
- plant for starch extraction and refining
- plant for clarifying of damped waters and fibre separation

Besides that a complex potato starch production factory may include other plants like: protein production plant from cell sap, starch de-watering and drying plant in order to obtain starch flour, de-watering and drying plant of waste pulp, plant for mechanical and biological waste water cleaning (where hydrocyclones are also applied). In some of these plants (22) the potatoes from the flumes or other transportation means are delivered to the washing machine "1" to which also fresh water is supplied, especially in the last zone of potatoes washing. Washed potatoes are further taken to the buffer vessel "2" from where by a dosemeter "3" they are delivered to the high-speed rusping machines "4". Next properly crushed potato pulp is undergoing separation in a juice centrifuge "5" from which water is directed for production of high quality protein. While the de-watered pulp is fed into the buffer vessel "6" equipped with a mixer in order to be properly diluted by fresh water or/and returns averaging the density.

This way the prepared pulp pump "7" delivers to de-sanding hydrocyclones "8" from which the underflow fraction diluted by returns, gravitationally feeds the battery of bi-hydrocyclones is recirculated to the buffer vessel "6" making a closed circuit, while the underflow containing in its mass mainly mineral impurities is led out from the technological system in a form of thickened waste. The overflow from the de-sanding hydrocyclones flows into the chamber I of buffer vessel "10". This vessel is divided into three chambers and its baffles are equipped with proper overflows.

From chamber I of the buffer vessel "10" pump "11" suppliers the de-sanded pulp to the first three stage multi-hydrocyclone assembly to which also washing water fresh or returned is fed. Intermediate products from this assembly are returned back to the beginning of the system i.e. to buffer vessel "10" from where by pump "12" it is delivered to the second assembly II to which also the washing water fresh or recirculated is led. The overflow fraction from assembly II flows to chamber III of buffer vessel "10" from where gravitationally or with pump "13" it is transported to the station of arc sieves in order to remove the fibres. While the underflow fraction from assembly II flows into chamber I and intermediate products to chamber II of buffer vessel "10".

Separated from the potato pulp starch and partly freed from fine fibres thickened starch milk which constitutes the underflow fraction from assembly I flows into chamber I of buffer vessel "14" also divided into three chambers. In this chamber the starch milk is diluted by returns and next by pump 15 pumped to assembly III from which the underflow fraction after washing by return supplied by pump "16" flows to chamber II, whereas the overflow fraction flows to chamber "0" while intermediate products flow to chamber I of the buffer vesel 14. From chamber II pump "17" delivers the starch milk to assembly IV to which also fresh water is supplied. The underflow from this assembly as a thickened starch milk is directed to a drum filter "18" in which dewatering takes place and next the wet starch is entering the drying plant from where a ready product is received in the form of starch flour.

The overflow fraction from assembly IV is returned to chamber I, whereas one of the intermediate products of this assembly becomes a washing medium of the assembly III 1st stage underflow and the second product i.e. the underflow fraction of the 2nd clarifying stages of assembly IV is returned to chamber II of buffer vessel 14.

To the overflow fraction of assembly III besides fine fibres also a significant amount of starch is also passing through which should be separated, especially in the case if the overflow partly or in whole is drained to the sewer. Therefore, in order to prevent loss of starch the installation was extended by an additional assembly V; which is equipped with hydrocyclones of 15mm diameter.

From here the diluted suspension of fine fibres and starch from chamber "O" pump "19" is delivering to the clarifying assembly V, from which the underflow fraction is returned to chamber I and intermediate products of separation to chamber "O", whereas the overflow fraction as a whole or partly is returned by pump "20" to the flow circuit as a diluting medium. The overflow from the clarifying assembly V practically contains the total amount of the fine fibre which together with the water is returned to the individual plant system units. In order to separate this fibre the overflow from assembly V can be directed e.g. into the arc sieves. It is however, more advantageous to dispose earlier the mass of fibre in particular after separation of the starch pulp in assembly I before directing it to the chamber I of buffer vessel "14".

Research on removal of fine fibre on arc sieves have been carried out before assembly III as well as after assembly V. Two types of arc sieves designed by the author have been used :

- pressure-gravitational arc sieves Fig. 5 (23)
- pressure arc sieve with vibrating sieves inserts Fig. 6 (24)

The carried out research have demonstrated that in case of fine fibre separation from starch milk before the assembly III more useful was the pressure-gravitational sieve in which the separated on the first sieve fibre has been undergoing washing on the second sieve. This way a fibre impoverished of starch was obtained which could be directly put together with the pulp separated in the arc sieve 21 originating from drains due to starch washing in assemblies I and II. While the arc sieve with vibrating sieve inserts was rather better suited for fibre removal from assembly V. This type of arc sieves were used only for fibre removal of the overflow fraction coming from assembly II which should be treated as waste water.

Considering the short lasting researches conducted at the end of the campaign no clear influence of fine fibres separation from starch milk before assembly III on the total instalation separation of starch production was found. Therefore it can be concluded that hydrocyclones are separating adequately the fibre from starch.

REFERENCES

01 Bednarski S.: Przem. Chem. 12, 303 (1956)

02 Bednarski S.: Chemik 9, 174 (1956)

03 Brzyski W.: Przemysk Spozywczy 10, 461 (1957)

04 Brzyski W.: Przemysk Spozywczy 11, 324 (1958)

05 De Laval Leaflet 1957

06 Bednarski: "Extruction and refining of potato pulp in the eight-stage system of hydrocyclones". Paper 4 3rd Int. Conf. on Hydrocyclones, Oxford, England, 30 Sept. - 02 Oct. 1987.

07 Bednarski: Przem. Chem. 13 73 (1957)

08 Bednarski: Przem. Chem. 13 312 (1957)

09 Bednarski: Hydrocyklony, Published by W.P.L.S. Warszawa 1967

10 Coransa A. Stärke 22, 27 (1970); 25, 27 (1973);

11 Van der Dorpel I.J. Stärke 25 143 (1953); 13, 407 (1961)

12 Van der Wal G.J.: Stärke 10, 277 (1958)

13 Lange A. Stärke 25 146 (1953)

14 Bednarski S. Kaliszon Z.: "Neue Erkenntnisse un der Anwendung der Hydrozyklone in der polnischen Stärke Industrie". Stärke Industrie". IV Stärkekolloqkim Proceedings, Pelhrimow CSSR, 8-12 Dec. 1986

15 Kurocichij C.K., Leberman A.L. Cholmianskij Ju.: Sgch. Prom. 11, 53 (1984)

16 Leaflet of Nivoba Firzu (1985)

17 "New designs of classifying hydrocyclones with high separation acuity" 2nd Int. Conf. on Hydrocyclones, Bath, England 19-21 Sept. 1984

18 Bednarski: Inz. i Apurat. Chem. 26 nr. 1, 6 (1987)

19 Patent RP nr P-275267 (Polish)

20 Patent RP nr 68099 (Polish)

21 Patent RP nr. P-290174 (Polish)

22 Bednarski: Stärke 15, 18, 171, 371 (1963)

23 Patent RP nr. 107195

24 Patent RP nr. 101289

Fig. 1 Cross-section of conicul-cylindrical hydrocyclone (a) with corrugated inner surface (b)

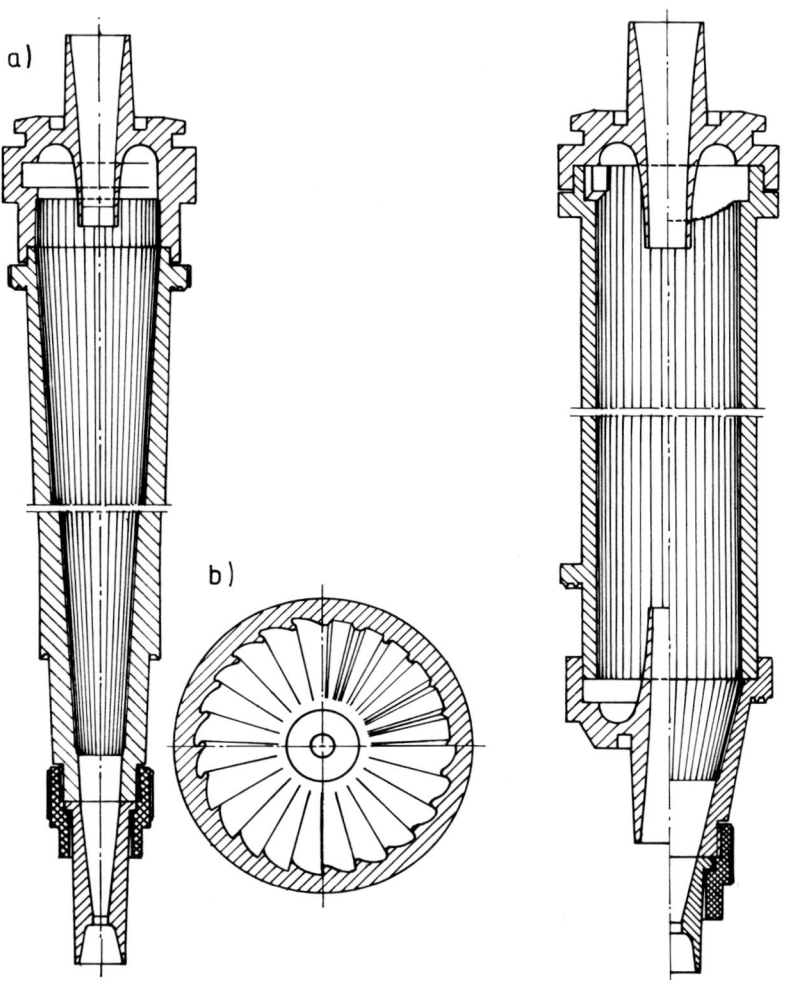

Fig. 2 Another type of hydrocyclone used during described research work

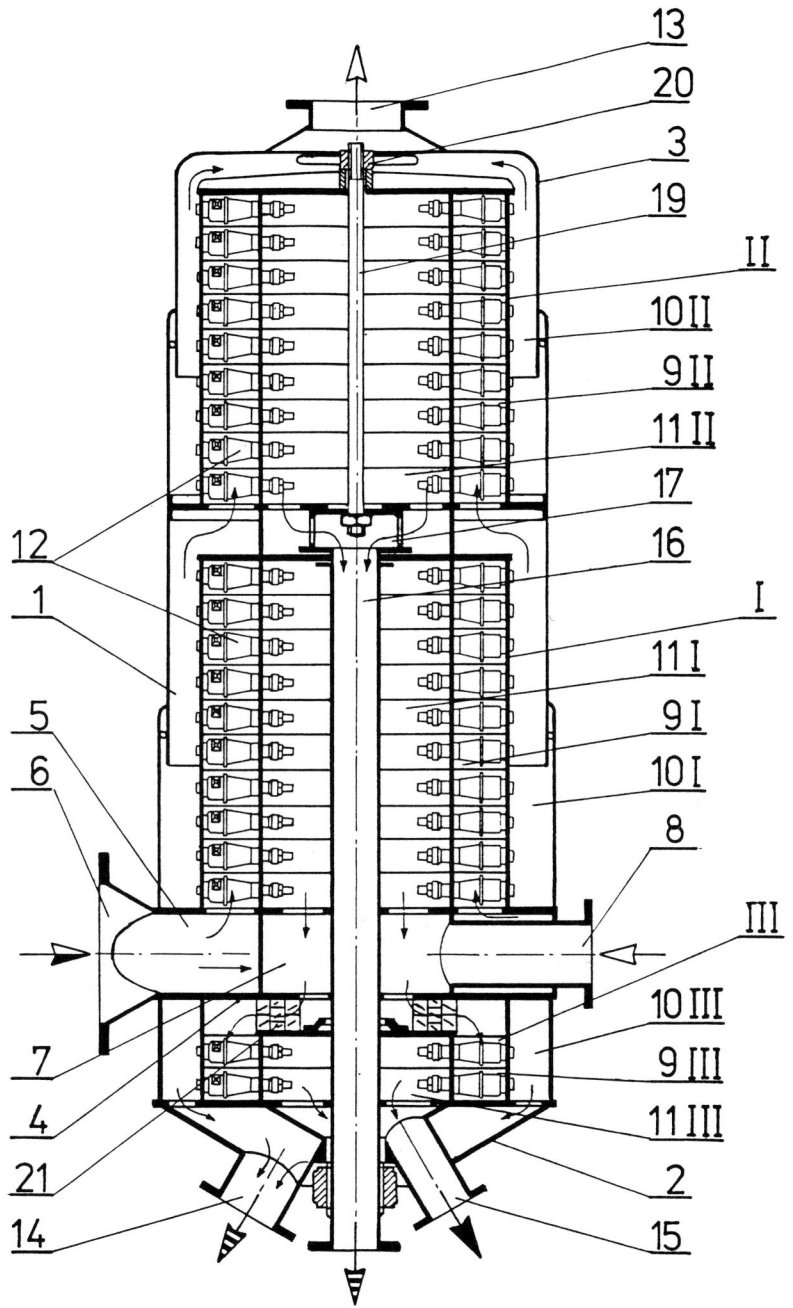

Fig. 3   Design of multiple multi-hydrocyclone

Stanisław Bednarski

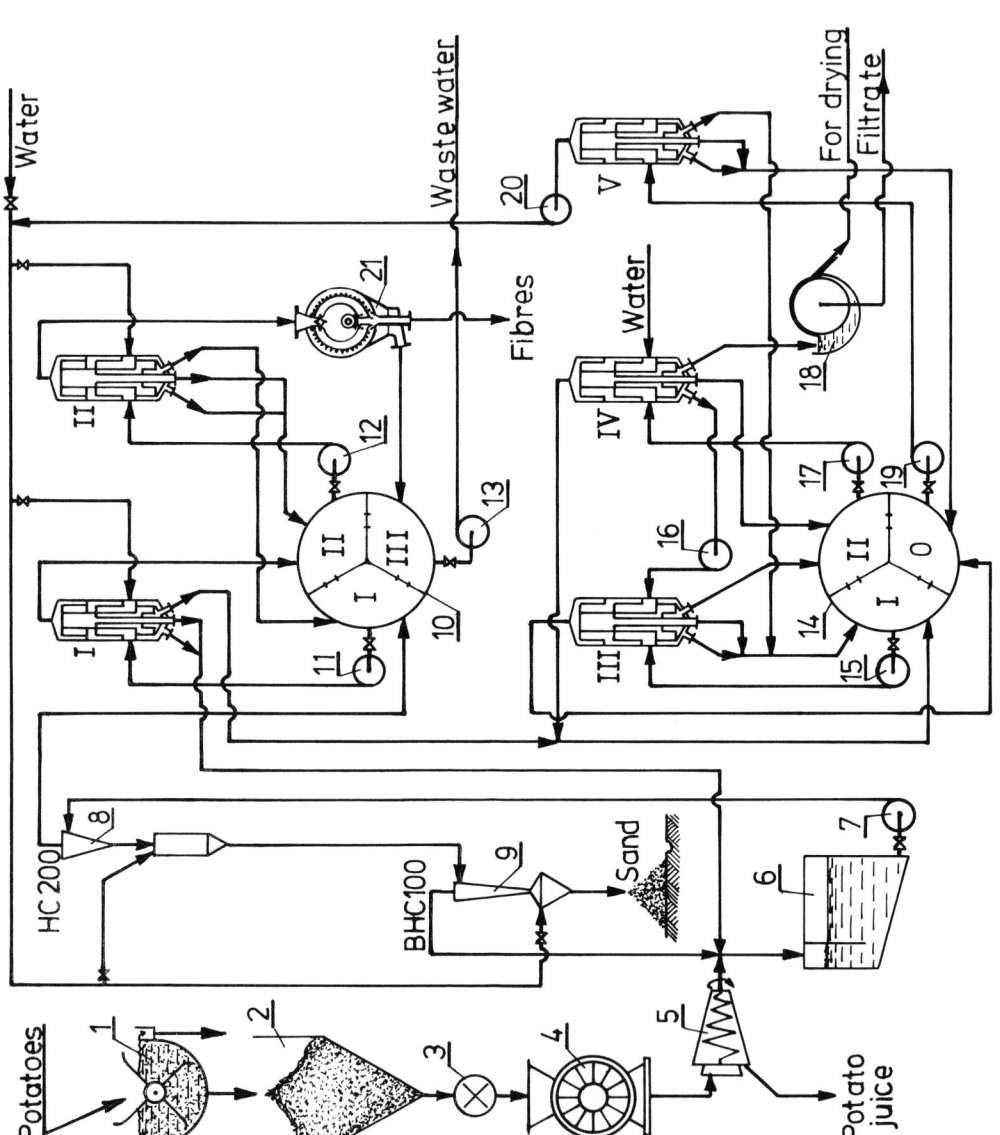

Fig. 4 Technological process flow sheet of potato starch production

325

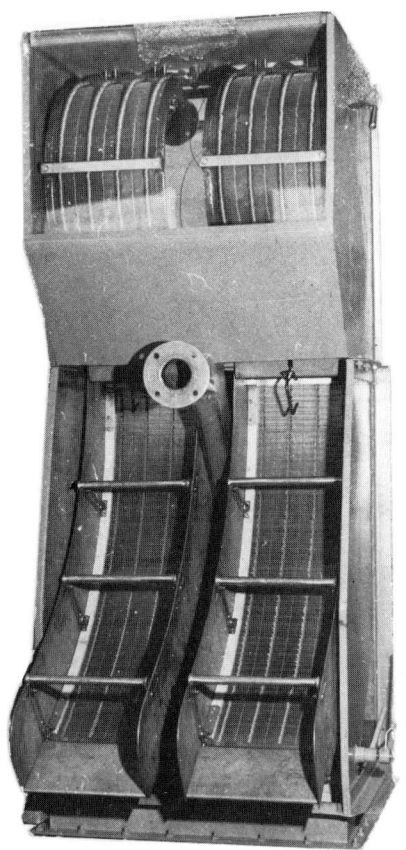

Fig. 5   Pressure-gravitational arc sieves

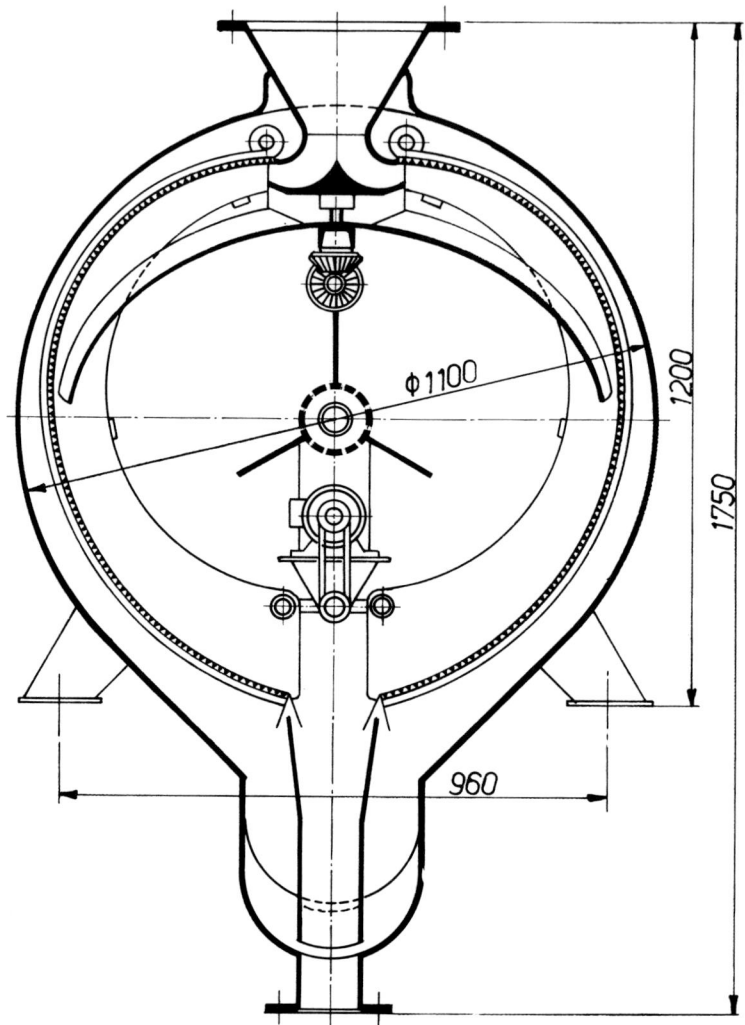

Fig. 6   Pressure arc sieve with vibrating sieve inserts

DEVELOPMENT

SEPARATION OF LIQUID-LIQUID SOLIDS MIXTURES IN A HYDROCYCLONE -COALESCER SYSTEM.

S. Bednarski
Technical University
Cracow-Poland
Institute of Chemical Eng.
and Ph. Chemistry

J. Listewnik
Maritime University
Szczecin-Poland
Marine Engineering
Department

SUMMARY

The hydrocyclone design presented at 3rd Int. Conference on Hydrocyclones in paper G-2 has been undergoing further thorough testing with the main aim of better optimization of its design and operational parameters.
Particular attention was concentrated on the usefulness of this apparatus in the process of separation emulsified oil-in water mixtures containing as well solid particles especially of mineral origin. The concentration level of oil during test varied from 0.1 to 3% (by mass) and solids from 0.01 do 10 g/dm^3.
The paper describes our continuing programme on the hydrocyclone type for light dispersions and solids separation where the influence of flow rate on separation efficiency as a function of inlet, outlet areas, length of cylindrical and conical part as well as pressure was examined.
Part of the experiments was also devoted to trials on separation of heavy liquid dispersions to check the suitability of our hydrocyclone as well for this purpose by only modifying the apertures but still retaining the basic structure of the hydrocyclone. Further research was carried out by connecting the hydrocyclone in series with special construction coalescers of ring shaped needle structures (patented)
The obtained global effects are described in the paper indicating very promising results to satisfy stringent IMO requirements on prevention of marine pollution by oil.

NOMENCLATURE

C_B	– coefficient of hydrocyclone orientation
D	– hydrocyclone major diameter
d_{50}	– mean drop diameter, classifying cut size
L_i	– total hydrocyclone length
l_i	– length of hydrocyclone conical part
l_c	– length of hydrocyclone cylindrical part
Δp	– pressure drops, $\Delta p = p_e - p_o$
p_e	– inlet pressure
p_o	– pressure at hydrocyclone overflow
d_o	– overflow nozzle diameter
d_u	– underflow nozzle diameter
d_e	– feed diameter
φ	– cone angle
ΔH	– head differential
H_e	– total head at the inlet
H_o	– total head at the overflow
n, m	– exponents in equation (2) and (4)
I	– split ratio $\dfrac{V_o}{V_u}$
V_u	– underflow volume flow rate
V_o	– overflow volume flow rate
V_e	– inlet volume flow rate
l_o	– vortex finder length
$\Delta \rho_a$	– differential density between continuous and dispersed material
μ_z	– dynamic viscosity (effective) of emulsion
ν	– kinematic viscosity

INTRODUCTION.

The generation of multi-phase systems is an intermediate or final product of a large number of technological processes being carried out in various branches of industry, quite frequently they are an undesired product, destructive to the environment and a burdensome ballast for the majority of plants.
The process of development and physical properties of these systems are generally very diversified as far as the method of their production and qualitative-quantitative composition of the contained phases is concerned.
The separation in hydrocyclones of two-phase systems such as liquid-liquid or gas-liquid and more so three-phases such as liquid-solid-liquid or gas-solid-liquid has not been investigated by many researchers.
The first separation attempt of a two-phase liquid-liquid system namely ether from water was carried out in 1943 by Tepe and Woods (1), but the achieved results should be considered as rather poor. Whilst the first patent application concerning the method of separating two immiscible liquids in a cyclone was made in Holland in 1948.
During the next decade several further reports on separation of various liquid-liquid mixtures have been reported (3,4,5,6,7,8,9,10,11,12).
In the beginning of 1960's a research program of quite an extensive nature have started investigating the possibilities of removing impurities in the form of water and solid particles mainly of organic origin from extraction naphtha used in extraction processes of vegetable oils(13).
Next in the years 1962-63 research work was carried out further on separation of water and mineral impurities as well as paraffin from crude oil and naphtha, further desalting of crude oil by applying a washing method and

separation of water and solids from diesel oil and petrol was also tried (14).

From later works in the field of heavier liquid dispersions separation from lighter continuous component other possibilities of hydrocyclones application have emerged namely the removal of solid particles and water from transformer oil as well as the separation of ammonia water and pitch mixture forming during the process of coal gasification. The ability of separating in hydrocyclones two immiscible liquids containing solid particles was also explored in the process of complex purification and dehydration of raw pitch using a method of water washing (15).

A large number of research results and numerous analytical and graphical relations have been given by Bohnet (16-18). He separated two-phase systems like compounded oil-water and spindle oil-water mixtures also with the presence of solid particles. Two more comprehensive reports discussing the separation of two insoluble liquids ought to be included also works of: Burrill and Woods (19), Sheng, Welker and Sliepcevich (20), Minart (21), Johnson, Gibson and Libby (22). The last decade have brought a large number of significant publications with very promising research results on the subject of liquid-liquid separation in hydrocyclones mainly thanks to the organized by BHRA already three International Conferences on Hydrocyclones (23-33).

THEORETICAL BASE.

Whereas the construction of the hydrocyclone is a simple one, the flow structure of liquid and solid particles for separation of which the hydrocyclone most frequently is applied is already a very complicated one (Fig.1). The individual spiral trajectories described by the circulating around the hydrocyclone axis liquid and solid particles have very complex and diversified directions, which are changing depending on the size and shape of the apparatus and its operating conditions.

The spiral trajectories are becoming much more complex and their orientations diversified if in the hydrocyclone flows a suspension consisting of two insoluble liquids of different densities, out of which one constitutes the dispersed and the other the continuous phase. The description as well as graphical presentation of flow directions of individual phases inside the hydrocyclone is much more difficult and practically impossible to realize. Even more difficult is to present the occuring fluctuations of that type of mixtures, if their contents includes additionally dispersed in them solids or a gaseous phase. In such case we have to deal with three or four-phases systems.

The theoretical formulation of separation mechanism of two immiscible liquids in a hydrocyclone is also a very difficult problem and so far not fully solved. The mentioned relations, especially the physical properties of the continuous medium containing particular dispersed phases as well as their quantitative ratio are changing in a wide range the qualitative-quantitative data of separation products.

Hence numerous obstacles frequently difficult to overcome in proper calculation and selection of a seemingly simple and easy to operate appliance.

The deepening of occuring difficulties is due to the fact that in most cases the dispersed phase is the lighter liquid and its volumetric quantity is many times smaller from the surrounding volume of the heavy liquid. Further more when the hydrocyclone operating conditions are such that one or both outlets are connected with the atmosphere, droplets of the lighter liquid may be subjected to aeration the results of it will be formation of foam what leads to the creation of flotation phenomenon. This phenomenon is often the cause that solid particles even several times heavier then the heavy liquid are being lifted by the lighter phase especially if they are better wetted by the less dense liquid than by the heavy. Hence the utilization of the hydrocyclone as a flotation machine for minerals enrichment.

The separation of solid particles contained in a mixture, of a three-phase system (liquid-solid-liquid) into one or both liquid fractions will depend not only on their density and particle size distribution but first of all on the ability of wetting the surface by the individual liquids. Thus the solids will go over to that liquid phase which wets it better and in quantities proportional to the degree the surface forces acting on them are resisting the action of centrifugal and buoyancy forces.

For the theoretical formulation of the droplet size limit during separation in the hydrocyclone, by analogy correlations can be used for defining the solids cut size d_{50}. But those correlations can only be related to mixtures in which the dispersed phase is the heavier liquid, and the lighter liquid is the continuous phase e.g. water-petrol, water-solids-edible oil or freon-water. Besides that to be able to adopt these correlations it is necessary to assume appropriate assumptions namely: the dispersed droplets of liquid are not subjected to break-up, flattening or elongation and don't flocculate or undergo the process of coalescence.

Out of tens of known equations defining the cut size d_{50} for solid particle separation an empirical formula given by (37) was chosen for adaptation to liquid-liquid separation case in a hydrocyclone designed initially for separation of lighter liquid dispersion and described in a paper given by the authors during the 3rd Int.Conference on Hydrocyclones (32). The described there hydrocyclone has been modified for separation of heavier liquid dispersions but basically retaining similar geometry.

The following adapted equation as mentioned before was verified during a 3 years trial period determining the equiprobable size d_{50} of heavier dispersion liquid droplets

$$d_{50} = C_B \sqrt{\frac{\mu_s}{\Delta \rho_a - 1}} \frac{d_e\, d_o}{d_u \sqrt{D}} \sqrt[4]{\frac{\mathrm{tg}\,(\varphi/4)}{\Delta p}} \left(\frac{D}{D+L}\right)^{0.4} \mu m \quad (1)$$

where:

- C_B — coefficient depending upon the hydrocyclone position during operation, the value of which at vertical position and assumed units remains within the limits 20-25 (average 22,5) and at horizontal position within limits 25-30 (average 27,5).
- μ_s — effective (apparent) viscosity of suspension (emulsion), cP
- $\Delta\rho_a$ — density difference between dispersed and continuous liquid phases, g/cm^3.
- $\Delta p = p_e - p_o$ — pressure drop in hydrocyclone (bar) where p_e — pressure at inlet, p_o — pressure at overflow; if $p_o = 0$ then $\Delta p = p_e = p$.
- $D, L, d_e, d_o, d_u, \varphi$ — hydrocyclone design data in succession — diameter, length of cylindrycal part, inlet overflow and underflow aperture diameters, cone angle.

Equation (1) includes nearly all more important hydrocyclone design parameters and reflects relations resulting from application of Stokes' Law i.e. for droplets of size d_{50} for which Reynolds Number is smaller then 1.

Further hydrocyclone operating data also very important from the point of view of process engineering calculations and their proper selection for concrete tasks is the feed flow rate and volume split.

For the feed flow rate calculations of the particular hydrocyclone design described in this paper application of an empirical equation given by (38) is suggested which characterizes the performance of this cyclone with a reasonable degree of accuracy.

The equation is

$$V_e = K_B \, d_e \, d_o \left[\left(1 - \frac{d_e}{D}\right)\left(1 - \frac{d_o}{D}\right) \right]^{0,5} \Delta H^n \qquad (2)$$

where:

D, d_e, d_o, L, φ as for equation (1)

$\Delta H = H_e - H_o$ - similar as for Δ_p but expressed in head [m].

n - exponent dependent on the shape of the cyclone inlet head has a value in the range 0.44 - 0.62 average 0.55.

K_B - coefficient dependent on many variables, which if the V_e is expresed in dm^3/min is calculated as follows:

$$K_B = \left[3.0 + \frac{0.3}{0.04 + tg\,\varphi/2}\right] \frac{D+20}{D+10} \left(1 - \frac{b}{D}\right)^{\frac{1}{4}} \left(\frac{D+L}{D+l_o}\right)^{\frac{1}{6}} \quad (3)$$

where:

l_o - length of vortex finder

b - inlet aperture breadth (see fig.2)

As mentioned equation (2) serves enough the purpose for technical calculations.

Much more difficult is a precise definition of the overflow and underflow rates. Tille (39) have given the possibility of an approximate determination of the volume split i.e. the volumetric flow ratios of underflow to overflow.

$$I = \frac{V_u}{V_o} = \frac{V_e - V_o}{V_o} = \frac{V_e}{V_o} - 1 \simeq K \left(\frac{d_u}{d_o}\right)^m \quad (4)$$

where:

K - coefficient = 1.13

m - exponent = 3.0

Detailed research work carried out by the author (40) has shown that the exponent m is strongly dependent on the total length of the hydrocyclone l_c and the angle of hydrocyclone inclination θ towards the level.

Generally its value is within large limits as for an exponent, namely in the range (2.2 - 3.50, whereas for hydrocyclones operating in a vertical position lower values should be assumed for larger l_c and larger values for smaller l_c.

From equation (4) it is possible to determine the volumetric flow ratio of the overflow or underflow diameter.

Rearranging this equation we receive:

$$V_u = \frac{V_e}{1 + K' \left[\dfrac{d_o}{d_u} \right]^m} \quad (5)$$

and

$$d_u = d_o \sqrt[m]{\frac{I}{K}} = d_o \sqrt[m]{I K'} = d_o \sqrt[m]{K' \frac{V_u}{V_o}} \quad (6)$$

Where

$$K' = \frac{1}{K} = 0.885$$

HYDROCYCLONE DESIGN.

Despite the existence of many known hydrocyclone design solutions applied for separation of non homogeneous mixtures of the type liquid-liquid, liquid-gas and liquid-solid-liquid, being different by design and even by working principle, on no one so far fully satisfactory efficiency has been reached.
This was causing the necessity of using multistage systems which from the operational point of view and frequently also for economical reasons were not widely accepted in the industry.
In the limited scope of this paper only the latest model of hydrocyclone design shall be discussed in which it is practically possible to separate with high efficiency the above mentioned multiphase systems [41]. Fig.2 ilustrates the design of the centrifugal separator with marked design parameters It consist of four basic elements: the preliminary swirling chamber 1, conical part 2, cylindrical part 3, solids collecting reservoir 4. In the upper part of the preliminary swirling chamber there are two diametrically opposed inlets situated tangentially on the chamber side surface and in the centre of the chamber top cover located is an opening serving the purpose of replacing the overflow aperture.

The solids collecting reservoir is equipped with a tangentially attached connector pipe. Hydrocyclone downstream end has been fitted out with an arrangement controlling the flow rate through the under-flow and with a disc shaped element preventing the passing through to the underflow some of the lighter dispersion which haven't reached yet the hydrocyclone axis forcing it at the same time to flow upwards to the overflow.
The working principle of the described appliance is similar to other types of hydrocyclones, with the difference that in the case of solid particles presence in the mixture they separate

from the rotating stream of underflow fraction and fall through a slot into the cylindrical settling vessel provided with a drain pipe for solid removal.

DISCUSSION OF SOME EXPERIMENTAL RESULTS.

The necessary conditions to achieve a good separation of multiphase suspension systems is the preservation of constant volumetric proportions of the mixture individual phases. The separation is occuring more easy the larger are the differences of densities between the continuous and dispersed liquid phase or solid if present.
The accuracy of this separation depends also on the interfacial tension value which in turn affects the size of the dispersed liquid droplets and the viscosity of the continuous phase.
The varying in time volumetric ratio of the individual liquid phases as well as of the solid of the mixture are creating large operational difficulties in carrying out their proper separation.
Similarly the degree of dispersion of the dispersed liquid and grain-size distribution of the solid particles as well as the ability of solids wetting by one or both liquids forming the non-homogeneous system have some time a decisive influence on the separation process.
Low interfacial tensions likewise don't promote the separation, since the occuring in the hydrocyclone shearing forces cause breaking-up instead of coalescence or coagulation of the dispersed liquid droplets. Especially in a hydrocyclone with a short cone in the zone of maximum acceleration, large shearing values do appear, which are the cause of dispersed phase increasing dispersion and in the presence of air this will be accompanied by the phenomenon of foam formation causing flotation of solids.
The easiest to separate are two-phase mixtures in which the dispersed phase is the heavier liquid and when the density difference between it and the continuous phase is larger than

0.05 while its volumetric quantity is not too small and not too large, best if it remains within 1-20% by volume. At lower contents e.g. of water in oil, efficiency of separation distinctly decreases and at higher contents especially within the range 1/3 - 2/3 volume of heavier phase emulsification occurs.

For example let's take a mixture solution of vegetable oil in solvent i.e. a miscella which through a long-lasting extraction of oil producing plants, particularly not sufficiently dried up gets contaminated by water and also by solids mainly of organic origin. With the increase of water content in the miscella the extraction effect is diminishing and thus the extraction time is significantly prolonged. By a single passage through the hydrocyclone of a miscella containing from 1 to 10% of water a following separation can be achieved: overflow fraction contains still 0.2 - 0.8% H_2O and the underflow fraction 0.06 - 0.3 % of solvent with which almost all solid particles present in the miscella are also flowing out. On a three-stage hydrocyclone installation the separation indices are much more beneficial and so with the same water content in solvent we have in the overflow, hardly 0.004 - 0.02% H_2O and in the underflow 0.0003 - 0.0005% of solvent what corresponds to an oil content in water from 30 - 50 ppm.

In a similar manner other two-phase liquid systems can be separated e.g. water-crude oil, water-kerosene, water-gasoline water-hexane, water-chloroform, water-isobutyl alcohol, water -freon etc. In general most difficult to take place is the separation in the hydrocyclone of two immiscible liquids when the dispersed phase is the lighter component constituting only a very small fraction of the whole mixture, typical example is an oil water emulsion, stearin-water and other.

It is possible to use the hydrocyclone for separating most of the mineral contaminants including water from crude oil. Besides that by adding a proper quantity of water it is also possible to carry out the leaching (extraction) of dissolved and suspended salts in crude oil.

The made first in this part of the paper general appraisal concerning the hydrocyclone performance when separating immiscible liquids and in them suspended solids is based on a series of carried out trials in a described already hydrocyclone (Fig.2) which by modifying the sizes of the overflow and underflow apertures was made suitable for effective separation of various liquid mixtures and solids removal.

In the following text the carried out tests shall be described. Actually the objectives of this further continued research work on liquid-liquid separation were:
- closer examination whether the hydrocyclone presented during 3rd Int.Conference on Hydrocyclones (Ref. 32) designed for removal of small percentage (typically from 0,1% to 3%) lighter liquid dispersions from heavy liquid continuous phase with a density difference $\Delta\rho$ above 140 kg/m^3 could also separate efficiently liquids with smaller density difference down to $\Delta\rho \sim 70$ kg/m^3 or so, simultaneously removing the highest possible percentage of solids,
- testing the mentioned type of hydrocyclone efficiency performance in an inverse situation i.e. separating heavy dispersions from lighter continuous phase retaining still its basic design features except for the overflow and underflow apertures sizes.

Here we describe in the following text the carried out tests. The process liquids used were mains water and three types of oils:
1. rape-oil ρ = 912 kg m^{-3} ; ν = 74 cSt @ 20°C
2. lubricating oil ρ = 882 kg m^{-3} ; ν = 8-10.5 cSt @ 100 $^{\circ}$C
3. cylinder lub.oil ρ = 933 kg m^{-3} ; ν = 20-36 cSt @ 100°C
4. extraction naphta ρ = 740 kg m^{-3} ν = 0,8 cSt @ 20°C
(gasoline solvent)

Oil-in-water (O/W) emulsions of following concentrations: 500 ppm, 1000 ppm, 2000 ppm, 4000 ppm, 8000ppm, 15000 ppm, and 30000 ppm, and water-extraction naphta (W/S) of following water concentrations in solvent: 1000 ppm, 2000 ppm, 5000 ppm; $1 \cdot 10^4$ ppm; $2 \cdot 10^4$ ppm, $5 \cdot 10^4$ ppm, $10 \cdot 10^4$ ppm have been tested.

Emulsification times of each emulsion were quite different and depended on the type of oil as well as its concentration. With the increase of oil concentration in water the emulsification time was increasing. The (O/W) emulsions were emulsified so long until all droplets were smaller then 50 μm on an average. Emulsions (O/W) of lower oil concentrations have shown higher stability, and out of the three tested oils the highest stability demonstrated the rape-oil next was the lubricating oil and the cylinder lub.oil had the lowest stability.

Emulsions of the W/S (water-extraction naphta) type have been prepared in two ways:
- by mechanical breaking-up using a centrifugal masher of special design,
- by condensation of steam in the solvent.

The emulsion produced by steam condensation in gasoline have shown significantly higher stability then the produced by mechanical mashing. The average and maximum size of water droplets diameters in case of achieved dispersion by condensation and mechanically were successively produced in several ranges 20 - 24 - μm, 31 - 36 μm and 28 - 33 μm, 75 - 90 μm respectively.

This way produced emulsions of O/W and W/S type have been subjected to separation in a hydrocyclone of 90 mm diameter in which following design parameters have been systematically varied:

- inlet area $f_e = \left(\dfrac{\Pi d_e^2}{4} \right)$,
- diameter of overflow nozzle d_e,
- diameter of underflow nozzle d_u,
- pressure drop Δ_p between inlet and overflow varying from 0,004 - 0,32 MPa.

It has to be assumed that during the flow of emulsions from the emulsifier to the hydrocyclone they have undergone partially the process of coalescence, especially those which are less stable, but this can hardly be avoided.

In table 1 some results of pressure effect on flow rate in the hydrocyclone and separation efficiency outcomes for a 3-phase emulsion composed of cylinder lub.oil ~ 4000 ppm, solids ~ 2000ppm and water have been compiled. This results were achieved on the hydrocyclone designed by the authors (Ref.32) operating in a vertical position illustrated in Fig.2 havingthe following characteristic dimensions (mm): $D = 90$, $L = 180$, equivalent inlet diameter $d_e = 18$, $d_o = 8$, $l_i = 700$ ($\varphi = 5^o$), $d_c = 35$, $l_c = 600$, $s = 20$, $D_s = 90$, $L_s = 250$, $d_s = 5$ the equivalent diameter d_u was being changed from 16 to 25. The total height of the hydrocyclone was $L_j = 1500$ mm.

In Fig.3 the individual parts of the experimental hydrocyclone with some interchangeable parts all made out of clear 'Perspex' are shown.

From the presented results it can be seen that with the increase of pressure drop in the hydrocyclone the efficiency for both oil and solids removal is clearly rising up to the pressure level of 0.28 MPa. Above this pressure to the highest tried i.e.0.32 MPa practically no deterioration of oil removal efficiency was observed, while the occuring differences resulted from measurements errors, whereas the efficiency of solid separation was still growing, what should be considered as normal.

The above statement relating the increase of oil removal efficiency with the increase of pressure is at variance with the majority of all works which have dealt with the problem of two immiscible liquids separation.

This phenomenon besides the proper choice of hydrocyclone design and its parameters ratios could be explained yet by other effects and dependences. In such a long hydrocyclone type the tangential velocities V_t having maximum values in its upper part are decreasing with the flow moving downstream causing the increase of axial velocities V_a. This is resulting from internal frictions occuring in the rotating motion of fluid and wall friction, also influenced by shape and orientation of hydrocyclones as well as the static head forcing the axial flow of the continuous medium.

In this manner through a proper shaping of the hydrocyclone and adequate selection of its design parameters for a specific task it is possible to achieve a large volumetric flow capacity V_e together with a high separation efficiency η_o and η_s. Where η_o the separation efficiency of two immiscible liquids during their process separation and is defined as follows $\eta_o = 1 - (c_c / c_i)$ with c volumetric concentrations c_i - inlet concentration c_c - concentration in the clean stream; η_s indicates the separation efficiency of solids defined as follows $\eta_s = V_u \cdot c_u / V_e \cdot c_i$ where:
c_u - a solid concentration in underflow
c_i - solid concentration in feed.
The presented above results constitute only a certain fragment of recently carried out research work, which due to their large number (over 1000 tests) and necessity of further data treatment can't be now made available but shall be published later. Parallel with above discussed research works, tests were also carried out on lowering the oil content in the water leaving the hydrocyclone underflow, as well as on the reduction of water in the oil separated from the oil-water emulsion. It has to be mentioned that the oil fraction leaving the hydrocyclone overflow frequently contains significant amount of water, which depending on the kind of oil is dispersed in it or surrounds droplets of highly concentrated emulsified oil i.e. fills the free space between not coalesced droplets into a uniform oil phase.
In order to clean further the leaving from hydrocyclone water from oil and if necessary from solids as well as dewater the oil the hydrocyclone have been coupled in turn with two proper construction coalescers. In Fig. 4a,b two types of coalescing devices are shown, which differ from each other by flow direction of the separated phases and by application. In coalescer a) (Fig.4) the flow of water in relation to the ascending oil separated from the water is a counter-current one, while in coalescer b) (Fig.4) both flows of water and oil are a parallel flow. Besides that, coalescer a) is only suitable for separation of clean emulsions such as

oil-water type or foaming emulsions as gas-oil-water (G/O/W), whereas coalescer b) can be also applied for treating clean or foaming emulsions, but mainly for separation 3 or even 4 phase system i.e. oil-solids-water (O/S/W) and gas-oil-solids-water (G/O/S/W) as well as for separation of two-phase systems water-oil (W/O) or water-extraction solvent (W/S).

In Fig.5a to 5g several types of the mentioned coalescing devices packings are shown, fig.5a illustrates an insert of ring-shaped leaflet structure (RLS) with a specific surface of 322 m^2/m^3 made out of any kind of metal; fig. 5b is a needle structure filling(NFS) with a specific surface of 600 m^2/m^3 made of plastics. In fig. 5c to 5g illustrated are varies needle type forms of inserts having specific surface from 360 to 480 m^2/m^3. As it was already on few occasions pointed out and is quite obvious from table 1 even on a well optimized hydrocyclone design it is not possible to obtain completly clean separation products. Depending on the type and concentration of oil in the emulsion subjected to separation in the hydrocyclone, the amount of oil still remaining in the water fraction was mostly within 35 - 160 ppm. On the other hand the amount of water passing into the oil fraction could be partly limited by the diameter of the overflow d_o. Generally through a proper selection of d_o and with optimum values of other design and operation data of a hydrocyclone it was possible to obtain oil fractions containing 1-3% of water. Most often the water content in the oil fraction was several times higher but it still was not a significant percentage when compared with the total flow rate.

Passing hydrocyclone underflow water fractions containing oil quantities quoted above through the described coalescers resulted in a distinct reduction of oil content in water as well as water and solids in a solvent.

By way of example at cylinder oil contents in hydrocyclone underflow fractions being: 23, 39, 76, 93, 128 ppm at the outlet of the coalescers these were reduced respectively to: 7,5,6,8,11 and 12 ppm. Practically these small values have created large difficulties by their analytical determination.

Much the same results were obtained during separation of other mineral and vegetable oils as for example the rape oil, which was giving very stable and strongly dispersed oil in water emulsions.

At concentrations of this oil equal 151 ppm in the hydrocyclone underflow water fraction and its further flow through the counter-current coalescer the oil level was reduced to 9 ppm of oil in water.

It should be added that the linear velocities of the flowing through coalescers media were contained within limits 0.32 - 0.97 mm/s (1.15 - 3.49 m/h) and in this range of velocities no significant change of oil separation efficiencies was observed.

Because the problem concerning the phenomenon of oil coalescence is outside this conference scope which obviously is devoted to hydrocyclones it shall be discussed more comprehensively on other occasion. But here some informations have been given which indicate that hydrocyclones can also be coupled with coalescer similarly as with other clarifying - thickening or classifying - enriching appliances in order to achieve finally a very high efficiency of liquid-liquid separation.

CONCLUSIONS.

1. A design of hydrocyclone for separating both light and heavy liquid dispersions has been discussed and shown incorporating features of the hydrocyclone presented at the previous 3rd Int.Conf. on Hydrocyclones.
2. It has been confirmed during tests that the hydrocyclone can be a universal unit i.e. used in various industries for deoiling or dewatering of liquids provided for each special duty the overflow and underflow hydrocyclone apertures are properly interchanged.
3. Equation (1) defining the cut size droplet for heavy liquid

dispersions removed in the described hydrocyclone gives within the range of tested flow capacities an accuracy of ± 20%.

4. To satisfy stringent IMO requirements of oil removal from ship's waters it is necessary to connect the hydrocyclone in series with a coalescer unit such as briefly described in the paper, in such a combined oil separator unit it is then possible to reduce the oil content of even highly dispersed oils below 10 ppm.

5. The authors are working already on a project of an oil sweeping barge where a battery of hydrocyclones with a pumping capacity between $5 \cdot 10^3 - 5 \cdot 10^4$ m^3/h and a mentioned type of coalescer of appropriate volume shall be installed. Carried out tests revealed that for a hydrocyclone output of 20 m^3/h a coalescer volume of 1 m^3 is required to reduce oil content below 10 ppm.

6. Based on done experimental work it is expected that a compact small compound oil separator consisting of a hydrocyclone(s) and coalescer(s) shall be an efficient system on board ships for water deoiling, while a large deoiling unit placed on a barge may be a very effective tool in case a disastrous oil spills at sea.

REFERENCES

1. Tepe J.B., Woods W.K.: U.S.Atomic Energy Comission Report.AECD. 2864 (Jan.1943).

2. Stamicarbon N.V. Dutch Pt. No.67244 (Nov. 1948).

3. Klein F.G.: M.S.Thesis, North Western Univ. (1951).

4. Ellefson R.R.: M.A.Thesis, North Western Univ. (1952).

5. Simkin D.J. Olney R.B.: A.I.Ch.E.J.2. 545 (1956).

6. Hitchon J.W.: United Kingdom Atomic Energy Authority Report,AERE CE/R 2777 (1958).

7. Molyneux F.: Chem.Proc.Eng. 43,10,502 (1962).

8. Sheng H.P.: Ph.D.Dissertation, University of Oklahoma (1968).

9. Burill K.A.: M.Eng.Thesis, McMaster University. Hamilton. Ont.Canada, 1967.

10. Holland C.D., McDonough, J.A. and Tomme, W.J.: A.I.Ch.E.J.6.(1960).

11. Regehr H.U.: Forsch-Gebiete Ingenieurwes. 28.11/27(1962).

12. Bednarski S.: Prace Naukowe Akademii Górniczo-Hutniczej,Nr 113/59, Kraków, 1959.(Scientific Works of Mining-Metallurgic Academy.Cracow 1959).

13. Bednarski S.: Sprawozdanie z pracy nt. "Rozdział misceli na hydrocyklonie" wykonanej na zlec.Instytutu Kwasów Tłuszczowych w Warszawie, 1960.(Report: "Separation of

miscell in a hydrocyclone" prepared for the Institute of Fatty Acids in Warsaw 1960).

14. Bednarski S.: Ogólnokrajowa Konferencja Naukowo-Techniczna nt. "Wydobycie i przeróbka Ropy Naftowej". Ref. pt. "Hydrocyklony w procesie uszlachetniania ropy i jej produktów". Kraków 28-29 wrzesień 1963. (Domestic Scientific-Technical Conference on the Subject "Crude Oil production and processing" Paper. "Hydrocyclone in the process of crude oiland its products refining. Kraków, September 1963).

15. Bednarski S.: Opracowanie Ośrodka Badawczo-Rozwojowego Urządzeń Chemicznych "CEBEA" w Krakowie nr 856, grudzień 1968. (Report of the Research-Development Centre for Chemical Appliances "CEBEA", Kraków nr 856, December 1968).

16. Bohnet M.: Chemie-Ing. - Techn. 41.5-6.381 (1969).

17. Bohnet M.: VDI Berichte Nr 121. Verfahrenstechnik 153 (1968).

18. Bohnet M.: MM-Industrie Journal Wurzburg, 77(1971) 53-VT60.

19. Burill K.A. , Woods D.R.: Ing. Eng. Chem. 9,4,545 (1970).

20. Sheng H.P., Welker J.R., Sliepcevich C.M.: Canadian J. of Chem. Eng. 52..487 (1974).

21. Minart P.: La Mouille Blanche, 5/6.325 (1975).

22. Johson R.A., Gibson W.E., Libby D.R.: Ind. Eng. Chem. Fundam. ,15.2.110 (1976).

23. D.A.Colman, M.T.Thew, D.R.Corney: "Hydrocyclones for oil/water separation"- 1st Int. Conf. on Hydrocyclones.

Cambridge, England , October 1980.

24. I. C. Smyth. M. T. Thew, P. S. Debenham and D. A. Colman "Small scale experiments on hydrocyclones for de-watering light oils" - <u>1st Int Conf. on Hydrocyclones, Cambridge,</u> England, October 1980.

25. D. A. Colman, M. T. Thew. "Hydrocyclone to give a highly concentrated sample of a lighter dispersed phase <u>1st Int. Conf. on Hydrocyclones, Cambridge,</u> England, October 1980.

26. Thew M. T. , Wright C. R. , Colman D. A. : R. T. D. characteristic of hydrocyclones for the separation of <u>light dispersions. 2nd International Conferences on Hydrocyclones, Bath,</u> England 19-20 September 1984. Paper, E1, p. 163.

27. Smyth I. C. , Thew M. T. , Colman D. A. : The effect of split ratio on heavy dispersion liquid-liquid separation in hydrocyclones". <u>2nd International Conference on Hydrocyclones.</u> Bath , England, Paper E2, p. 177.

28. Listewnik J. : "Some factors influencing the performance of deoiling hydrocyclones for marine applications". <u>2nd International Conference on Hydrocyclones. Bath,</u> England 19-21 September 1984. Paper E3, p. 191.

29. King N. W. , Purfit G. L. : "Experiments on oil/gas separation in helical passages". <u>2nd International Conference on Hydrocyclones, Bath.</u> England, 19-21 September 1984, Paper F1 , p. 205.

30. Colman D. A. , Thew M. T. . D. Lloyd. : "The concept of hydrocyclones for separating light dispersions and a comparison of field data with laboratory work". <u>2nd International Conference on Hydrocyclones. Bath, England,</u> 19-20 September , 1984. Paper F2, p. 217.

31. Nezhati K., Thew M.T.: Aspects of the performance and scaling of hydrocyclones for use with light dispersions". 3rd International Conference on Hydrocyclones. Oxford, England, 30 September - 2 October 1987. Paper G1, p.167.

32. Bednarski S., Listewnik J.: "Hydrocyclones for simultaneous removal of oil and solid particles from ship's oily waters". 3rd International Conference on Hydrocyclones. Oxford, England,1987. Paper G2, p.181.

33. Smyth I.C., Thew M.T.: "A comparison of the separation of heavy particles and droplets in a hydrocyclone". 3rd International Conference on Hydrocyclones. Oxford, England, 30 September - 2 October 1987. Paper G3, p.193.

34. Bednarski S.: "Hydraulika układów wielofazowych typu ciecz-ciało stałe-ciecz i gaz-ciecz w hydrocyklonach". II Ogólnopolska Konferencja nt. "Przepływy wielofazowe". Gdańsk 9-12 października 1989. Mat.Konferencji.
("Hydraulic multiphase systems of type liquid-liquid, liquid-solid-liquid and gas -liquid in hydrocyclones" 2nd-All-Polish Conference on Multiphase Flows, Gdańsk, 9-12 October 1989. Conference Papers).

35. Bednarski S.: "Oddzielanie wody i cząstek stałych z misceli". V Konferencja Naukowo-Techniczna nt." Budowa i eksploatacja maszyn w przemyśle spożywczym". Poznań, 8-9 października 1990. Mat. Konferencji, str. 73.
("Separation of water and solid particles from miscella" 5th Scientific-Technical Conference on Construction and Design of Machines in food industry. Poznań, 8-10 Oct.1990).

36. Bednarski S.: "Oddzielanie olejów i tłuszczów ze ścieków". I Ogólnopolska Konferencja Naukowa nt." Inżynieria

procesowa w ochronie środowiska". Warszawa - Jachranka, wrzesień 1991. Mat. Konferencji, str. 258.
("Separation of oils and fats from waste waters". 1st All-Polish Scientific Conference on Process Engineering in Environmental Protection. Warsaw, September 1991. Conference Papers.

37. Bednarski S.: Hydrocyklony. WPLiS. Warszawa. 1967. (Hydrocyclones).

38. Bednarski S.: Chem. Techn. 20.1.12 (1968).

39. Tille R.: Ann. Techn. Belgigue 50.6.788 (1951).

40. Bednarski S.: Starke. 15.2.18; 5.171 i 10.371 (1963).

41. Patent PRL nr P 271532.

Table 1

Test No.	Δp m H$_2$O	Feed V$_e$ dm^3/min	Oil Fraction		Water Fraction			Solid Reservoir Fraction V$_s$ dm^3/min	Efficiency	
			V$_o$ dm^3/min	Oil Content g/dm^3	Oil Content ppm	Solid Cont. *	V$_u$ dm^3/min		η$_o$ %	η$_s$ %
1	4	89.3	0.72	445.0	424.6	331.0	86.66	1.92	89.70	83.9
2	8	117.52	0.96	446.2	196.4	215.5	114.40	2.16	95.22	89.5
3	10	142.02	1.42	390.0	103.6	156.0	138.20	2.40	97.48	92.4
4	15	178.98	1.30	542.0	63.3	117.8	175.18	2.50	98.45	94.2
5	20	205.92	1.44	565.2	48.8	82.5	201.80	2.70	98.80	96.0
6	24	230.93	1.48	617.3	44.9	60.3	226.50	2.95	98.90	97.0
7	28	251.62	1.52	656.9	32.6	46.5	247.00	3.10	99.20	97.7
8	32	267.95	1.60	661.9	48.9	38.9	263.10	3.25	98.80	98.1

* – ppm by mass

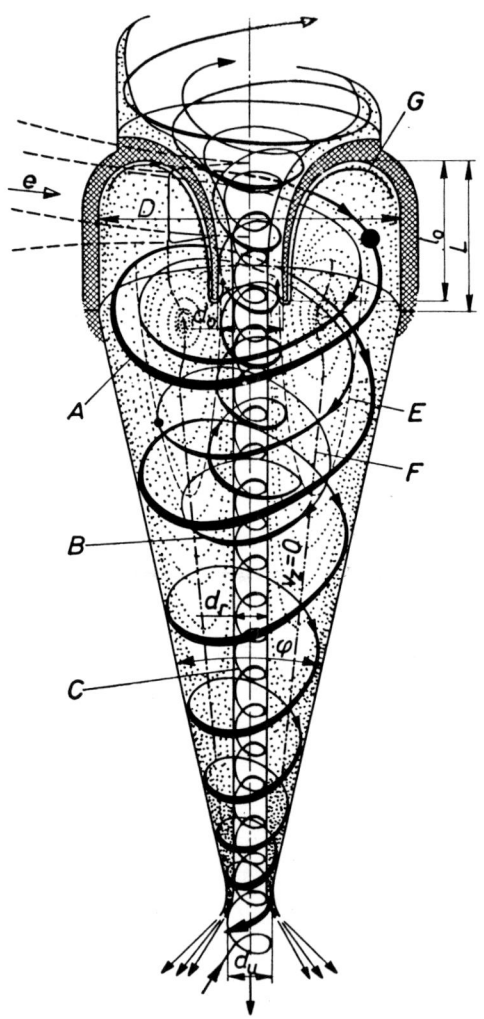

Fig. 1. Visualization of flow in a conventional hydrocyclone.

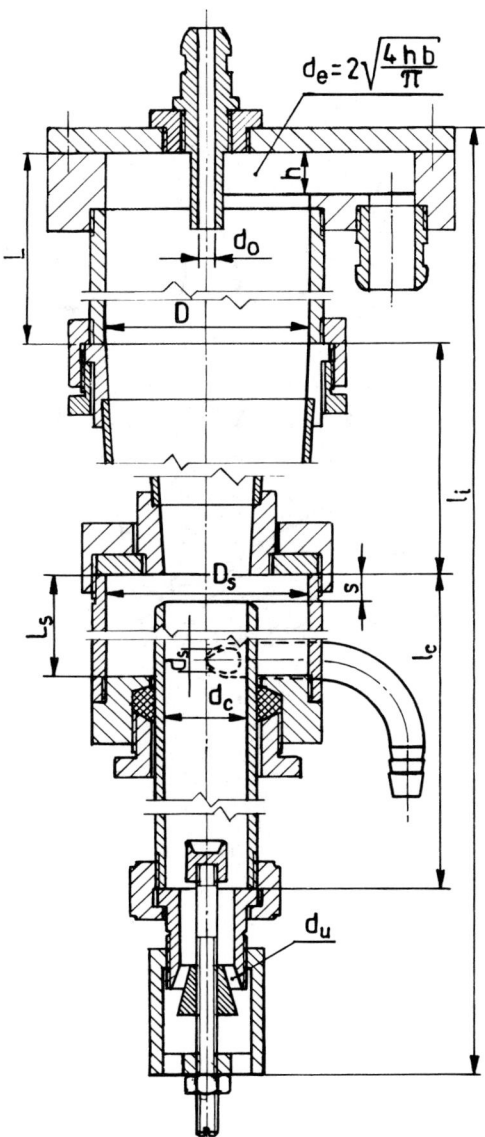

Fig. 2. Hydrocyclone design for liquid-liquid-solids separation.

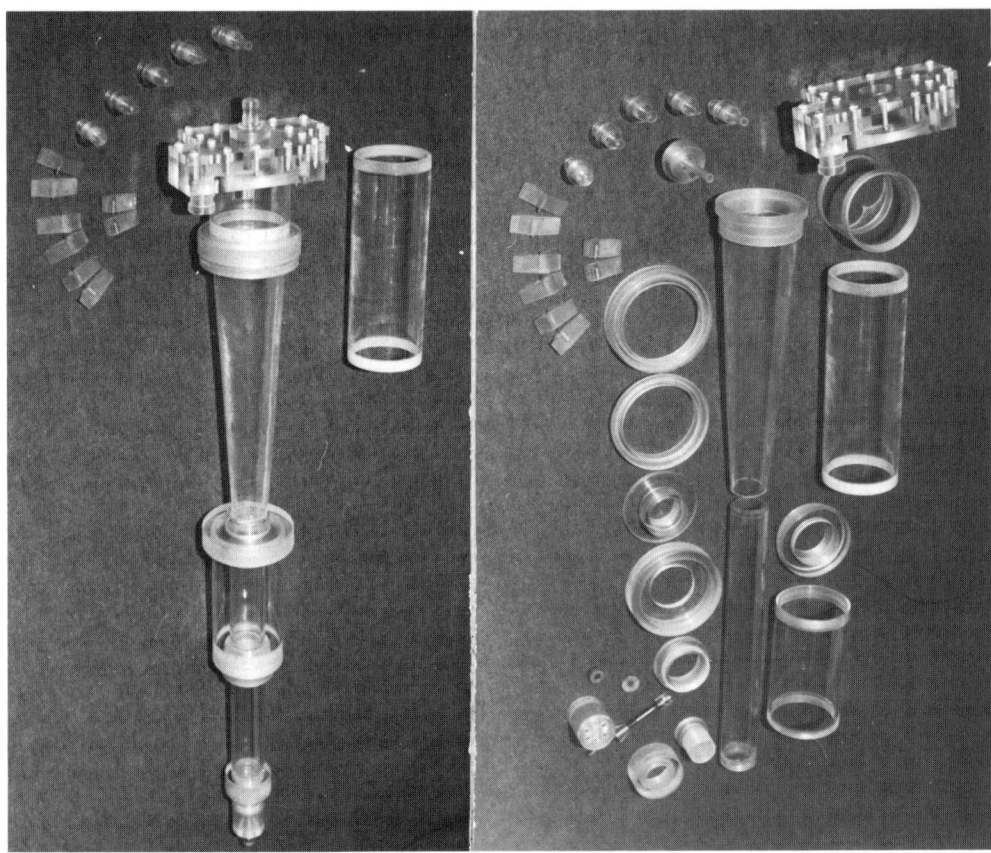

Fig. 3. View of laboratory testing hydrocyclone and its single parts made of transparent material.

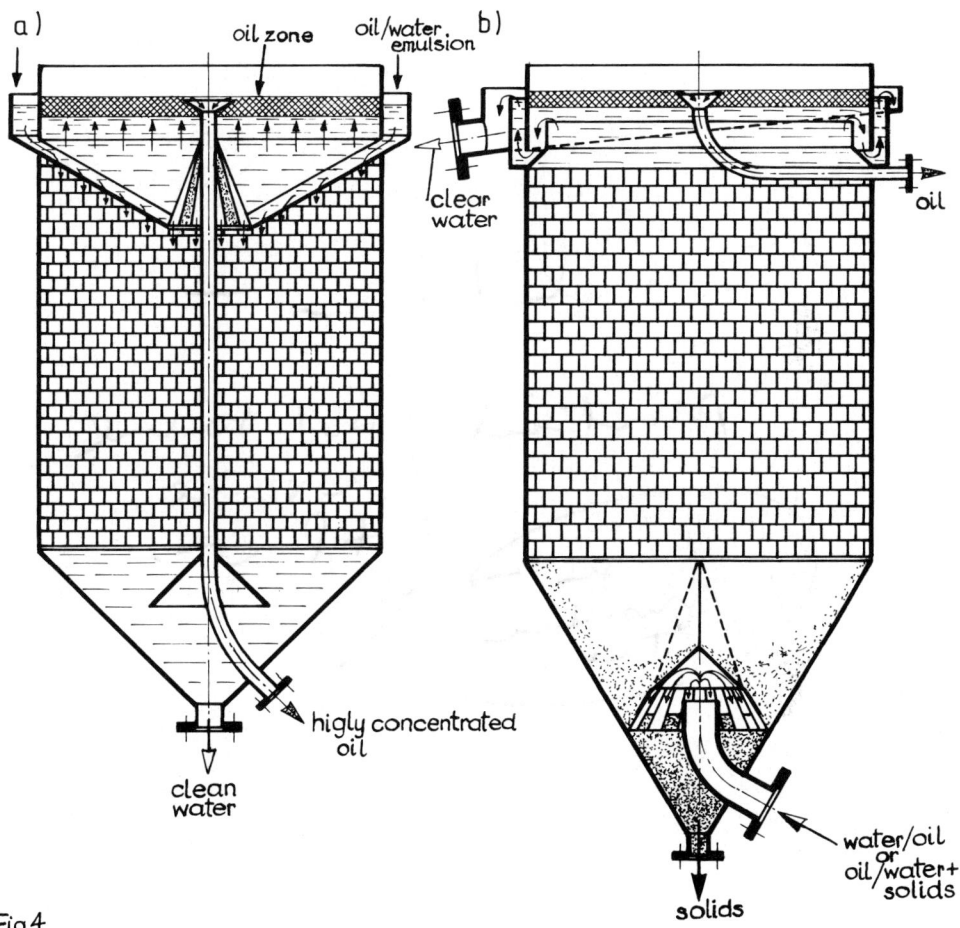

Fig.4.

Fig. 4. Cross-sections of coalescers operating in series with the hydrocyclone during tests.

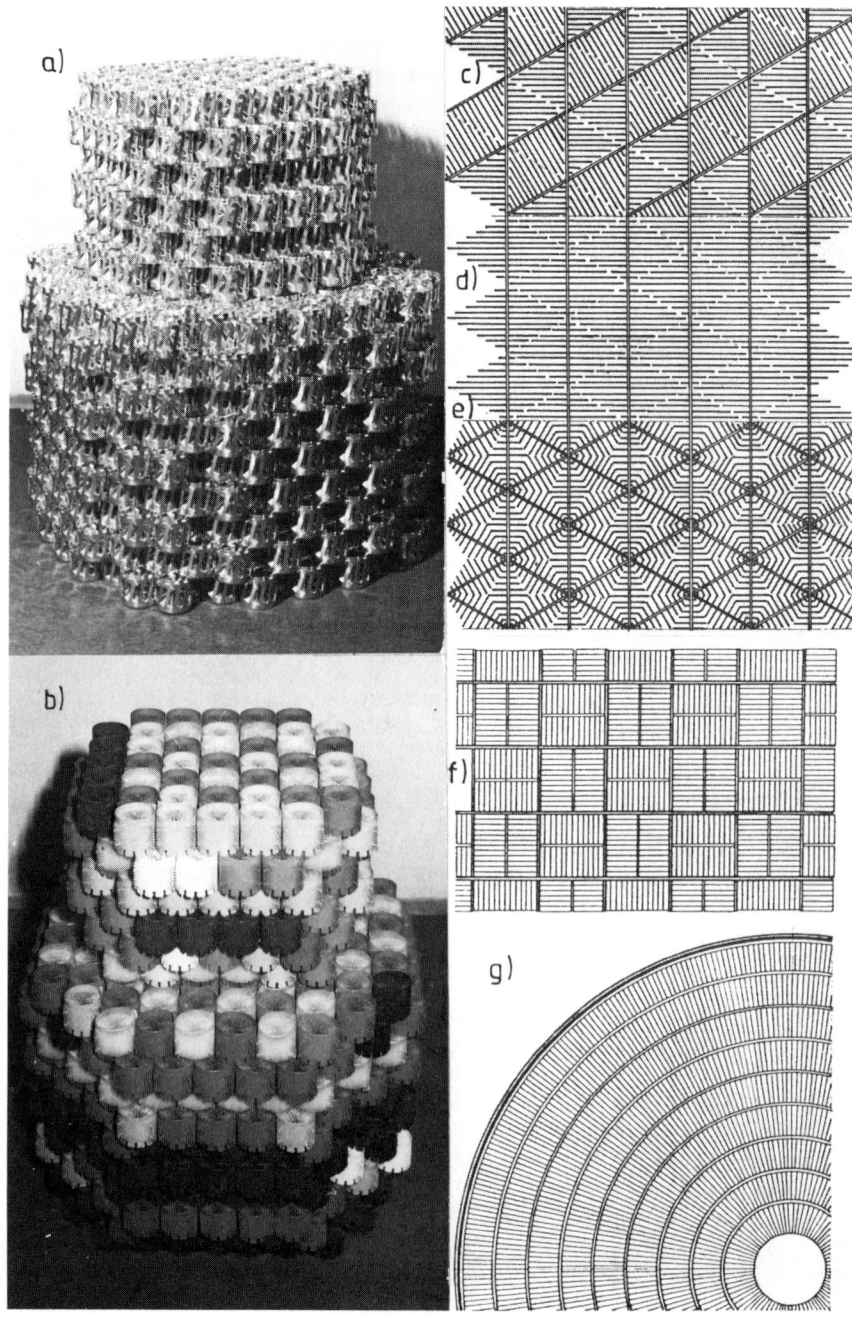

Fig. 5. Various forms of coalescers inserts applicable for highly dispersed oil separation.

LDV MEASUREMENTS IN THE VORTEX FLOW CREATED BY A ROTATING WALL DEWATERING CYCLONE

P. SCHUMMER & P. NOE (ACB-CERG, France)
M. BAKER (BP Engineering, Great Britain)

Summary

A series of LDV[1] measurements enable the investigation of axial and tangential velocity fields inside the cylindrical separation chamber of a new rotating wall cyclone aimed at oil/water separation in offshore applications (dewatering process).

Experiments are undertaken with a special pure transparent cosmetic oil of high viscosity to simulate the laminar/turbulent flow conditions of crude oil. The experiment is described.

Velocity measurements show the influence of rotation speed and flow rate on the resulting acceleration field (more than 900 gees achieved) and reveal non uniform axial profiles with high velocities near the wall and the core while at mid-radius an axial velocity reduction and even reversed flow is observed at low throughputs.

Tangential profiles are barely affected by flow withdrawal through an annular downstream exit, enabling potentially good separation and extraction of water droplets from oil. Thanks to the rotating wall there is no swirl decay ; very low turbulence and reduced shear effects are expected due to special design of the rotary cyclone.

Nomenclature

- D : diameter of cylindrical vortex separation chamber (m)
- d : diameter of a droplet (m)
- L : length of cylindrical vortex separation chamber (m)
- n : exponant of vortex law : Vr^n = constant
- Ro : radius of the cyclone's inner wall : Ro = D/2 (m)

---

[1] Laser Dopppler Velocimetry

r : radius *(m)*
$\vec{u}$: flow velocity vector *(m/s)*
u_r, u_θ, u_x: velocity components (radial, tangential and axial) *(m/s)*
U : axial velocity *(m/s)*
V : tangential velocity *(m/s)*
Vo : tangential wall velocity *(m/s)*
Vmax : maximum of tangential velocity *(m/s)*
W : migration radial velocity of droplets *(m/s)*
X : abscissa along axis of vortex separation chamber *(m)*, X ranging from 0 to L
γ : intensity of acceleration field *(m/s²)*
Γ : circulation (swirl intensity of vortex) *(m²/s)*
Θ : angle *(rd)*
µ : viscosity of continuous phase *(kg/ms)*
$\delta\rho$: specific gravity difference or density difference *(kg/m³)*

1. INTRODUCTION

Hydrocyclones are increasingly used for liquid-liquid separation especially in oil industry : their low mean residence time (a few seconds vs. several minutes for gravity settling tanks) reduces the space and weight required on offshore platforms. Their separation efficiency is also very attractive in comparison with conventional treatment units.

Hydrocyclones with special design have been developped for the treatment of produced oily water to respect legislation before discharge to the sea : the oil in water content has to be reduced to less than 40 ppm (North Sea). Separation can be obtained with conventional hydrocyclones *(Ref. [1])* but higher efficiencies were obtained with rotary hydrocyclones *(Ref. [2], [3] and [4])*.

The use of hydrocyclones may also be considered in the reverse process which consists of dewatering the produced crude oil before transport into pipelines, especially for heavy oils of high viscosity. Up to now, very few studies have been carried out on that subject *(scarce datas in Ref. [3])*. The present study is thus included in a development program on dewatering rotating hydrocyclones.

Inside a de-oiling hydrocyclone the oily droplets migrate towards the axis of the vortex chamber, the continuous phase being water ; with a dewatering cyclone, the continuous phase is crude oil and the water droplets migrate towards the wall. The migration velocity W can be estimated according to Stokes'law :

$$W = \frac{\delta\rho \cdot \gamma d^2}{18\mu}$$

in which :

- $\delta\rho$: density difference between oil and water (kg/m^3)
  (typically $\delta\rho$ = 100 to 200 kg/m^3)

- γ : intensity of acceleration field (m/s^2)

- d : mean drop diameter (m)
  (typically d ranges from 10 to 50 microns)

- µ : viscosity of continuous phase $(kg/m.s)$
  (typically : µ = 0.001 kg/m.s with water
   µ ⩾ 0.020 kg/m.s with crude oil)

As viscosity of oil may be 20 times higher than viscosity of water, a dewatering cyclone has to produce a very fast swirl to get the same migration velocity and by this way the same separation efficiency as in the de-oiling application. An acceleration field of several hundred gees is required to achieve good performance.

With conventional hydrocyclones, this swirl results from a tangential jet inside a conical vessel : high tangential velocities are needed to achieve high accelerations. But strong turbulence distributions appear especially on the radial and axial flows (Ref. [5]). Turbulence and wall friction (thick boundary layers) are partially counterbalanced by the conical shape of the separation chamber avoiding a rapid swirl decay. Nevertheless turbulent shear stress and energy dissipation in the entry affect the size of the droplets : large water droplets in oil are stretched by turbulent vortices and broken up into smaller particles (Ref. [6]) much more difficult to separate. Moreover high turbulence rates induce diffusion and modify the ideal settling trajectory of a small droplet inside the separation chamber (Ref. [7] and [8]). That's why conventional hydrocyclones are badly suited for dewatering of oil.

On the contrary, rotary hydrocyclones are specially designed to overcome above limitations :

- rotating the walls of the separation chamber reduces wall friction and limits turbulence,

- axial entry through rotating blades (semi-axial pump) at low velocity avoids drop breakup and enables the swirl to be independent from the flow rate, giving more operating flexibility.

The separation chamber is rotated by an external electrical motor. Some new advantages arise also, such as the use of a cylindrical vessel which is easier to manufacture and lower pressure drop requirements as some head is produced by the inlet pump design.

Rotary de-oiling hydrocyclones have already proven their superiority, leading to very low absolute cut sizes and good separation efficiencies over a great range of operating conditions (Ref. [2] and [4]).

This paper presents some measurements of the axial and tangential flow fields inside the vortex chamber of a rotary cyclone designed for dewatering crude oils. *Figure 1* shows a general sketch of the cyclone with the following main features :

- diameter of vortex chamber D : 100 mm
- length of vortex chamber L : 1500 mm (15D)
- axial entry in the upper part of the cyclone
- induced rotation of the fluid by inlet blades and ducts and by the rotation of the walls
- axial main outlet of the purified crude oil through a downstream reduced section orifice (1/3 of cyclone's radius)
- axial secondary outlet to withdraw the settled water in an annular downstream section close to the wall (main flow and extraction are co-current).

The knowledge of the flow field will enable the validation of a mathematical model of the rotary hydrocyclone (to be published soon) and also to set the best working conditions (optimal rotation, flowrate and reject ratio) before testing the equipment in an oil treatment process.

2. **EXPERIMENTAL SET UP**

The best way to measure the flow field velocities inside the vortex chamber of a rotary cyclone seems to be Laser Doppler Velocimetry (LDV) since it is a non intrusive technique. Use of dyes *(like in Ref. [9], [10] and [11])* is not easy because of the rotation of the walls and large particle seeding and following up (Cf. PIDV[(2)] techniques) is not compatible with the high swirls achieved.

As the medium needs to be transparent for LDV measurements, the cylindrical separation chamber is made of high quality altuglass perspex (thickness of walls = 20 mm) perfectly well machined. The process fluid (crude oil) is replaced by a pharmaceutic synthetic oil

---

(2) **P**articule **I**mage **D**isplacement **V**elocimetry

used for cosmetic purpose (ARAL, AUTIN PL). This oil is very pure and transparent (it looks as clear as water), has a viscosity of 27 centistokes at ambient operating temperature, a specific gravity of 855 kg/m^3 and a measured refractive index of 1.48, very close to the index of perspex (1.49). The main reason for using such a costly oil is that the viscosity is in the same range as for crude oils and so the laminar/turbulent flowing conditions are pretty well represented (low Reynolds numbers for axial and radial flows).

A powerful 5 watts argon dual-beam DANTEC laser-Doppler velocimeter is used to investigate the flow field. The probe head linked with the laser source by fiber optics has a focal length of 310 mm and is spacially positioned by a three motor traverse assembly (X, Y and Z axis). Since there is no air core inside the vortex chamber of the rotary cyclone (controlled flow conditions through downstream valves), it is possibe to use forward scattering alignment : more light is collected on the photomultipliers and a better accuracy is reached compared to backscatter alignment. A Bragg Cell frequency shifter permits the detection of flow direction as well as the measurement of low velocity flow.

Classical LDV measurements through cylindrical or conical walls use a special set up : the hydrocyclone is immersed in a water-filled parallelepipedic jacket to minimize the complex optical refraction of the laser beams *(Ref. [12])*. Again because of the rotation of the walls that idea is difficult to apply : water tightness would require a special seal along the rotating wall of the cyclone and increased power for the electrical driving motor.

The set up remains therefore very easy as the cylindrical vortex chamber is used as it is. The tube wall curvature is taken into account to calculate precisely the optical trajectories of the laser beams and to locate the focus point at the right place knowing the different refractive indexes : there is in effect a position shift of the probe volume along the LDV axis due to refraction but only for tangential velocity measurements. That's why solely the green beam of the argon laser is used : it is not possible to measure at the same time both axial and tangential velocities since the focus points are not the same.

The optical axis of the LDV intersects the tube axis (symetrical case) to limit aberration problems, which are not large enough to cause serious degradation of the LDV signal *(Ref. [13])*.

The cylindrical tube is perfectly well aligned and balanced to limit vibrations due to the rotating walls : the focus point doesn't move and precise measurements of mean velocities can be done. Nevertheless accurate measurements of the turbulence rate of the flows are not possible since micro-vibrations would affect the results.

The fluid has to be seeded to provide a maximum of signal to the Doppler signal processor : the frequency shift is than converted into flow velocity. A very fine powder of pearl lustre pigment called IRIODIN 110 fine satin is used to seed the flow. The mean size of the particules is lower than 7 microns ; as their density is very close to 1, the particules don't have time to settle inside the vortex chamber and thus are assumed to follow quite well the streamlines of the flow. The proper trajectory of a particle would affect first the radial velocity (not measured during the tests) but is of less importance and thus negligable compared with axial and tangential velocities.

The experimental test rig consists of the rotary cyclone, a sump with smoothing grid, a pump, a heat exchanger to keep temperature at a precise level, a coarse filter and some instrumentation (flowmeters, pressure gauges, ...). *Figure 2* presents the set up with the LDV system. Two control valves placed downstream enable to control pressure and flow rates of main flow and annular withdrawal.

3. **EXPERIMENTAL RESULTS AND COMMENTS**

3.1 **Tangential velocities**

Inside a perfect irrotational free vortex flow, the theory tells us that circulation Γ (swirl intensity of vortex) remains constant along a closed loop around the axis.

With $\overrightarrow{\mathrm{rot}\, \vec{u}} = \vec{0}$

$$\Gamma = \int_0^{2\pi} u_\Theta \, r \, d\Theta = \text{constant}$$

thus :

$$\boxed{V = u_\Theta = \frac{\Gamma}{2\pi r} = \frac{V_o R_o}{r}}$$

where :

. u_Θ or V represent the tangential velocity (V_o tangential velocity at the wall)

. r represents the radius (R_o wall radius)

A lot of work on conventional hydrocyclones has shown that the tangential velocity is rather proportional to r^{-n} where n is an empirical exponent lower than 1 :

$$V \cdot r^n = \text{constant}$$

There are discrepancies about the value of n depending on the geometrical design of the cyclones : n is ranging from values close to 1 *(Ref. [14])*, to more commonly 0.8 and 0.7 *(Ref. [12] and [17])* and even 0.4 *(Ref. [15])*.

Figure 3 shows a typical tangential velocity profile measured on the dewatering rotary cyclone rotating at 1250 RPM with a throughput of 8 m^3/h. Maximum of velocity is reached at a radius of 17 mm approximately equal to Ro/3 at a value of :

$$V_{max} = 1,4 \cdot V_o$$

The profile is correlated with a mean value of n equal to 0.4 from the wall to the radius Ro/3. At a lower radius, in the core of the vortex flow, a solid body rotating liquid column is encountered as classically described in litterature.

According to Escudier et al. *(Ref. [9])*, the location and magnitude of the maximum swirl velocity and thus also the dimensions of the solid body rotating core are linked to the size of the exit hole. Withdrawing some flow through the secondary annular outlet of the dewatering cyclone close to the wall might then affect the vortex flow at high reject rates. But *figure 4* shows little influence of the split ratio even at 20 % : the solid body rotating core remains at Ro/3, the maximum of tangential velocity stands now at 1.28 Vo and the value of n is about 0.3. Extraction of settled water in a layer close to the wall is thus possible without affecting to any great extent the separation power of the cyclone.

A change in total flow rate has more drastic effects : *figure 5* compares results at 8 m^3/h and 4 m^3/h. The shape of the tangential velocity profile is nearly flat at 4 m^3/h, the velocity remaining constant from the wall to the solid body rotating core at Ro/3 thus giving an exponant n equal to zero. The performance is still better than with centrifuge separation (at the same rotation speed) with solid body rotation but nevertheless the resulting acceleration field is very poor compared to the one reached at 8 m^3/h or more. Therefore there is a benefit in increasing the throughput of rotary dewatering cyclones ; but at the same time the residence time of droplets inside the vortex field is decreased, lowering the separation power of the unit. An optimal flow rate should be found.

Figure 6 shows two tangential velocity profiles measured at two different axial positions along the cylindrical vortex chamber : the first cross section is placed at 500 mm downstream of the inlet that is at 1/3 of the total length of the chamber (X/L = 1/3) while the second cross section is placed at 1000 mm downstream at a location of 2/3 of the total length (X/L = 2/3). No swirl decay is noticed over the 500 mm length corresponding to these measurements, proving the great benefit of rotating the walls of the cyclone. These results introduce the possibility of increasing the length of the separation chamber to a much greater extent, the only limitation being technological : these would thus extend the residence time of droplets inside the vortex field and thus enhance the separation efficiency.

The resulting acceleration fields are plotted on *figure 7* at three different rotation speeds corresponding to 1250, 2500 and 3000 RPM at a flow rate of 8 m^3/h and a 20 % split ratio (reject rate). Maximum acceleration reaches easily more than 900 times the gravity at a radius of Ro/3 enabling a good migration velocity (according the Stokes'law) even for small water droplets in viscous oil.

3.2 Axial velocities

Conventional hydrocyclones are designed with an overflow outlet (vortex finder) and an underflow outlet (called spigot or apex) so that mainflow and extraction are counter-current. Therefore, there is an axial flow reversal with downward flow along the walls and upward flow along the axis. Between both flows, there is a zero velocity surface (called mantle) shown to remain cylindrical even in a conical vessel at approximately mid-radius of the cylindrical upper part *(Ref. [10], [11] and [12])* ; some publications show a rather large annular zone of very low axial velocity at mid-radius, depending on design of the vortex finder *(Ref. [15] and [16])*.

Since in a rotary cyclone both main flow and extraction are co-current (downstream exits), a smooth unidirectional axial profile should be expected.

Figure 8 presents a classical axial velocity profile measured at 2500 RPM and 8 m^3/h with 20 % split ratio :

- there is a high axial velocity near the rotating wall of the cyclone together with a thin boundary layer (zero velocity on the wall),

- a high axial velocity zone is also encountered inside the solid body rotating core,

. in the annular zone situated around mid-radius the axial flow is very low building a kind of "mantel" as in conventional hydrocyclones.

These results are barely affected by the extraction rate (split ratio) through the annular exit and profiles remain roughly the same.

Figure 9 shows the influence of a decrease of the total flow rate : the mid-radius low axial velocity zone turns negative at low flow rates (4 m^3/h). A recirculating annular shell with reverse flow appears inside the vortex chamber. Again with low throughputs the rotary cyclone presents flow patterns that may hinder a good separation of water in oil droplets : residence time is not uniform across a section and settling probability depends therefore a lot on radial position of each droplet.

Influence of the rotation speed on axial velocity is presented in *figure 10* again at 8 m^3/h with 20 % split ratio. At 1250 RPM, there is less flow along the walls than at 2500 RPM and therefore axial velocities are increasing at mid-radius and in the core of the vortex ; the general shape of the axial profile is characterized by two annular peaks (located at 0.2 Ro and 0.8 Ro), while the velocity on the axis remains surprisingly the same as at 2500 RPM. Similar axial flow patterns with annular peak were measured in a cylindrical vortex chamber without wall rotation *(see Ref. [9])*. The risk of flow reversal at mid-radius is decreasing with the rotation speed of the walls and the axial profiles become less disturbed with better radial flow distribution.

Measurements at different axial positions confirm that there is a radial redistribution of the flow towards the center of the cyclone in the downstream sections : the axial velocity along the axis, in the whole core of the vortex flow is increasing with the abscissa. That effect is due to the exit contraction of the flow that forces the streamlines to converge towards the axis *(see figure 11)*.

4. **CONCLUSIONS**

The Laser Doppler Velocity measurements undertaken on a rotating wall dewatering cyclone lead to a great amount of datas of paramount importance to the understanding of the flow patterns inside the vortex chamber. Main results are as follows :

- Annular flow withdrawal with an extraction rate up to 20% doesn't affect to any great extent the tangential velocities ; the profiles are correlated by the law : Vr^n = constant with n ranging from 0.4 to 0.3.

- The rotation speed of the walls of the cyclone induce the swirl and resulting acceleration fields are more especially strong as this rotation speed increases ; nevertheless with high swirls the axial velocity increases along the walls of the separation chamber and flow reversal may occur at mid-radius at low flow rates.

- Counter-current flow is in effect the main disadvantage of low throughputs associated with flat tangential velocity profiles leading to poor acceleration fields. With a 100 mm rotary cyclone the minimum flow rate should stand above 6 m^3/h, an optimum being linked to the effective residence time of droplets inside the separation chamber.

- Finally, there is no swirl decay in the cylindrical vortex chamber as wall friction is reduced to a minimum thanks to the rotating wall ; long separating tubes may be achieved according to only some technological limitation.

The special design of the dewatering rotary cyclone should limit turbulence and shearing effects (especially at the inlet). The measured velocity profiles and acceleration fields are promising especially at high flow rates and medium rotation speeds. Good separation efficiencies should be expected on water in oil emulsions, to be checked with on site crude oil tests.

5. ACKNOWLEDGMENTS

This research was supported by British Petroleum Engineering.

The authors thank Y. LECOFFRE and J. WOILLEZ for having initiated work on rotary hydrocyclone on behalf of TOTAL/CFP and ALSTHOM/NEYRTEC.

6. REFERENCES

[1] D.A. COLMAN, M.T. THEW & D.R. CORNEY
   "*Hydrocyclones for oil/water separation*"
   Proc. Int. Conference on Hydrocyclones, paper 11, page 143, BHRA 1980

[2] J.C. GAY, G. TRIPONEY, C. BEZARD & P. SCHUMMER
"*Rotary cyclone will improve oily water treatment and reduce space requirement/weight on offshore platforms*"
Society of Petroleum Engineers, 1987, SPE 16571

[3] J. WOILLEZ, P. SCHUMMER & Y. LECOFFRE
"*A new high efficiency liquid/liquid separator*"
Proc. 4th Int. Conf. on Multiphase Flow, BHRA 1989, Chapter 8, pp. 117-132

[4] P.S. JONES
"*A field comparison of static and dynamic hydrocyclones*"
Society of Petroleum Engineers, 1990, SPE 20701

[5] XU JIRUN, LUC QIAN & QIU JICUN
"*Studying the flow field in a hydrocyclone with no forced vortex - Part II : Turbulence*"
Filtration & Separation - Sept/Oct. 1990, page 356

[6] J.O. HINZE
"*Fundamentals of the hydrodynamic mechanism of splitting in dispersion process*"
AIChE J. Vol. 23 - n° 4, 1977

[7] KUBIE
"*Settling velocity of droplets in turbulent flows*"
Chem. Eng. Sci., Vol. 35 - 1980

[8] Y. LECOFFRE
"*Turbulence et séparation*"
Session SHF des 18/19 mars 1987, IVe Demi-journée : La Turbulence dans l'Industrie

[9] M.P. ESCUDIER, J. BORNSTEIN & N. ZEHNDER
"*Observations and LDA measurements of confined turbulent vortex flow*"
J. Fluid Mech. (1980), Vol. 98, part 1, pp. 49-63

[10] P. BHATTACHARYYA
"*The flow field inside a conventional hydrocyclone*"
Proc. 2nd International Conference on Hydrocyclones BHRA, 1984, paper H2, page 323

[11] D. BRADLEY
"*The hydrocyclone*"
Int. Series of Monographs in Chemical Engineering, Pergamon Press, 1965

[12] K.T. HSIEH & R.K. RAJAMANI
"*Mathematical model of the hydrocyclone based on physics of fluid flow*"
AIChE Journal, May 1991, Vol. 37, No 5, page 735

[13] R.P. DURRETT, R.D. GOULD, W.H. STEVENSON & H.D. THOMPSON
"*A correction lens for Laser Doppler Velocimeter measurements in a cylindrical tube*"
AIAA Journal, Vol. 23, No 9, Sept. 1985, page 1387

[14] GU FANGLU & LI WENZHEN
"*Measurement and study of velocity field in various cyclones by use of laser doppler anemometry*"
Proc. 3nd International Conference on Hydrocyclones - BHRA, 1987, paper C2, page 65

[15] XU JIRUN, LUO QIAN & QUI JICUN
"*Studying the flow field in a hydrocyclone with no forced vortex - Part I : Average velocity*"
Filtration & Separation - July/August 1990, page 276

[16] B. DABIR & C.A. PETTY
"*Laser Doppler Anemometry measurements of tangential and axial velocities in a hydrocyclone operating without an air core*"
Proc. 2nd International Conference on Hydrocyclones, BHRA, 1984, paper A2, page 15

[17] J. WOILLEZ
"*Collection efficiencies of liquid/liquid hydrocyclones*"
Advances in Water Treatment and Environmental Management
BHR Group - Elsevier Applied Science, 1991

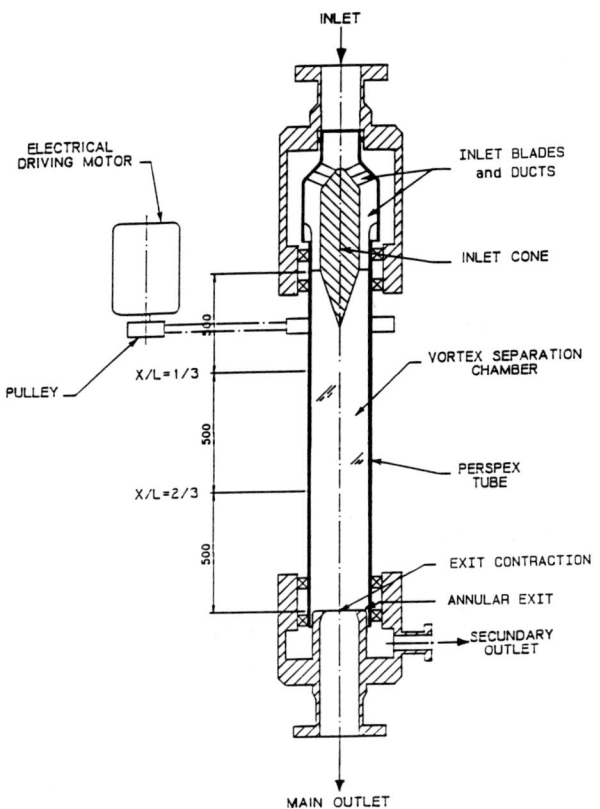

Figure 1
Schematic representation of a cross section of the dewatering rotary cyclone

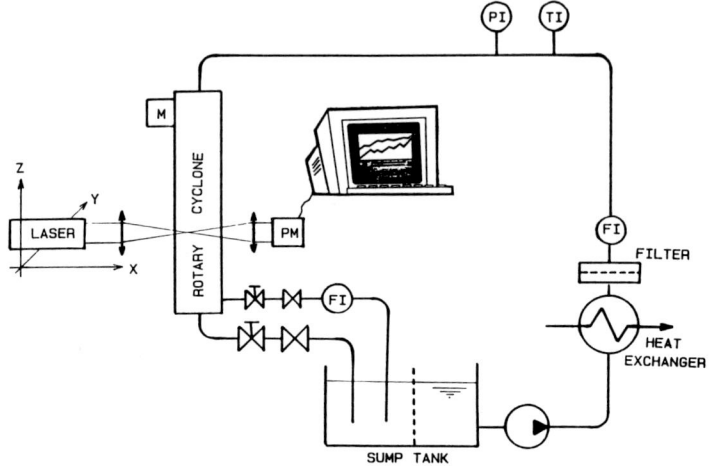

Figure 2
Schematic diagram of experimental set up

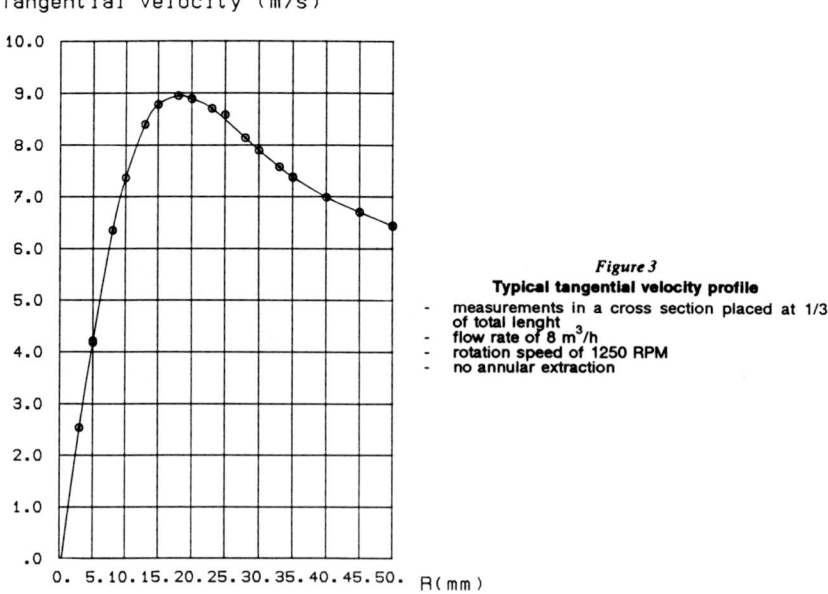

Figure 3
Typical tangential velocity profile
- measurements in a cross section placed at 1/3 of total lenght
- flow rate of 8 m^3/h
- rotation speed of 1250 RPM
- no annular extraction

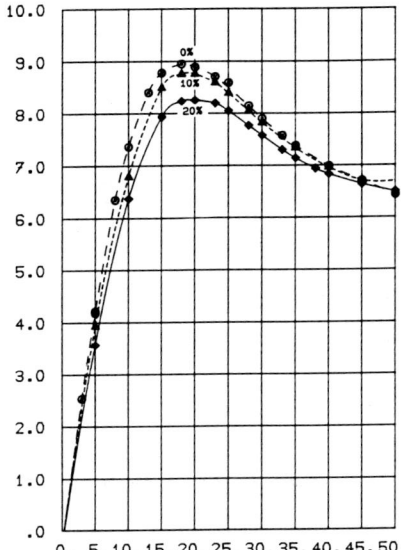

Figure 4
Influence of annular extraction on tangential velocity profiles with extraction rates of 0 %, 10 % and 20 %
- measurements in a cross section placed at 1/3 of total lenght
- flow rate of 8 m^3/h
- rotation speed of 1250 RPM

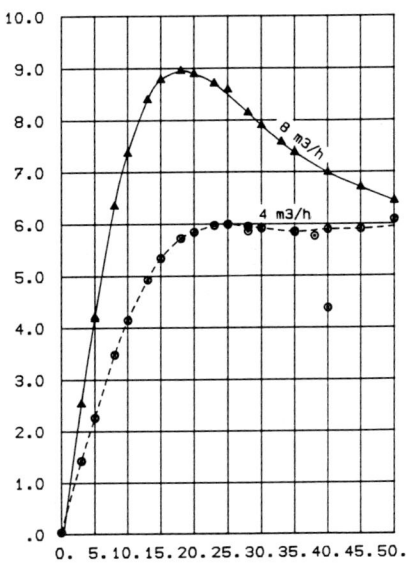

Figure 5
Influence of flow rate on tangential velocity profiles with flow rates of 4 m^3/h and 8 m^3/h
- measurements in a cross section placed at 1/3 of total lenght
- rotation speed of 1250 RPM
- no annular extraction

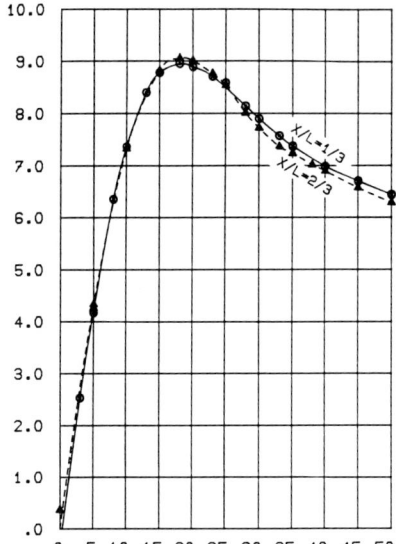

Figure 6
Axial influence on tangential velocity profiles with measurements in two different cross sections places at 1/3 and 2/3 of total lenght
- flow rate of 8 m^3/h
- rotation speed of 1250 RPM
- no annular extraction

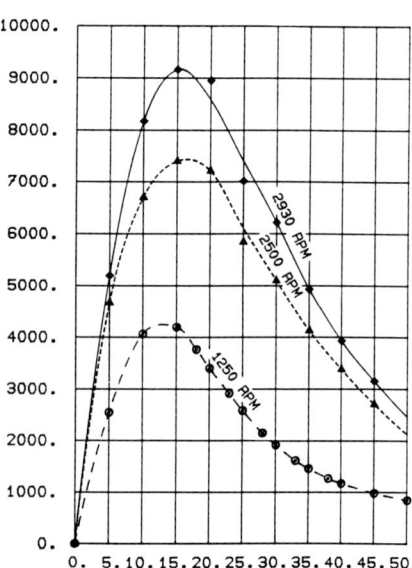

Figure 7
Acceleration fields at three different rotation speeds of the walls : 1250, 2500 and 2930 RPM
- measurements in a cross section placed at 1/3 of total lenght
- flow rate of 8 m^3/h
- annular extraction rate of 20 %

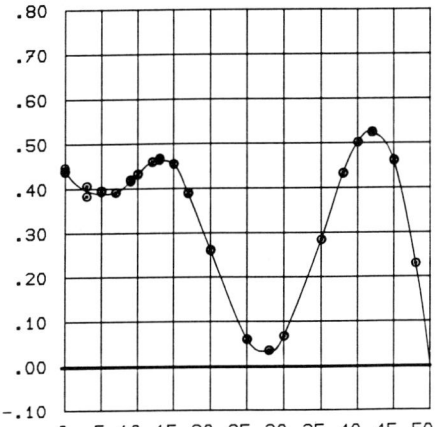

Figure 8
Typical axial velocity profile
- measurements in a cross section placed at 1/3 of total lenght
- flow rate of 8 m³/h
- rotation speed of 2500 RPM
- annular extraction rate of 20 %

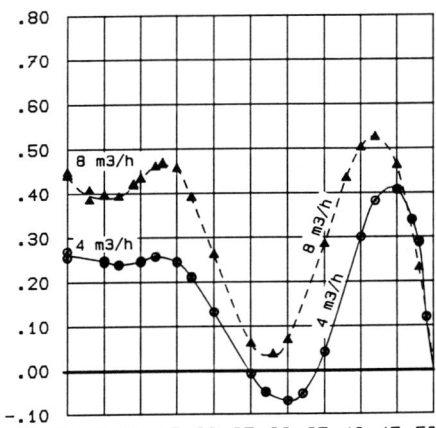

Figure 9
Influence of flow rate on axial velocity profiles with flow rates of 4 m³/h and 8 m³/h
- measurements in a cross section placed at 1/3 of total lenght
- rotation speed of 2500 RPM
- annular extraction rate of 20 %

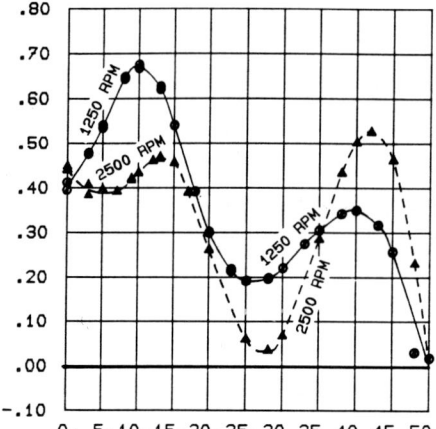

Figure 10
Influence of rotation speed of the walls on axial velocity profiles at 1250 and 2500 RPM
- measurements in a cross section placed at 1/3 of total lenght
- flow rate of 8 m^3/h
- annular extraction rate of 20 %

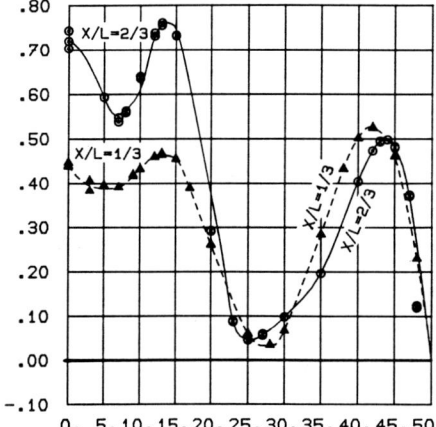

Figure 11
Axial influence on axial velocity profiles with measurements in two different cross sections placed at 1/3 and 2/3 of total lenght
- flow rate of 8 m^3/h
- rotation speed of 2500 RPM
- annular extraction rate of 20 %

IN-LINE FREE VORTEX SEPARATOR
USED FOR GAS/LIQUID SEPARATION
WITHIN A NOVEL TWO-PHASE PUMPING SYSTEM

Dr E G Arato and Dr N D Barnes
BHR Group Limited
Cranfield, Bedford, U.K.

SUMMARY

An In-line Free Vortex Separator was designed to separate various quantities of pure liquid from a gas/liquid mixture. This device was installed downstream of a rotodynamic pump and its function was to facilitate recirculation of a certain proportion of the liquid through the pump, via an ejector or jet pump located in the suction pipe. The increase in the liquid flow rate through the pump combined with the action of the ejector in dispersing and compressing the gas phase at the pump entry allowed the rotodynamic pump to operate efficiently up to an observed volumetric gas void fraction of 37% (ie based on the suction side). This limit was imposed by the pressure rating of the experimental facility rather than by the overall performance of the arrangement.

Efficient operation of the two-phase pumping system could only be achieved if the separator functioned efficiently, so that the return fluid passing through the ejector was essentially pure liquid. This fact was confirmed by visual observations through the transparent return pipe.

The "straight" (or axial) pressure losses produced by the device amounted to about 35 times dynamic pressure (or velocity head) at the inlet and were not considered excessive for the initial stage of development of the separator. The pressure losses however from the inlet to the tangential outlet were much smaller than the "straight" pressure losses, reflecting the effect of the centrifugal pressure gradient inside the separator.

1. INTRODUCTION

Pumping gas/liquid mixtures is a frequent requirement in offshore oil production. Here, the liquid phase generally contains both oil and water, while the gas phase includes a variety of hydrocarbons and impurities. In addition to the above, the mixture to be pumped also contains a small amount of a third phase in the form of highly abrasive silt or sand.

Clearly, positive displacement pumps would not be ideal for the above application, as they would require frequent maintenance and possibly suffer breakdowns caused by the presence of abrasive material. Rotodynamic pumps however, are more flexible and less susceptible to abrasive damage or failure, although they are only capable of handling a relatively small proportion of gas in the mixture, beyond which the pump will suddenly de-prime. Typically, the gas handling capability of a centrifugal pump can vary between about 5% and an upper limit of about 15% concentration by volume, just upstream of the pump. The actual proportion of gas in the mixture which can be handled by a given pump without de-priming however, can be increased (within the above limits) by increasing the speed of rotation, and/or the flow rate and improving the homogeneity of the mixture.

A collaborative research project is being undertaken at BHR Group with the specific objective of improving the gas handling capability of conventional centrifugal pumps by means of liquid recirculation via an ejector or jet pump located in the suction line of the pump. This system, in addition to increased liquid flow rate through the pump is also instrumental in reducing the volumetric concentration of gas (due to pressure rise) and improving the homogeneity of the mixture (due to momentum exchange).

The in-line free vortex separator discussed in the paper forms a major part of this multi-phase pumping system. Its function is to separate part of the liquid from the bulk of the gas/liquid mixture at the discharge side of the pump and return it via an ejector to the suction line.

In order to ensure efficient operation of the ejector and ultimately the whole pumping system, the recirculated liquid needs to be practically free of the gas fraction.

2. THE TWO PHASE PUMPING SYSTEM

2.1 Description of the arrangement

The liquid re-circulation system, which is a novel method of pumping gas/liquid mixtures consists of an in-line free vortex separator just downstream of the pump, a jet pump or ejector located in the suction line of the main centrifugal pump and a return line from the separator to the ejector supplying the high pressure primary or drive fluid to the latter. A diagrammatic arrangement of the pumping system, showing the locations of measuring points, is illustrated in Fig. 1.

By recirculating part of the liquid through the pump the volumetric proportion of the gas (ie gas fraction) in the mixture entering and leaving the pumping loop is reduced. In addition, by passing the liquid through the ejector the gas fraction is further reduced at the pump inlet, in proportion to the pressure rise produced by the ejector. The intense mixing (or momentum exchange) which takes place in the mixing chamber of the ejector will create a practically homogeneous gas/liquid mixture and this aspect will further improve the gas handling capability of the centrifugal pump.

2.2 Operational features

On the experimental facility the increase in the quantity of air injected in the suction side was accompanied by an increase in the flow rate of water recirculated via the ejector. The latter was achieved by throttling the valve near the outlet of the test rig. (See Fig.1.) Typical static pressure profiles at key positions along the rig are illustrated in Figs. 2 and 3. These graphs clearly show the substantial increase in the static pressure just upstream of the pump (ie from 1.4 to 2.2 bar abs.) due to the increased rate of recirculation through the ejector, necessitated by increasing the gas void fraction (GVF) from 20.8 to 30.7%.

The variation of flow rates of the mixture and individual phases within the pumping loop are shown in Fig. 4. Here, the recycled liquid flow rate increases with gas fraction, while the suction liquid upstream of the ejector reduces with the same. The cross over point between the two flow rates occurs at about 30% gas fraction, beyond which the recirculated liquid flow will be greater than the liquid delivery down the pipeline.

The graphs in Figs. 2 to 4 serve to illustrate the overall operation of the proposed two-phase flow pumping system of which the IFV Separator is a major constituent.

3. THE IFV SEPARATOR
3.1 Principle of operation

The flow rotation is produced by a four-start helical guide which is designed to impose a full 360° turn on the flow within one pipe diameter length. The actual separation takes place in the cylindrical settling chamber, located just downstream of the helical guide, where a free vortex type flow is expected to develop. Although at this stage of the project no direct measurements have been carried out to confirm this, visual observations of the gas core (through the transparent settling chamber) and pressure measurements upstream and downstream of the device and in the tangential off take indicate free vortex type of radial velocity and pressure distribution in the settling chamber.

The pressure variation in the radial direction for a rotating flow can be expressed as:

$$\frac{dp}{dr} = \frac{\rho V^2}{r} \qquad (1)$$

for a free vortex flow $vr = c$ (2)

where;
    v = tangential/ velocity
    p = pressure
    r = radius of rotation
    ρ = liquid density
    c = constant

After substituting (2) into (1) and integrating the resulting equation from an inner (r_1) to an outer radius (r_2) we obtain,

$$p_2 - p_1 = \frac{1}{2}\rho c^2 \left[\frac{1}{r_1^2} - \frac{1}{r_2^2} \right] \qquad (3)$$

Assuming free vortex type of flow regime in the separator, the value of the constant "c" is a measure of the intensity of the free vortex and for a given geometry of the helical passage it will only depend on the incoming axial flow velocity.

Bubbles of gas (of negligible mass) dispersed in the rotating liquid outside the gas core will be displaced by the heavier fluid and propelled rapidly inwards under the effect of the pressure gradient expressed by equation (3). During their passage, the volume of the bubbles increase in direct proportion with the radial pressure drop and eventually they form a continuous gas core at the centre. The size of the latter will depend on the concentration of the gas in the liquid and on the intensity of the vortex. For a typical operating condition, the conditions in the settling chamber, including the gas core are shown in Fig. 5.

The gas core and some of the surrounding liquid is discharged through a vortex finder, the diameter of which is designed according to the requirements of the separation. The vortex finder is joined to the downstream pipeline by a diffuser designed to provide a pressure recovery, thereby reducing the axial pressure losses produced by the separator. The pure liquid is discharged tangentially via an "involute" arrangement where the flow rate is determined by the back pressure and the physical size of the tangential outlet. The overall arrangement of the separator is shown in Fig.5a.

3.2 Pressure loss performance of the separator

Since the pressure losses across the device represent a reduction in useful pumping output it is of prime importance to minimise these losses, while maintaining high separation efficiency. The latter was actually achieved under the full range of operating conditions studied on the arrangement.

An indication of the static pressure variation along the whole pumping system for two particular operating conditions can be seen in Figs. 2 and 3 which were discussed in section 2.1. Specifically for the separator, operating over a range of flow rates, the pressure drop between upstream and downstream of the separator against the total flow rate at the pump exit (or separator inlet) is plotted in Fig. 6 and the variation of pressure drop between the inlet and tangential outlet versus total flow is given in Fig. 7.

The pressure drop in Fig. 6 increases with total flow rate, while in Fig. 7 it shows the opposite trend. These trends illustrate the operating mode of the system where the reduction of the total flowrate is achieved by throttling the outlet valve of the loop, thereby increasing the rate of liquid recirculation. The increase of the pressure drop between separator inlet and tangential outlet versus recirculated flow rate, mainly due to friction losses in the return pipe, is illustrated in Fig. 8.

In order to obtain more general information regarding the pressure loss characteristics of the IFV Separator, the pressure drops illustrated in Figs. 6 and 7, were expressed in terms of loss coefficients and plotted against the ratios of recirculated to total flowrates.

For instance, ignoring the relatively minor effects of differences in elevations and dynamic pressures at the relevant sections, the "straight" or axial loss coefficient is defined as:

$$K_{1,2} = \frac{\Delta p_{1,2}}{\frac{1}{2}\rho_1 V_1^2} \tag{4}$$

while, the loss coefficient of the tangential off take is defined as:

$$K_{1,3} = \frac{\Delta p_{1,3}}{\frac{1}{2}\rho_1 V_1^2} \tag{5}$$

where;
    Δp = pressure drop
    ρ = mixture density
    v = velocity
subscripts refer to :
    1 : section upstream of separator
    2 : section downstream of separator
    3 : section in off-take, upstream of ejector nozzle

The definition of the above loss coefficients follow the method used for "dividing" flows in "T" junctions, see Ref.1.

Both loss coefficients are based on the dynamic pressure at a section upstream of the separator and the straight losses were approximately constant, within the range tested and the individual values did not show excessive scatter around the average value of 35 (see Fig 9.). This means that at any flow rate the head or pressure loss across this device in the axial direction is 35 times the velocity head or dynamic pressure of the flow. This value is not comparable to the loss coefficients obtained in some typical hydrocyclones, (See Ref. 2.) due to the basically different geometric configurations. However, the actual head losses produced by this device are similar to those occurring in some hydrocyclones.

The loss coefficient of the tangential off take increases approximately linearly with the ratio of recirculating to total flow rate (see Fig 10). The substantially lower value of this loss coefficient compared to the previous one, illustrates the effect of the centrifugal pressure rise in maintaining relatively high pressures at the ejector nozzle. Due to friction losses in the off-take pipeline the pressure losses will gradually increase with increasing proportion of recirculating flow. However, at a 50% split (ie $Q_{rec}/Q_{tot} = 0.5$) the off take losses still amount to only about 70% of the "straight" losses.

The effectiveness of the ejector in compressing the gas at the pump inlet, facilitated by efficient separation combined with relatively high pressures prevailing in the off take line due to the centrifugal effect, is illustrated in Fig 11.

3.3 Special features of the IFV Separator

Although the tests were carried out on air/water mixtures, due to the very large "g" forces (ie amounting to several hundred times "g") being generated in the device, the results obtained are considered to be broadly applicable to gas/oil mixtures.

In addition to gas/liquid separation, this device is thought to be adaptable for a variety of liquid/solid and liquid/liquid separation duties, presently performed by hydrocyclones.

Special features of the IFV Separator which represent distinct advantages over conventional hydrocyclones are:

(i) can be installed into pipelines as an in-line device.

(ii) the flow emerging from the tangential off take (ie product line) is effectively free from rotation, therefore throttling of this outlet should not affect separation performance.

(iii) due to the absence of flow rotation combined with elevated pressures in the tangential off-take this device is suitable for multi-stage operation (ie several devices in series).

4. CONCLUSIONS

(1) The effectiveness of the gas/liquid separation was confirmed directly by visual observations through the transparent tangential off-take and indirectly by the efficient operation of the overall pumping system.

(2) Although the pressure losses across the separator were not considered excessive, there is a scope for pressure reductions by improving the geometry of the helical passage and for increasing the pressure recovery by improving the outlet diffuser.

5. REFERENCES

1. Miller, D S   "Internal Flow Systems, Design and Performance Prediction" BHRA 1978, 1990

2. Arato, E G   "Reducing Head or Pressure Losses Across a Hydrocyclone" Filtration and Separation 21, 3, May/June 1984, pages 181-2

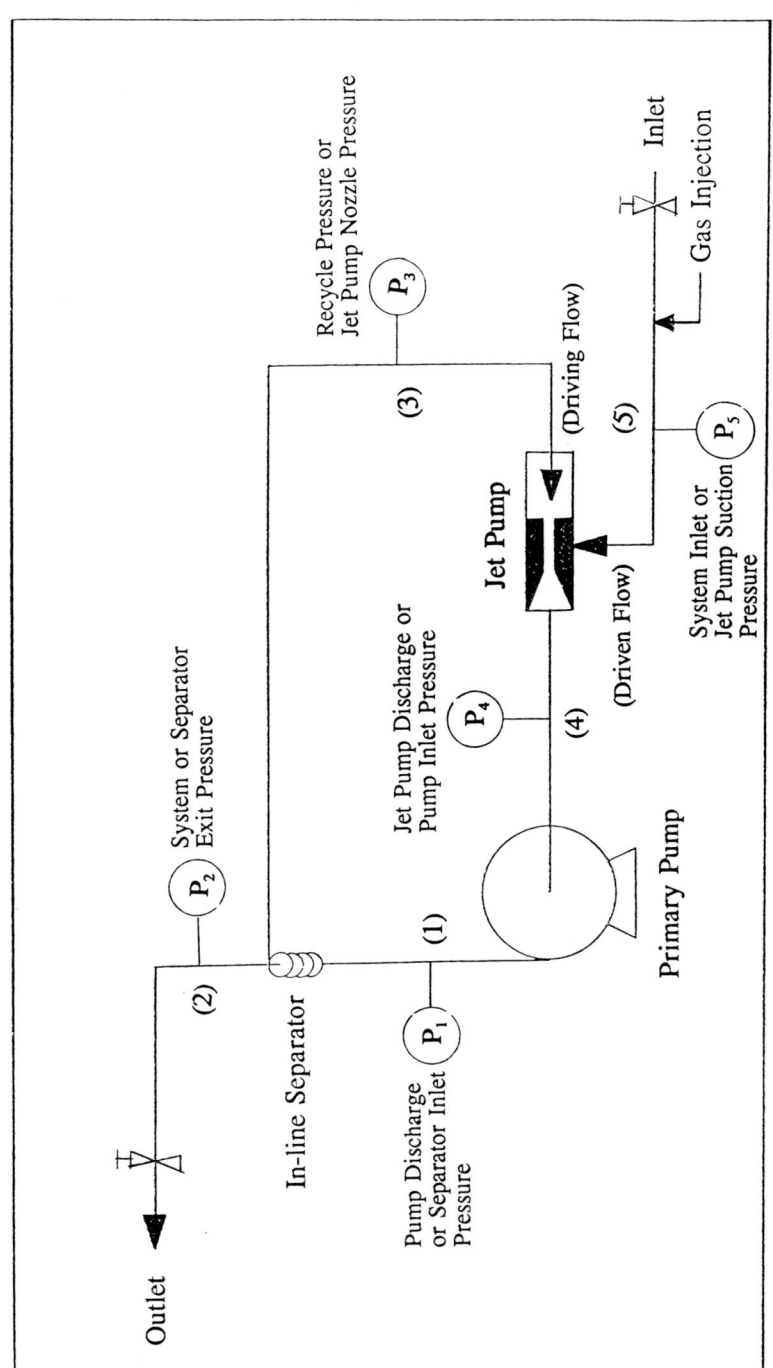

Figure 1 : Schematic Diagram Showing Locations of Pressure Transducers

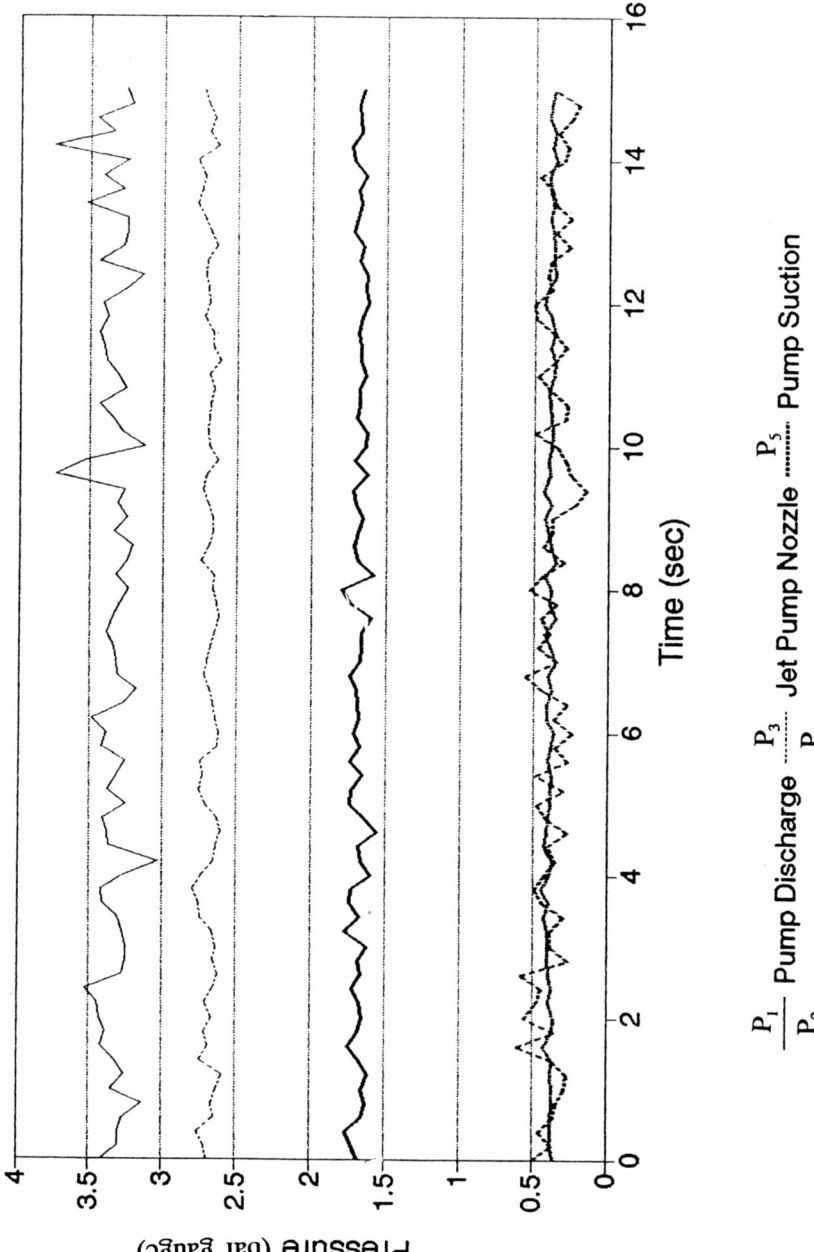

Figure 2 : Pressure Fluctuations Over Sampling Period (GVF = 20.8%)

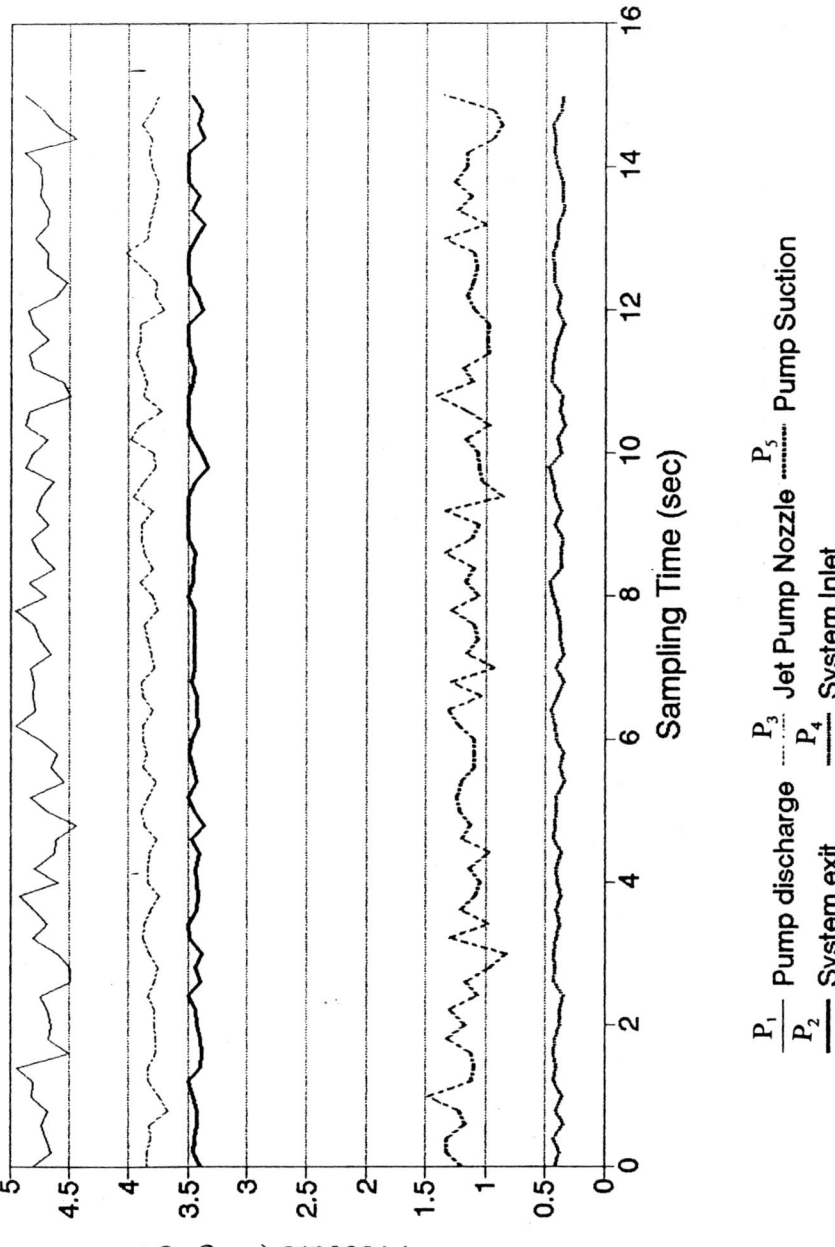

Figure 3 : Pressure Fluctuations Over Sampling Period (GVF = 30.7%)

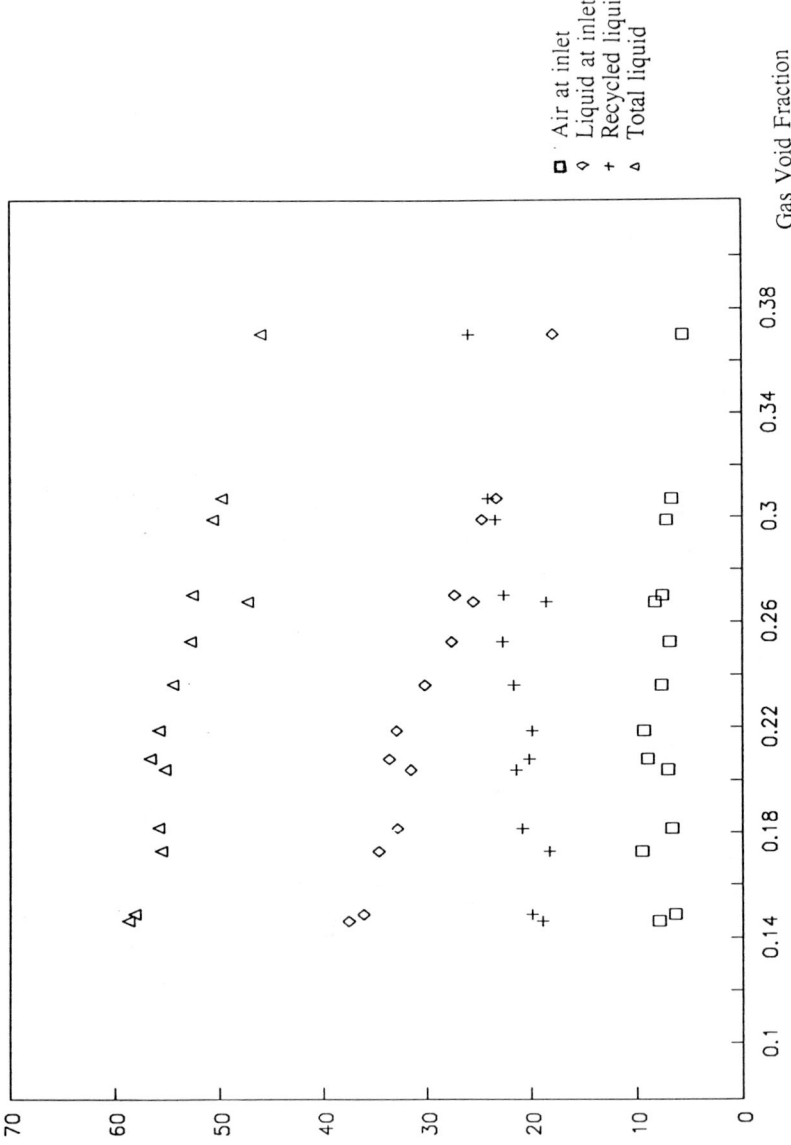

Figure 4 Flow Rates versus Gas Void Fraction At System Inlet

Figure 5 : Photograph of Phase Separation in In-Line Free Vortex Separator

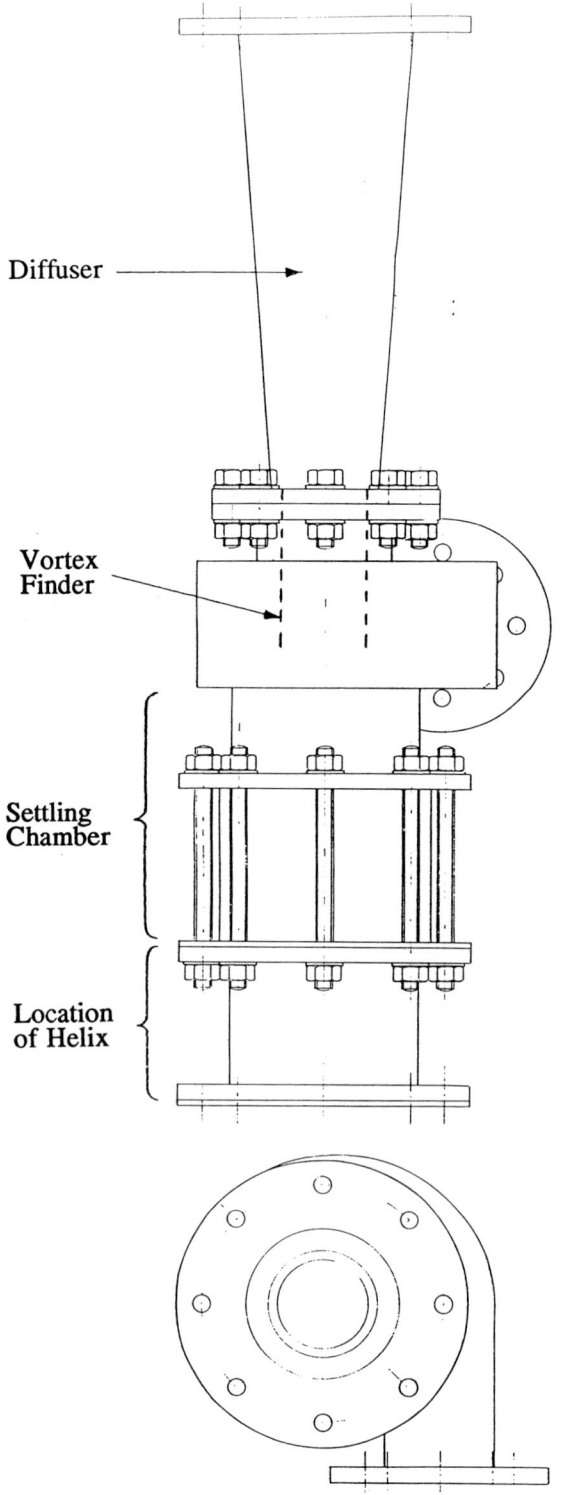

Figure 5A: In-Line Free Vortex Separator

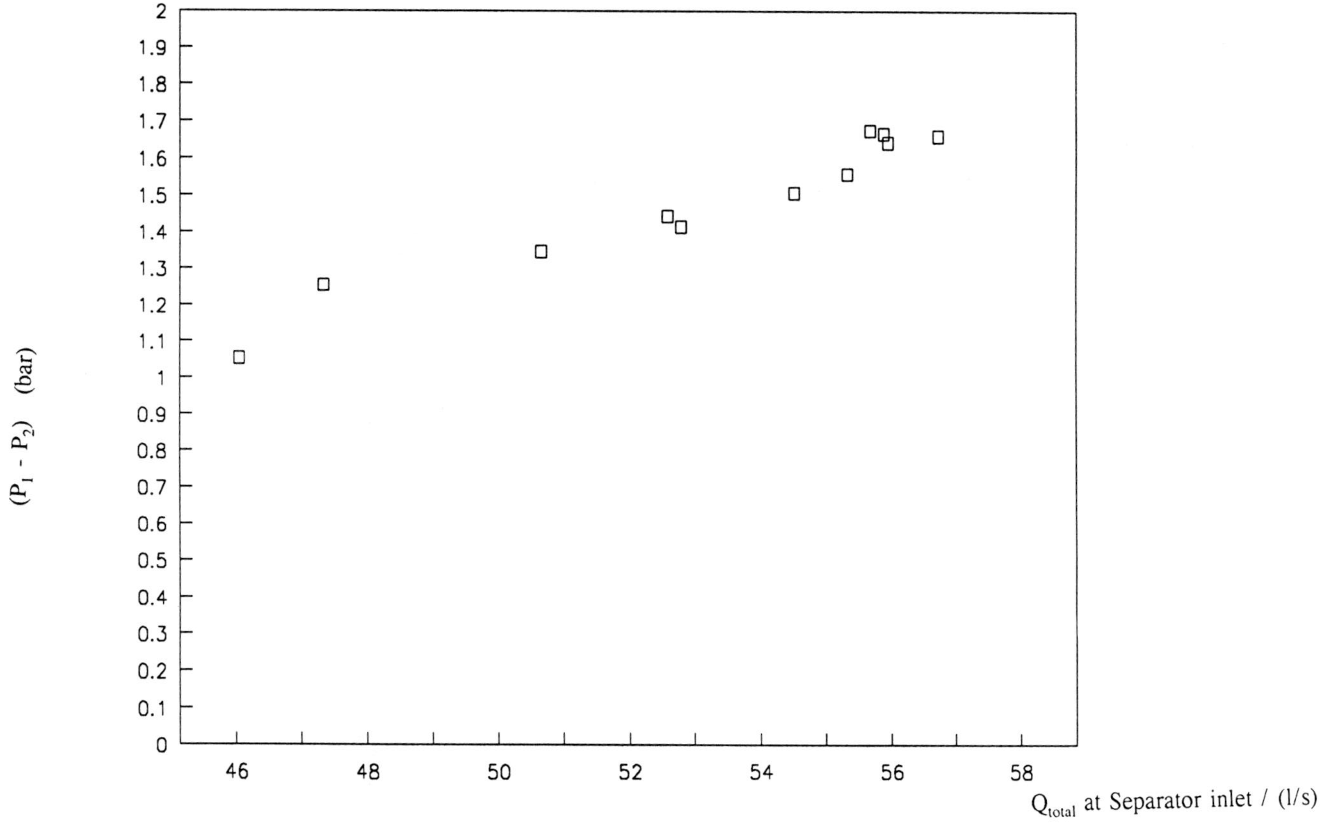

Figure 6   Pressure Drop Across Separator versus Total Flow At Separator Inlet

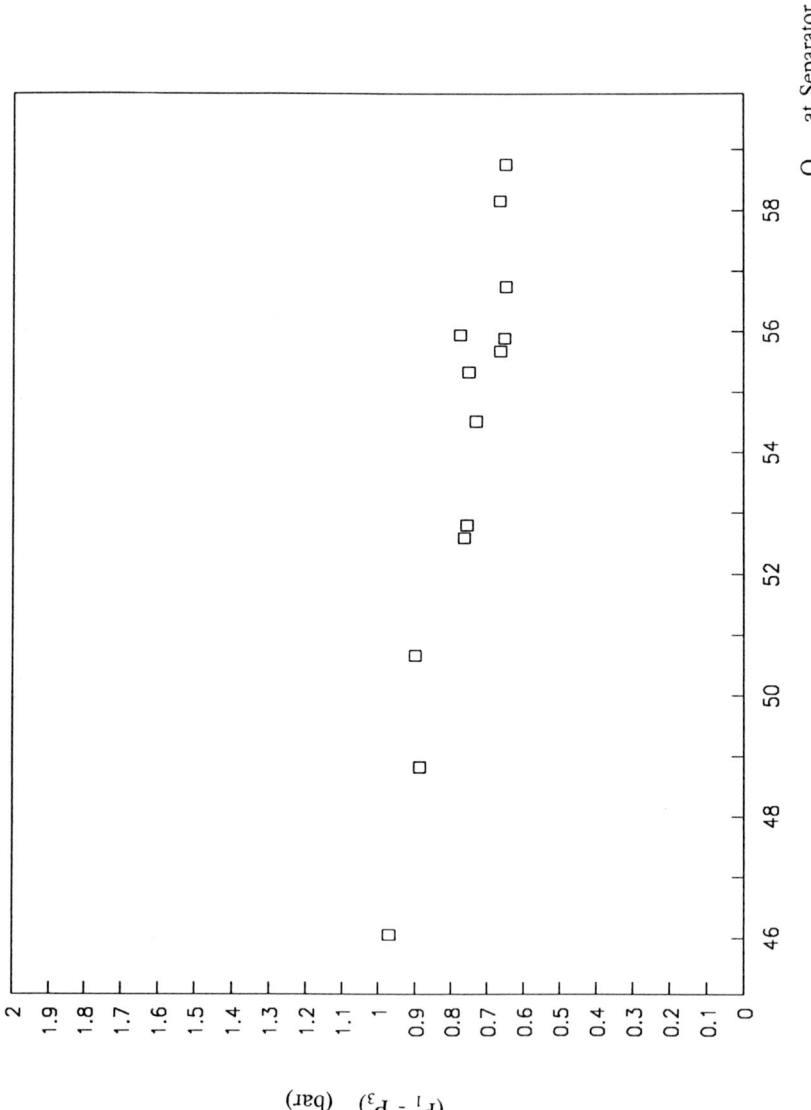

Figure 7   Pressure Drop From Separator To Ejector versus Total Flow At Separator Inlet

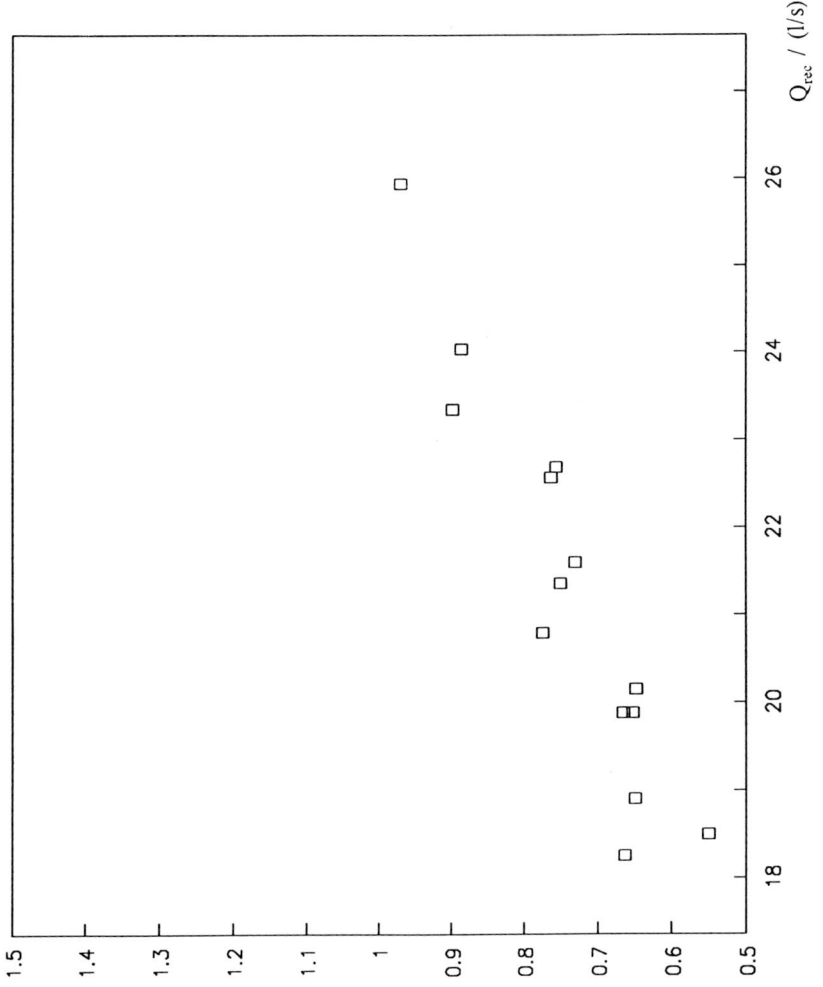

Figure 8 Pressure Drop From Separator To Ejector versus Recirculating Flows

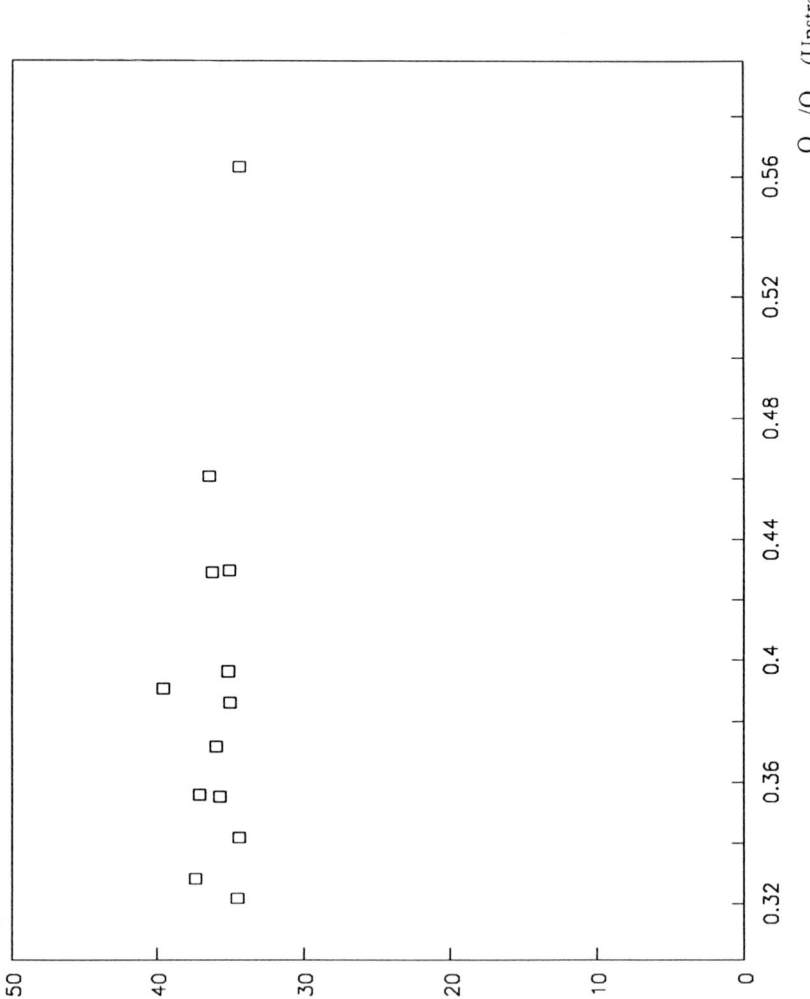

Figure 9 "Straight" Loss Coefficient versus Recirculating Total Flow Ratios

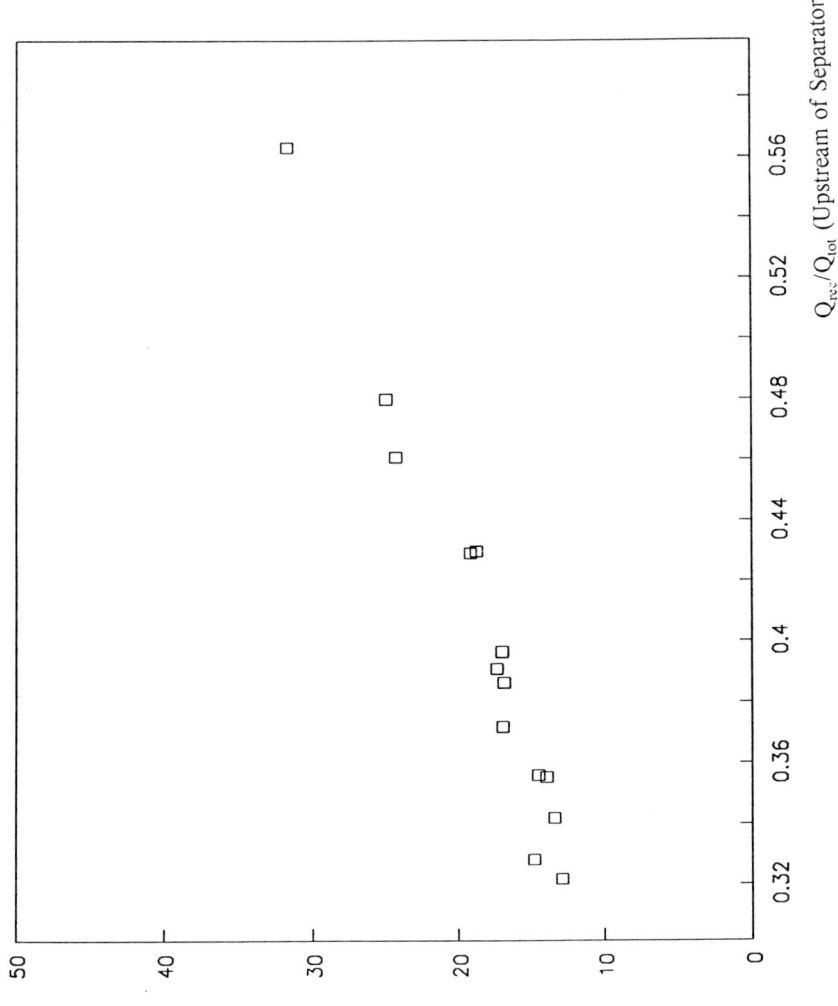

Figure 10 Offtake Loss Coefficient versus Recirculating Total Flow Ratios

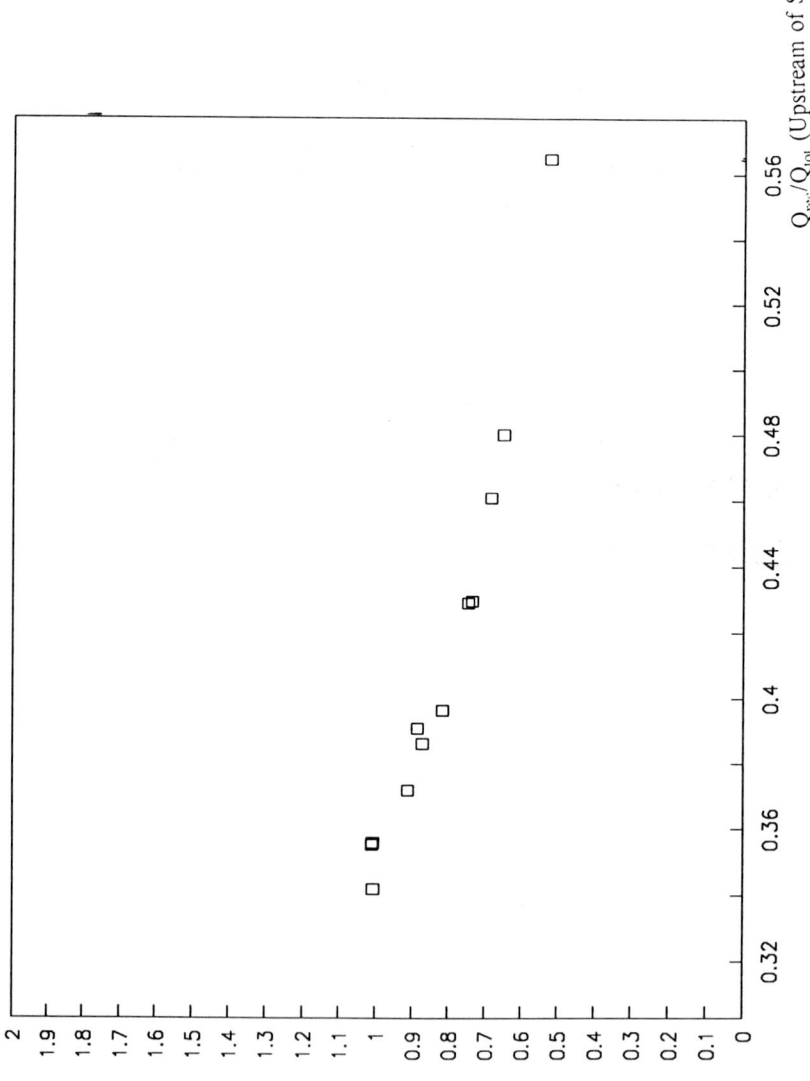

Figure 11 Air Flow Ratios versus Recirculating To Total Flow Ratios

THREE - PRODUCT HYDROCYCLONE FOR SIMULTANEOUS SEPARATION OF SOLIDS BOTH HEAVIER AND LIGHTER THAN LIQUID MEDIUM

STANISŁAW BEDNARSKI
Technical University of Cracow, Poland

ABSTRACT

A new construction of the three-product hydrocyclone for separation (in one device) of solids both heavier and lighter than the liquid medium has been presented. There is a possibility for application of this type of hydrocyclone for separation of three-phase systems i.e. liquid-solid-liquid, gas-liquid-liquid and gas-solid-liquid. Some operating variables as well as the performance of this type of hydrocyclone have been discussed. The instructions for both designing and technological calculations as well as the directions for the selection have been given, especially in application to the industrial processes (petroleum, food, pulp and paper).

INTRODUCTION

In many technological processes of such industries as refinery - petrochemical, paper, chemical or food, numerous operations can be encountered in which it can be necessary to separate the suspensions forming multi-phase systems. Variety of these systems, often occuring in close cycles, causes several problems related to their separation.
 The majority of operations consists in the separation of two-phase suspension systems like solid-liquid or liquid-liquid. In some technologies also suspension systems consisting of three phases can appear like e.g. liquid-solid-liquid or gas-solid-liquid as well as suspension systems consisting of two or more solids, some of which as heavier than the surrounding medium sink in it and the others, as lighter, rise to the surface.
 Devices basing on the action of various forces are applied for separation of the latter systems and different physical properties of solids and suspensions are improved. Efficiency of these devices and of the applied separation method for this kind of suspensions (in general multi-stage processes) depends on many factors which affect to a high degree the profitableness of the process.
 In the main, the old methods, basing only on the action of gravitation, demand the construction of expensive objects which occupy a great area, are heavy and troublesome in maintenance and, as a result, their efficiency is low. More modern methods however, basing mainly on

centrifugal separators, not always can accomplish a task. Moreover their investment cost is high and they require studious and highly qualified maintenance.

All these aspects were the reason for which solutions were searched basing on other devices, equally efficient and much more economical. Three-product hydrocyclones appeared to be this kind of devices which distinguish themselves by high effectiveness and process intensity and moreover, which is very advantageous, they are simpler in construction and maintenance and cheaper in purchase and exploitation. It is safe to say that these devices became nowadays basic apparatus in many technological operations of the modern works, in particular paper mills which will serve as an example described below.

DESCRIPTION OF OPERATION AND CONSTRUCTION OF A THREE - PRODUCT HYDROCYCLONE

The scheme and operation principle of a three-product hydrocyclone was presented in Fig.1, and its construction in Fig.2. As it can be seen from Fig.1, operation principle of this device is similar to that of a conventional hydrocyclone. Tangentially introduced suspension raises the circulating movement for which flow rate at the strictly selected constructional parameters of a device is dependent upon pressure drop between the feed inlet and overflow outlet. Due to the circulating mevement of the suspension in the device, displacement of solid particles of different density takes place.

Particles heavier than the surrounding medium make toward the internal surface of the device. Moving on a three-dimensional helix they are by the way thickened and directed to the underflow hole. Particles lighter than the surrounding medium flow with it towards the vortex and all along the hydrocyclone they are pushed out into the vortex, from where they are taken out through the vortex finder passing in the center of the overflow nozzle.

The surrounding medium (paper pulp) released from the heavy and light impurities, like in a normal hydrocyclone, flows through the overflow nozzle and leaves the device through a helical connector pipe. It is advantageous that the outlet flows of both upper fractions are respectively throttled or they have to overcome the hydrostatic pressure of a column of 0.5 - 1.5 m, which to a high degree prevents from plugging of the underflow hole and eliminates the air core in the hydrocyclone axis.

In the case where a three-product hydrocyclone is used for separation e.g. of an oil-water emulsion polluted with solid particles, the flow of particular components of a suspension is similar to that described above whereas the lighter fraction (separated oil) flows through the vortex finder. To prevent the excess of separated oil from getting to the overflow and underflow, appropriate screens are applied in a form of cone, spherical segment or paraboloidal segment.

Such protections are necessary for they automatically control the flow rate of the separated oil at the outlet, particularly when its concetration in the oil-water emulsion is variable and the suspension density and the content of solid particles in it changes. When the amount of oil in the feed increases, the rotating oil core becomes thicker i.e. its diameter increases and the rotating heavier liquid sets itself against this phenomenon and causes the displacement of the lighter liquid and increases its flow rate.

This phenomenon, from the practical point of view, is very

advantgeous because it leads up to the self-regulation of the amount of oil flowing out, depending on its contents in the emulsion. Quantitative range of the oil flow rate at the outlet from minimum to maximum can achieve the values differing from each other several times. For example: if in the process of separation of an emulsion containing 2 % vol. of oil the appropriate diameter of the vortex finder is selected, the maximum amount of oil in the emulsion can increase even 2 - 4 times without leaking back to the deoiled water. When oil concetration in the emulsion decreases, its lower limit at a stable run of the hydrocyclone can be practically equall zero.

Taking into account its construction, the three-product hydrocyclone does not differ considerably from the conventional hydrocyclone and after the lockout of the vortex finder or its removal it works like two-product hydrocyclone. Essential parts of this device are the head and the cone ended with an underflow nozzle. The head can be made in one piece including the feed and overflow nozzles and the vortex finder. The upper part of the head can also be made separately and the overflow nozzle and the vortex finder can be separate parts joined with the head respectively.

The conical part of this hydrocyclone is almost identical with the cone of a normal hydrocyclone with only one difference, that at the lower outlet there is an element in the shape of a bowl which prevents oil from mixing with the underflow fraction. At the upper part of the hydrocyclone head, around the vortex finder, there is a second limiter in the shape of a bell or a piece of pipe with a bottom, which prevents oil from mixing with the overflow fraction. The conical part can be made of one or several segments and its convergence or apex angle can have the constant value or variable on the certain segments or on the whole lenght.

Three-product hydrocyclones can work separately or in a parallel system with the common feed and overflow chambers and the oil chamber (or for the floating particles) as well as the underflow chamber. Besides, the hydrocyclones of this kind can work in both vertical and horizontal positions and also under any angle related to the horizontal level.

Three-product hydrocyclones can work within the wide range of pressures. When they are applied to the sorting of waste paper pulp, where three-stage systems in series are installed, the pressure value before each stage should be 0.2 - 0.3 MPa, whereas in the process of oil and solids separation from water the pressure should be from the range 0.1 - 0.2 MPa. This operation at a lower pressure is little effective and a higher pressure can cause the foam generation, especially when the separated oil-water suspensions were aerated.

INVESTIGATION DESCRIPTION AND DISCUSSION OF THE RESULTS

Investigations were carried out with two different suspensions i.e. waste paper pulp with the addition of wood pulp containing ca 1 % d.s. (dry solids) with the impurities of mineral and organic origin and oil-water emulsion polluted with metallic and non-metallic solid particles from a forge, hydraulic press, scrap moulding and machining. Waste paper pulp with the wood pulp was polluted with the disintegrated bark, splinters, printer's ink, thermoplastics, latex, rubber, binder, coal, resins, polymers, silica and fine metallic elements passing through the sieve mesh of 3 mm.

This kind of impurities can be encountered nearly in every waste paper pulp and except for the bark and splinters also in a wood-free waste paper pulp. Investigations with the paper pulp were carried out in a single and a three-stage three-product hydrocyclone battery installed after the hydropuplers, initial classifiers and grinders. Pulp obtained after these devices was diluted to the concetration ca 1.1 % d.s. and with an impeller pump of a capacity $Q \approx 100$ m^3/h and pressure H = 35 m of water column was directed to a battery of three-product hydrocyclones.

Altought in one stage of three-product hydrocyclones very exact paper pulp purfication from mineral and organic pollutants can be achieved, but together with the separation products removing these pollutants a considerable amount of the paper raw material (fibre) was also discharged.

Mass amount of fibres getting through to the heavier fraction (underflow), of which volumetric flow rate referred to the feed was 5 - 6 %, was equal 91±3 %. The stream of fibres led out through the vortex finder together with the floating impurities (lighter than the paper pulp) with the volumetric flow rate referred to the feed was 4.2 - 4.7 % and its mass amount was ca 93±2 %.

These experimental results obtained in a single-stage three-product hydrocyclone battery were the reason, for which other solutions were searched to reduce the fibre losses. So great quantity of fibres penetrating with the impurities to waste water would be inadvisable. Therefore investigations were carried out in a three-stage system of three-product hydrocyclones for fibre separation from lighter and heavier wastes.

In Fig.3 the scheme has been presented of a three-stage waste paper wood pulp purification in three-product hydrocyclones from which the pulp separated from the light and heavy impurities, fraction with light impurities and fraction with heavy impurities are obtained. Such a system demands a pump before each stage; the biggest one in stage I and the smallest in stage III. Capacity of pumps in stages II and III were respectively 20 and 10 % of the capacity of the pump in stage I. In stage I ten hydrocyclones have been installed. Vortex finder diameter of them as well as of the hydrocyclone in stage III was increased by 20 % and the underflow nozzle diameter by 25 % related to the both hydrocyclones in stage II which had been investigations in a single-stage system. This increase of diameters was very advantageous for the quality of the purified pulp. In the presented scheme flow rates for particular fractions have been given as well as the solid concetration in them and the pump characteristics.

It follows from the values for the overflow fraction in stage III and the underflow fraction in stage II that about 38.5 kg d.s. per hour penetrates to the wastes. It makes ca 3 % of the solids introduces with the pulp to the system. Together with the light and heavy wastes also the fibrous pulp as their carrier is taken away. The amount of fibre coming to the waste light impurities can be from a wide range, its mean value is 45 % d.s. and the fibre losses are about 2.5 kg/h, whereas the amount of lost fibre with the waste heavy impurities is several times greater, in spite of its concetration in the wastes which is less than half of the value for light fraction and makes about 18 - 21 % d.s. Mass losses are equal to 6.5 kg/h.

Consequently the total fibre losses amount to 9 kg/h and referred to the quantity introduced to the separating system it makes ca 0.66 %. The amount of lost fibre depends significantly on the kind of separated waste paper and its pollution degree as well as on wood concetration in the pulp.

Investigations of the oil-water emulsion separation polluted with grease and solid particles like fine metallic fillings and non-metallic elements from the rubber abrasive disks have been carried out in one three-product hydrocyclone installed directly on the waste water output channel. Concetration of the dispersed oil in water was about 3.5±1.5 g/dm^3. Oil properties and the composition of solid impurities changed during the run within a wide range because of the unstable run of machines and devices and the application of various kinds of emulsion.

The run was carried out in the conditions which did not admit the formation of an air core inside the vortex for its presence caused flotation. This phenomenon affected even the transport of metallic fillings to the oil fraction. Pressure value before the hydrocyclone was 0.2±0.01 MPa and the counter - pressure caused by a liquid column of 1.2 m referred to the feed level resulted in the vanishing of the air core. In these feed conditions the total hydrocyclone capacity was equal ca 13.2 m^3/h and the underflow rate 0.583 m^3/h which makes 4.42 % of the feed, whereas the volumetric amount of the oil fraction was dependent on the oil concetration in water.

At the average oil contents in water equal ca 3.5 g/dm^3 flow rate of the oil fraction was from the range 0.125 - 0.150 m^3/h and the separation effect of the oil mixture with density 0.92 - 0.95 was equal ca 97.6 % which means that still about 84 mg/dm^3 remained in water. Solid pollutants were separated in the range 95 - 98 % and were thickened even up to 350 mg/dm^3 i.e. about 20 - 30 times in relation to the feed.

It can be seen from the presented data that considerable amount of water penetrated to the oil fraction; its concetration in this fraction was from the range 50 - 80 %. In this investigations, despite the screens used in the upper and lower part of the hydrocyclone, the maximum oil concetration in the oil fraction was about 50 %. The reason for this was a high dispersion degree of oil in water and the presence of a small amount of grease to which solid particles and air bubbles were sticked. Therefore in spite of the apparent elimination of the air core, foam generation always occured and this phenomenon was the reason of water transport to the oil fraction.

Also small amount of oil and grease penetrate to the underflow fraction together with the collected solid particles. Grease was occluded on the surface of the sediment. Grease and oil removal from the sediment was possible by its washing with hot water, steam or an appropriate solvent. Concetration of solids in the underflow fraction could be changed practically at will in the range from some scores to ca 800 g/dm^3 or even more.

Sediment thickening to the form of irrotational or line flow from the underflow nozzle could be easily achieved, particularly when the solids concetration in the feed was up to 10 -20 g/dm^3. It has been stated, however, that the thickening of underflow to a semi-fluid form leads to the aggravation of pulp purification from bark and splinters. These impurities, as lighter than mineral and metallic particles, are displaced to the main stream of pulp. Therefore it is inadvisable to increase the sediment contents over 200 - 300 g/dm^3.

SELECTION OF CONSTRUCTIONAL PARAMETERS OF A THREE - PRODUCT HYDROCYCLONE

Selection of the diameter and other constructional parameters of a three-product hydrocyclone depends on its application. For the simultaneous separation of solid particles sinking and floating, the cylindrical part diameter should be from the range D = 100 - 150 mm. In the case where this hydrocyclone is used for emulsion separation containing solids it is recommend to apply the cylindrical part diameters from the range D = 70 - 120 mm.

Three-product hydrocyclones can be cylindrical-conical or only cylindrical. In the case of a cylindrical-conical hydrocyclone the apex angle of the conical par should be from the range $\varphi = 4 - 8°$ where smaller angles should be used in hydrocyclones of smaller diameters and larger angles in hydrocyclones of larger diameter. For a cylindrical hydrocyclone the height of the cylindrical part should be from the range $L = (8 - 12 D)$.

Recommended diameters of the nozzles and vortex finder are within the ranges:

feed nozzle $d_o = (1/4 - 1/6)D$, advantageously $\approx 0.2\,D$

overflow nozzle $d_p = (1/3 - 1/4)D$, advantageously $\approx 0.3\,D$

undeflow nozzle $d_w = (1/8 - 1/15)D$, depending on the suspension properties

vortex finder $d_c = (1/10 - 1/20)D$, for a core-free flow

VOLUMETRIC FLOW RATE

It follows from the measured flow rates that the most suitable formula for a three-product hydrocyclone capacity is the author's equation in the form:

$$V_n = K_B d_o d_p [(1-d_o/D)(1-d_p/D)]^{0.5} \cdot \Delta p^n \qquad [1]$$

where:

$$K_B = \left[0.21 + \frac{0.018}{0.04\,tg\varphi/2} \right] \frac{D+20}{D+10} \left(1 - \frac{s}{D} \right)^{0.5} \left(\frac{D+L}{D+l_p} \right)^{1/6} \qquad [2]$$

$\Delta p = p_o - p_p$ — pressure drop between the pressure before the feed p_o and the pressure in the overflow connector p_p of a hydrocyclone

s — width of the feed nozzle
l_p — immersion depth of the overflow nozzle

$$n = B\,\xi\,(D/d_o)^m \qquad [3]$$

where:
B = 0.467 - constant
ξ — shape factor of the hydrocyclone head which for the discussed construction is $\xi=1$ and for all the others $\xi<1$
m — coefficient depending on the roughness of the internal parts of a hydrocyclone, for smooth surfaces m=0.12, for rough m>0.12

GENERAL CONCLUSIONS

Presented construction of the three-product hydrocyclone is an offshoot of a conventional hydrocyclone. This hydrocyclone can be applied for separation of two- and three phase systems i.e. heterogenous mixtures of the kind: solid-liquid, liquid-liquid, gas-solid-liquid, liquid-solid-liquid and solid-solid-liquid. It is a unique device which is the simplest, the cheapest, the most effective and the lightest, in which simultaneous separation of a three-phase mixture can be performed and which occupies very little space.

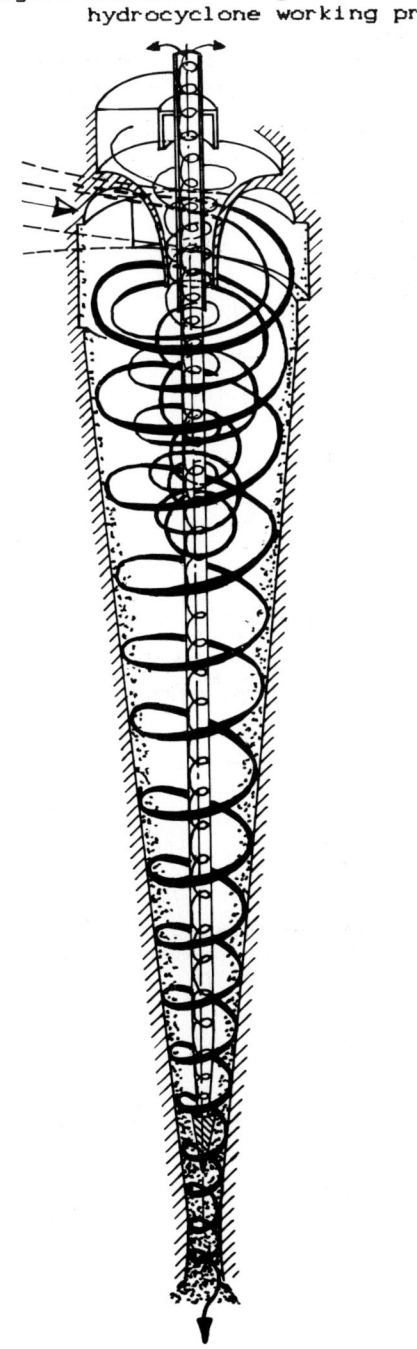

Fig.1. Schematic diagram of a three-product hydrocyclone working principle.

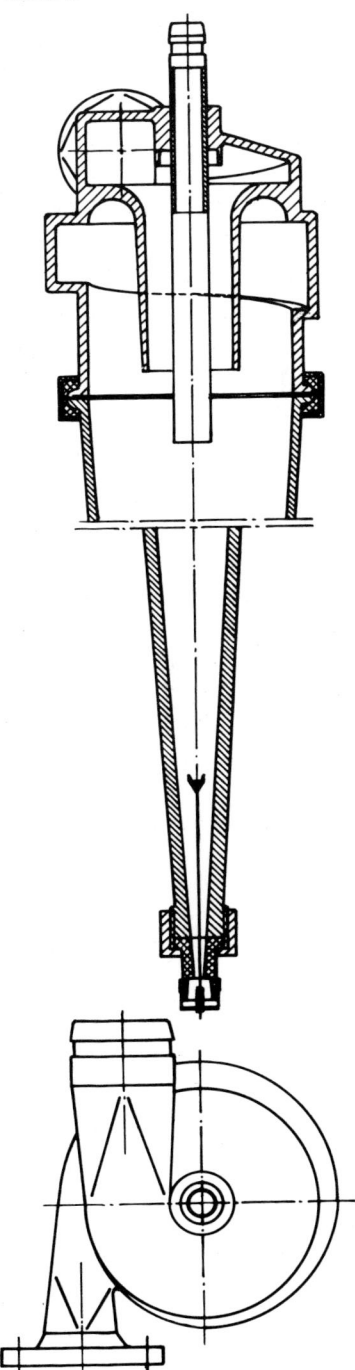

Fig.2. Three-product hydrocyclone design used during research work.

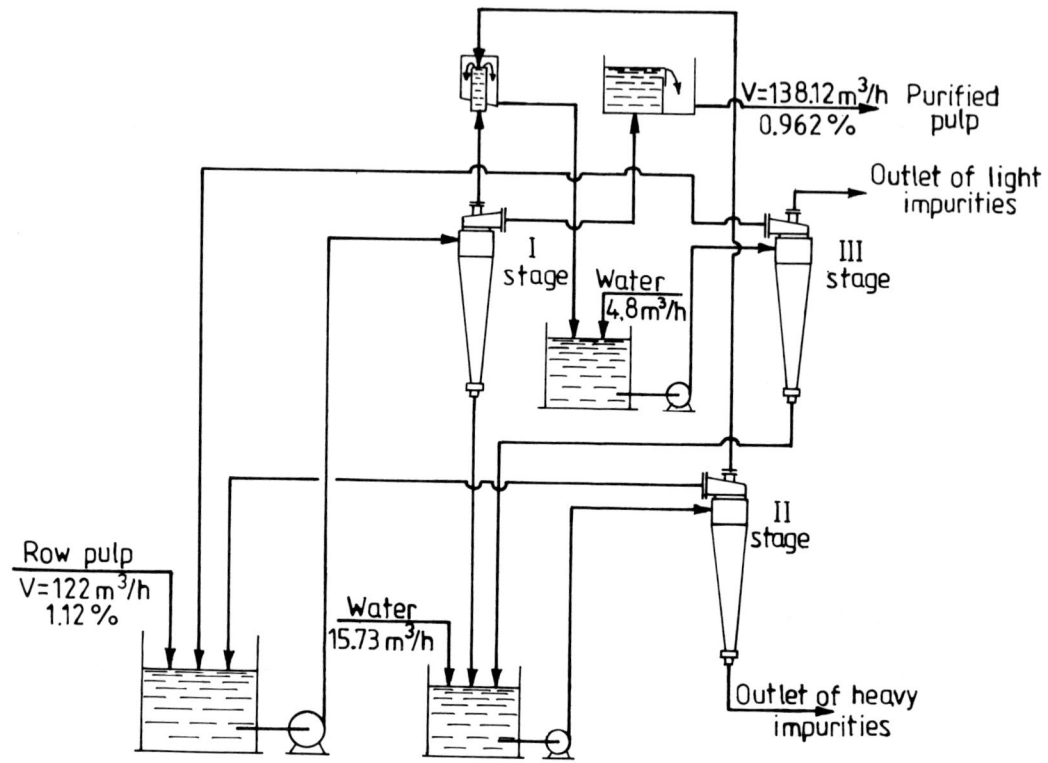

Fig. 3. Diagram of three stage installation for separation of lighter and heavier solid impurities from paper pulp.

OPERATIONAL PARAMETERS OF HYDROCYCLONES IN THE UREA CRYSTALS THICKENING

STANISŁAW BEDNARSKI
Technical University of Cracow, Poland

ABSTRACT

Effect of selected constructional parameters of a hydrocyclone on the process of separation and thickening of urea crystals obtained directly from a crystallizer has been investigated. Concentrated sludge of urea crystals has been dewatered in a pulse centrifuge type Sharples. Introduction of a crystal washing operation has improved significantly urea quality.
It has been stated that 50 % of used centrifuges can be shut down without production decrease. Improvement of crystallization and filtration processes has been noted. Electrical energy consumption has been considerably reduced.

INTRODUCTION

Urea synthesis from NH_3 and CO_2, independently from the applied technological variant of the process, leads mainly to its generation in a form of a solution which yields in the process of evaporation and afterwards crystallization a suspension of urea crystals in mother liquor.
 For separation of urea crystals, which are next directed to drying, centrifuges are used, among others pulsatory multistage e.g. type Sharples. Both, purchase and operating costs of centrifuges are generally high, they are particularly energy consuming. Therefore to solve the problem of energy consumption reduction as well as the costs of centrifuge repairs it has been proposed to thicken the urea crystals suspension before directing it to the centrifuges. Thus the number of operational centrifuges could be reduced. Investigations of urea crystals thickening have been carried out in hydrocyclones and sieve bends. The results obtained on sieve bends will be published in the Polish technical periodicals.
 In this paper only some results will be discussed, those obtained from experiments in which hydrocyclones colaborated with centrifuges.

EXPERIMENTAL INSTALLATION AND INVESTIGATION RANGE

Experimental installation has been constructed in the chemical works on a shunt of the line transporting urea for thickening and centrifuging as shown in Fig.1.

From the lower part of the crystallizer **1** the impeller pump **2** of capacity 150 m^3/h situated on the level 0.0 feeds the urea crystals suspension through the collector **3** to the thickener **4** placed on the level 12.5 m. Under the thickeners there was the hydrocyclone battery **5** consisting of 6 pieces of 200 mm diameter. This battery was fed through the by-pass **6** after the opening of the valve **7** and the closure of the valve **8**. On this by-pass pipe pressure and temperature were measured. Hydrocyclone battery as well as all the collectors were properly insulated, and in case of plugging of the valves **7** and **8** with crystals, steam **9** could be directed to them. Overflow streams from the hydrocyclones were fed through the collector **10** to an evaporator due to gravity, whereas underflow streams from the hydrocyclones were distributed, also due to gravity, through the collector **11** to the pulsatory centrifugal filters type Sharples.

Besides of the different urea crystals concentrations in the suspension and its temperature, which were independent of the investigators and resulted from the technological system, the pressure and some constructional parameters of hydrocyclones were changed in the experiments.

Two centrifuges were involved in the cooperation with the hydrocyclones. The other two were added only when the previous system was used for which four centrifuges were rquired. Hydrocyclones were connected directly with the central feed distributor and each of them could be shut down. The part of mother liquor from hydrocyclone overflows through the collector 13 was directed straight to the crystallizer.

Investigations were carried out in several series for which constructional parameters of hydrocyclones were changed. Moreover during the nearly continuous run of the station in half-year period test runs were carried out (one or two weeks each) for the most advantageous hydrocyclone parameters. As the urea crystals in hydrocyclone underflows were thickened to the significant degree before dewatering in one centrifuge they were washed with pure water in order to decrease the biuret concentration.

RESULTS DISCUSSION AND CONCLUSIONS

The results of urea crystals thickening in hydrocyclones are presented in the table inserted below, in which the range of investigated constructional parameters and the inlet pressure have been given in the following denotation: $200/d_e/d_o/d_u/\varphi$-p. Flow rate was designed V and urea crystals concentration in mass percentage as c, whereas subscript "e" refers to the feed stream, "o" - to the overflow and "u" to the underflow. In the last column (12) there is an efficiency of urea crystals separation from the solution i.e. their yield in the underflow fraction, defined as:

$$\eta = \frac{V_e c_e - V_o c_o}{V_e c_e} \cdot 100 = \frac{V_u c_u}{V_e c_e} \cdot 100 \ \%$$

In this definition volumetric flow rate was used instead of mass flows because of the difficulties with the mass flow rate measurements.

Therefore the values of this coefficient are not real but can be used for comparison.

In this table only some experimental results are gathered for which urea crystals concentration in the feed stream was from the range 16.6-17.4 % mass. In reality, however, mass amount of urea crystals after the crystallizer varied in the range 13.7-32 %, mainly 18+3 %.

From the data collected in the table various dependences can be presented graphically and appropriate conclusions can be drawn which can be the base for the selection of constructional and operational parameters. It follows from the data that te most advantageous operation can be achieved for the high hydrocyclones equipped with the conical part of small apex angles φ and for higher pressures p. The height of cylindrical part of the investigated hydrocyclone was equal 140 mm.

Pressure increase over 0.20 MPa did not cause the proportional increase of neither the underflow concentration nor the amount of separated urea crystals. It has been also stated that the optimum concentration of urea crystals in the hydrocyclone underflow stream, which feeds the centrifuge should be 41-43 %. Thickening of urea crystals over 45 % may cause the plugging of the pipeline feeding the centrifuge and causes the uneven distribution which can lead to the centrifuge vibration.

Washing of the thickened urea crystals provided for the significant decrease of the biuret content in the dewatered product to as low value as 0.5 %. Heavy metal concentration was also decreased, but they were cumulated and as a result they achived the state of equilibrium.

Mother liquor recycling to the evaporator and partly directly to the crystallizer provided for the crystal growth and facilitated the filtration process. Apart from many technological advantages the main economical profits were obtained by the shut down of 50 % of centrifuges at the same production capacity. Due to the removal of large thickeners the free space was obtained and the exchange of existing pumps for the pumps of a higher lift increased the power consumption only by 60 %.

In the thickening-dewatering system due to the urea crystals thickening up to 40-42 %, electrical power consumption was reduced by 30-35 %. Moreover, after the automatic control of this system has been introduced, technical staff was reduced to a minimum and the service was limited to periodical supervision.

p	φ	Constructional parameters			Feed		Overflow		Underflow		Effi-ciency
		d_e	d_o	d_u	$V_e \cdot 10^{-3}$	c_e	$V_o \cdot 10^{-3}$	c_o	$V_u \cdot 10^{-3}$	c_u	η
MPa	Deg	mm			$\frac{m^3}{min}$	%	$\frac{m^3}{min}$	%	$\frac{m^3}{min}$	%	%
1	2	3	4	5	6	7	8	9	10	11	12
0.10	8	33.3	40	20	311.9	16.80	221.0	10.01	90.9	33.85	58.17
		40.0	50	25	435.5	17.30	313.5	10.61	122.0	34.50	55.87
		50.0	65	30	613.8	17.21	458.0	10.72	155.8	36.30	53.54
0.15	8	33.3	40	20	373.6	17.08	266.2	8.14	107.4	38.88	65.44
		40.0	50	25	519.3	17.20	378.2	8.46	141.1	40.62	64.17
		50.0	65	30	750.0	17.05	565.0	8.65	185.0	42.70	61.78
0.20		33.3	40	20	434.5	16.96	311.0	7.03	123.5	41.97	70.33
		40.0	50	25	598.0	16.78	437.0	7.18	161.0	42.85	68.75
		50.0	65	30	861.0	16.92	647.0	7.73	214.0	44.70	65.66
0.10	12	33.3	40	25	275.4	17.36	201.2	10.85	74.2	35.00	54.32
		40.0	50	30	378.4	16.88	286.3	10.82	92.1	35.72	51.50
		50.0	65	35	546.0	16.69	426.0	10.90	120.0	37.25	49.05
0.15	12	33.3	40	25	333.6	17.13	245.4	9.11	88.2	39.45	60.89
		40.0	50	30	461.2	17.25	352.0	9.77	109.2	41.37	56.78
		50.0	65	35	663.2	16.97	523.0	9.97	140.2	43.10	53.69
0.20		33.3	40	25	385.9	16.75	286.0	7.91	99.9	42.05	65.00
		40.4	50	30	539.4	17.12	413.0	8.94	126.4	43.85	60.02
		50.0	65	35	764.0	17.10	602.0	9.61	162.0	44.93	55.71
0.10	20	33.3	40	30	240.0	17.25	182.0	11.25	58.0	36.08	50.55
		40.0	50	35	328.0	17.15	256.0	11.48	72.0	37.32	47.77
		50.0	65	40	483.5	17.18	391.4	12.14	92.1	38.60	42.80
0.15	20	33.3	40	30	295.9	16.68	224.9	9.33	71.0	39.96	57.48
		40.0	50	35	401.5	16.72	316.5	9.97	85.0	41.85	52.99
		50.0	65	40	585.8	16.70	478.0	10.65	107.8	43.52	47.16
0.20		33.3	40	30	342.0	17.29	260.0	9.02	82.0	43.46	60.27
		40.0	50	35	473.0	17.16	374.5	9.88	98.5	44.83	54.40
		50.0	65	40	690.0	17.38	563.0	10.99	127.0	45.72	48.42

p	φ	Constructional parameters			Feed		Overflow		Underflow		Efficiency
		d_e	d_o	d_u	$V_e \cdot 10^{-3}$	c_e	$V_o \cdot 10^{-3}$	c_o	$V_u \cdot 10^{-3}$	c_u	η
MPa	Deg	mm			$\frac{m^3}{min}$	%	$\frac{m^3}{min}$	%	$\frac{m^3}{min}$	%	%
1	2	3	4	5	6	7	8	9	10	11	12
0.10	8	33.3	40	20	311.9	16.80	221.0	10.01	90.9	33.85	58.17
		40.0	50	25	435.5	17.30	313.5	10.61	122.0	34.50	55.87
		50.0	65	30	613.8	17.21	458.0	10.72	155.8	36.30	53.54
0.15	8	33.3	40	20	373.6	17.08	266.2	8.14	107.4	38.88	65.44
		40.0	50	25	519.3	17.20	378.2	8.46	141.1	40.62	64.17
		50.0	65	30	750.0	17.05	565.0	8.65	185.0	42.70	61.78
0.20	8	33.3	40	20	434.5	16.96	311.0	7.03	123.5	41.97	70.33
		40.0	50	25	598.0	16.78	437.0	7.18	161.0	42.85	68.75
		50.0	65	30	861.0	16.92	647.0	7.73	214.0	44.70	65.66
0.10	12	33.3	40	25	275.4	17.36	201.2	10.85	74.2	35.00	54.32
		40.0	50	30	378.4	16.88	286.3	10.82	92.1	35.72	51.50
		50.0	65	35	546.0	16.69	426.0	10.90	120.0	37.25	49.05
0.15	12	33.3	40	25	333.6	17.13	245.4	9.11	88.2	39.45	60.89
		40.0	50	30	461.2	17.25	352.0	9.77	109.2	41.37	56.78
		50.0	65	35	663.2	16.97	523.0	9.97	140.2	43.10	53.69
0.20	12	33.3	40	25	385.9	16.75	286.0	7.91	99.9	42.05	65.00
		40.4	50	30	539.4	17.12	413.0	8.94	126.4	43.85	60.02
		50.0	65	35	764.0	17.10	602.0	9.61	162.0	44.93	55.71
0.10	20	33.3	40	30	240.0	17.25	182.0	11.25	58.0	36.08	50.55
		40.0	50	35	328.0	17.15	256.0	11.48	72.0	37.32	47.77
		50.0	65	40	483.5	17.18	391.4	12.14	92.1	38.60	42.80
0.15	20	33.3	40	30	295.9	16.68	224.9	9.33	71.0	39.96	57.48
		40.0	50	35	401.5	16.72	316.5	9.97	85.0	41.85	52.99
		50.0	65	40	585.8	16.70	478.0	10.65	107.8	43.52	47.16
0.20	20	33.3	40	30	342.0	17.29	260.0	9.02	82.0	43.46	60.27
		40.0	50	35	473.0	17.16	374.5	9.88	98.5	44.83	54.40
		50.0	65	40	690.0	17.38	563.0	10.99	127.0	45.72	48.42

p	φ	Constructional parameters			Feed		Overflow		Underflow		Efficiency
		d_e	d_o	d_u	$V_e \cdot 10^{-3}$	c_e	$V_o \cdot 10^{-3}$	c_o	$V_u \cdot 10^{-3}$	c_u	η
MPa	Deg	mm			$\frac{m^3}{min}$	%	$\frac{m^3}{min}$	%	$\frac{m^3}{min}$	%	%
1	2	3	4	5	6	7	8	9	10	11	12
0.10	8	33.3	40	20	311.9	16.80	221.0	10.01	90.9	33.85	58.17
		40.0	50	25	435.5	17.30	313.5	10.61	122.0	34.50	55.87
		50.0	65	30	613.8	17.21	458.0	10.72	155.8	36.30	53.54
0.15		33.3	40	20	373.6	17.08	266.2	8.14	107.4	38.88	65.44
		40.0	50	25	519.3	17.20	378.2	8.46	141.1	40.62	64.17
		50.0	65	30	750.0	17.05	565.0	8.65	185.0	42.70	61.78
0.20		33.3	40	20	434.5	16.96	311.0	7.03	123.5	41.97	70.33
		40.0	50	25	598.0	16.78	437.0	7.18	161.0	42.85	68.75
		50.0	65	30	861.0	16.92	647.0	7.73	214.0	44.70	65.66
0.10	12	33.3	40	25	275.4	17.36	201.2	10.85	74.2	35.00	54.32
		40.0	50	30	378.4	16.88	286.3	10.82	92.1	35.72	51.50
		50.0	65	35	546.0	16.69	426.0	10.90	120.0	37.25	49.05
0.15		33.3	40	25	333.6	17.13	245.4	9.11	88.2	39.45	60.89
		40.0	50	30	461.2	17.25	352.0	9.77	109.2	41.37	56.78
		50.0	65	35	663.2	16.97	523.0	9.97	140.2	43.10	53.69
0.20		33.3	40	25	385.9	16.75	286.0	7.91	99.9	42.05	65.00
		40.4	50	30	539.4	17.12	413.0	8.94	126.4	43.85	60.02
		50.0	65	35	764.0	17.10	602.0	9.61	162.0	44.93	55.71
0.10	20	33.3	40	30	240.0	17.25	182.0	11.25	58.0	36.08	50.55
		40.0	50	35	328.0	17.15	256.0	11.48	72.0	37.32	47.77
		50.0	65	40	483.5	17.18	391.4	12.14	92.1	38.60	42.80
0.15		33.3	40	30	295.9	16.68	224.9	9.33	71.0	39.96	57.48
		40.0	50	35	401.5	16.72	316.5	9.97	85.0	41.85	52.99
		50.0	65	40	585.8	16.70	478.0	10.65	107.8	43.52	47.16
0.20		33.3	40	30	342.0	17.29	260.0	9.02	82.0	43.46	60.27
		40.0	50	35	473.0	17.16	374.5	9.88	98.5	44.83	54.40
		50.0	65	40	690.0	17.38	563.0	10.99	127.0	45.72	48.42

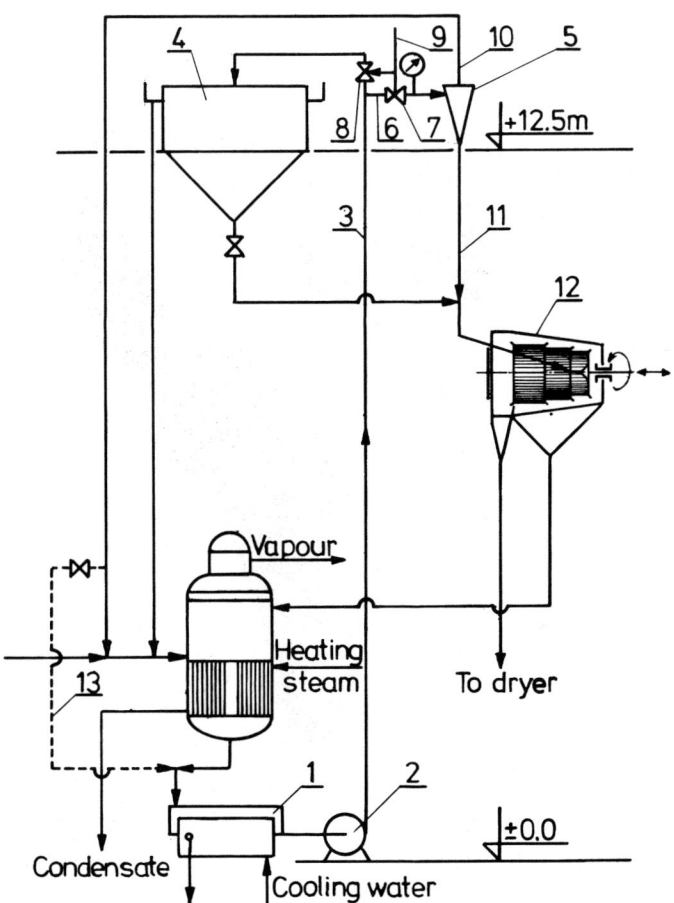

Fig.1. Scheme of experimental installation.

Fig. 2. View of a hydrocyclone of 200 mm diameter.

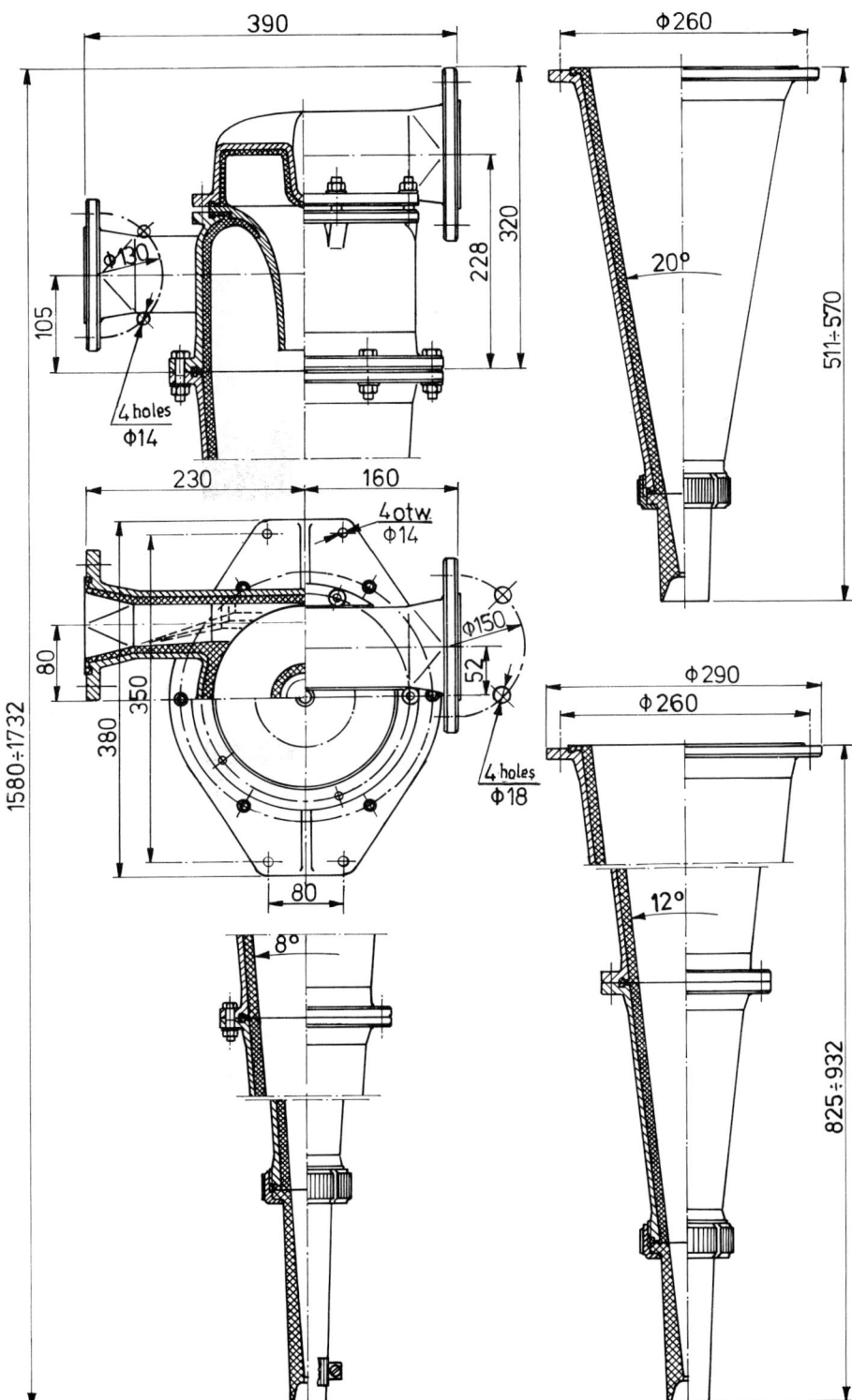

Fig.3. Design of hydrocyclone with interchangeable conical parts used during research work.

Mechanics

FLUID MECHANICS AND ITS APPLICATIONS

Series Editor: R. Moreau

Aims and Scope of the Series

The purpose of this series is to focus on subjects in which fluid mechanics plays a fundamental role. As well as the more traditional applications of aeronautics, hydraulics, heat and mass transfer etc., books will be published dealing with topics which are currently in a state of rapid development, such as turbulence, suspensions and multiphase fluids, super and hypersonic flows and numerical modelling techniques. It is a widely held view that it is the interdisciplinary subjects that will receive intense scientific attention, bringing them to the forefront of technological advancement. Fluids have the ability to transport matter and its properties as well as transmit force, therefore fluid mechanics is a subject that is particularly open to cross fertilisation with other sciences and disciplines of engineering. The subject of fluid mechanics will be highly relevant in domains such as chemical, metallurgical, biological and ecological engineering. This series is particularly open to such new multidisciplinary domains.

1. M. Lesieur: *Turbulence in Fluids*. 2nd rev. ed., 1990 ISBN 0-7923-0645-7
2. O. Métais and M. Lesieur (eds.): *Turbulence and Coherent Structures*. 1991
   ISBN 0-7923-0646-5
3. R. Moreau: *Magnetohydrodynamics*. 1990 ISBN 0-7923-0937-5
4. E. Coustols (ed.): *Turbulence Control by Passive Means*. 1990 ISBN 0-7923-1020-9
5. A. A. Borissov (ed.): *Dynamic Structure of Detonation in Gaseous and Dispersed Media*. 1991 ISBN 0-7923-1340-2
6. K.-S. Choi (ed.): *Recent Developments in Turbulence Management*. 1991
   ISBN 0-7923-1477-8
7. E.P. Evans and B. Coulbeck (eds.): *Pipeline Systems*. 1992 ISBN 0-7923-1668-1
8. B. Nau (ed.): *Fluid Sealing*. 1992 ISBN 0-7923-1669-X
9. T.K.S. Murthy (ed.): *Computational Methods in Hypersonic Aerodynamics*. 1992
   ISBN 0-7923-1673-8
10. R. King (ed.): *Fluid Mechanics of Mixing*. Modelling, Operations and Experimental Techniques. 1992 ISBN 0-7923-1720-3
11. Z. Han & X. Yin: *Shock Dynamics*. 1992 ISBN 0-7923-1746-7
12. L. Svarovsky and M.T. Thew (eds.): *Hydrocyclones*. Analysis and Applications. 1992
    ISBN 0-7923-1876-5
13. A. Lichtarowicz (ed.): *Jet Cutting Technology*. 1992 ISBN 0-7923-1979-6

Kluwer Academic Publishers – Dordrecht / Boston / London

Mechanics

SOLID MECHANICS AND ITS APPLICATIONS
Series Editor: G.M.L. Gladwell

Aims and Scope of the Series

The fundamental questions arising in mechanics are: *Why?*, *How?*, and *How much?* The aim of this series is to provide lucid accounts written by authoritative researchers giving vision and insight in answering these questions on the subject of mechanics as it relates to solids. The scope of the series covers the entire spectrum of solid mechanics. Thus it includes the foundation of mechanics; variational formulations; computational mechanics; statics, kinematics and dynamics of rigid and elastic bodies; vibrations of solids and structures; dynamical systems and chaos; the theories of elasticity, plasticity and viscoelasticity; composite materials; rods, beams, shells and membranes; structural control and stability; soils, rocks and geomechanics; fracture; tribology; experimental mechanics; biomechanics and machine design.

1. R.T. Haftka, Z. Gürdal and M.P. Kamat: *Elements of Structural Optimization*. 2nd rev.ed., 1990   ISBN 0-7923-0608-2
2. J.J. Kalker: *Three-Dimensional Elastic Bodies in Rolling Contact*. 1990   ISBN 0-7923-0712-7
3. P. Karasudhi: *Foundations of Solid Mechanics*. 1991   ISBN 0-7923-0772-0
4. N. Kikuchi: *Computational Methods in Contact Mechanics*. (forthcoming)   ISBN 0-7923-0773-9
5. Y.K. Cheung and A.Y.T. Leung: *Finite Element Methods in Dynamics*. 1991   ISBN 0-7923-1313-5
6. J.F. Doyle: *Static and Dynamic Analysis of Structures*. With an Emphasis on Mechanics and Computer Matrix Methods. 1991   ISBN 0-7923-1124-8; Pb 0-7923-1208-2
7. O.O. Ochoa and J.N. Reddy: *Finite Element Modelling of Composite Structures*. (forthcoming)   ISBN 0-7923-1125-6
8. M.H. Aliabadi and D.P. Rooke: *Numerical Fracture Mechanics*.   ISBN 0-7923-1175-2
9. J. Angeles and C.S. López-Cajún: *Optimization of Cam Mechanisms*. 1991   ISBN 0-7923-1355-0
10. D.E. Grierson, A. Franchi and P. Riva: *Progress in Structural Engineering*. 1991   ISBN 0-7923-1396-8
11. R.T. Haftka and Z. Gürdal: *Elements of Structural Optimization*. 3rd rev. and exp. ed. 1992   ISBN 0-7923-1504-9; Pb 0-7923-1505-7
12. J.R. Barber: *Elasticity*. 1992   ISBN 0-7923-1609-6; Pb 0-7923-1610-X
13. H.S. Tzou and G.L. Anderson (eds.): *Intelligent Structural Systems*. 1992   ISBN 0-7923-1920-6
14. E.E. Gdoutos: *Fracture Mechanics*. An Introduction   ISBN 0-7923-1932-X
15. J.P. Ward: *Solid Mechanics*. An Introduction. 1992   ISBN 0-7923-1949-4

Kluwer Academic Publishers – Dordrecht / Boston / London

Mechanics

From 1990, books on the subject of *mechanics* will be published under two series:
FLUID **MECHANICS AND ITS APPLICATIONS**
  Series Editor: R.J. Moreau
SOLID **MECHANICS AND ITS APPLICATIONS**
  Series Editor: G.M.L. Gladwell

Prior to 1990, the books listed below were published in the respective series indicated below.

MECHANICS: DYNAMICAL SYSTEMS
Editors: L. Meirovitch and G.Æ. Oravas

1. E.H. Dowell: *Aeroelasticity of Plates and Shells*. 1975 ISBN 90-286-0404-9
2. D.G.B. Edelen: *Lagrangian Mechanics of Nonconservative Nonholonomic Systems.* 1977 ISBN 90-286-0077-9
3. J.L. Junkins: *An Introduction to Optimal Estimation of Dynamical Systems.* 1978 ISBN 90-286-0067-1
4. E.H. Dowell (ed.), H.C. Curtiss Jr., R.H. Scanlan and F. Sisto: *A Modern Course in Aeroelasticity.* Revised and enlarged edition see under Volume 11
5. L. Meirovitch: *Computational Methods in Structural Dynamics.* 1980 ISBN 90-286-0580-0
6. B. Skalmierski and A. Tylikowski: *Stochastic Processes in Dynamics.* Revised and enlarged translation. 1982 ISBN 90-247-2686-7
7. P.C. Müller and W.O. Schiehlen: *Linear Vibrations.* A Theoretical Treatment of Multi-degree-of-freedom Vibrating Systems. 1985 ISBN 90-247-2983-1
8. Gh. Buzdugan, E. Mihăilescu and M. Radeş: *Vibration Measurement.* 1986 ISBN 90-247-3111-9
9. G.M.L. Gladwell: *Inverse Problems in Vibration.* 1987 ISBN 90-247-3408-8
10. G.I. Schuëller and M. Shinozuka: *Stochastic Methods in Structural Dynamics.* 1987 ISBN 90-247-3611-0
11. E.H. Dowell (ed.), H.C. Curtiss Jr., R.H. Scanlan and F. Sisto: *A Modern Course in Aeroelasticity.* Second revised and enlarged edition (of Volume 4). 1989 ISBN Hb 0-7923-0062-9; Pb 0-7923-0185-4
12. W. Szemplińska-Stupnicka: *The Behavior of Nonlinear Vibrating Systems.* Volume I: Fundamental Concepts and Methods: Applications to Single-Degree-of-Freedom Systems. 1990 ISBN 0-7923-0368-7
13. W. Szemplińska-Stupnicka: *The Behavior of Nonlinear Vibrating Systems.* Volume II: Advanced Concepts and Applications to Multi-Degree-of-Freedom Systems. 1990 ISBN 0-7923-0369-5
    Set ISBN (Vols. 12–13) 0-7923-0370-9

MECHANICS OF STRUCTURAL SYSTEMS
Editors: J.S. Przemieniecki and G.Æ. Oravas

1. L. Frýba: *Vibration of Solids and Structures under Moving Loads.* 1970 ISBN 90-01-32420-2
2. K. Marguerre and K. Wölfel: *Mechanics of Vibration.* 1979 ISBN 90-286-0086-8

Mechanics

3. E.B. Magrab: *Vibrations of Elastic Structural Members.* 1979 ISBN 90-286-0207-0
4. R.T. Haftka and M.P. Kamat: *Elements of Structural Optimization.* 1985
   Revised and enlarged edition see under Solid Mechanics and Its Applications, Volume 1
5. J.R. Vinson and R.L. Sierakowski: *The Behavior of Structures Composed of Composite Materials.* 1986 ISBN Hb 90-247-3125-9; Pb 90-247-3578-5
6. B.E. Gatewood: *Virtual Principles in Aircraft Structures.* Volume 1: Analysis. 1989
   ISBN 90-247-3754-0
7. B.E. Gatewood: *Virtual Principles in Aircraft Structures.* Volume 2: Design, Plates, Finite Elements. 1989 ISBN 90-247-3755-9
   Set (Gatewood 1 + 2) ISBN 90-247-3753-2

MECHANICS OF ELASTIC AND INELASTIC SOLIDS
Editors: S. Nemat-Nasser and G.Æ. Oravas

1. G.M.L. Gladwell: *Contact Problems in the Classical Theory of Elasticity.* 1980
   ISBN Hb 90-286-0440-5; Pb 90-286-0760-9
2. G. Wempner: *Mechanics of Solids with Applications to Thin Bodies.* 1981
   ISBN 90-286-0880-X
3. T. Mura: *Micromechanics of Defects in Solids.* 2nd revised edition, 1987
   ISBN 90-247-3343-X
4. R.G. Payton: *Elastic Wave Propagation in Transversely Isotropic Media.* 1983
   ISBN 90-247-2843-6
5. S. Nemat-Nasser, H. Abé and S. Hirakawa (eds.): *Hydraulic Fracturing and Geothermal Energy.* 1983 ISBN 90-247-2855-X
6. S. Nemat-Nasser, R.J. Asaro and G.A. Hegemier (eds.): *Theoretical Foundation for Large-scale Computations of Nonlinear Material Behavior.* 1984 ISBN 90-247-3092-9
7. N. Cristescu: *Rock Rheology.* 1988 ISBN 90-247-3660-9
8. G.I.N. Rozvany: *Structural Design via Optimality Criteria.* The Prager Approach to Structural Optimization. 1989 ISBN 90-247-3613-7

MECHANICS OF SURFACE STRUCTURES
Editors: W.A. Nash and G.Æ. Oravas

1. P. Seide: *Small Elastic Deformations of Thin Shells.* 1975 ISBN 90-286-0064-7
2. V. Panc: *Theories of Elastic Plates.* 1975 ISBN 90-286-0104-X
3. J.L. Nowinski: *Theory of Thermoelasticity with Applications.* 1978
   ISBN 90-286-0457-X
4. S. Łukasiewicz: *Local Loads in Plates and Shells.* 1979 ISBN 90-286-0047-7
5. C. Firt: *Statics, Formfinding and Dynamics of Air-supported Membrane Structures.* 1983 ISBN 90-247-2672-7
6. Y. Kai-yuan (ed.): *Progress in Applied Mechanics.* The Chien Wei-zang Anniversary Volume. 1987 ISBN 90-247-3249-2
7. R. Negruţiu: *Elastic Analysis of Slab Structures.* 1987 ISBN 90-247-3367-7
8. J.R. Vinson: *The Behavior of Thin Walled Structures.* Beams, Plates, and Shells. 1988
   ISBN Hb 90-247-3663-3; Pb 90-247-3664-1

Mechanics

MECHANICS OF FLUIDS AND TRANSPORT PROCESSES
Editors: R.J. Moreau and G.Æ. Oravas

1. J. Happel and H. Brenner: *Low Reynolds Number Hydrodynamics*. With Special Applications to Particular Media. 1983 ISBN Hb 90-01-37115-9; Pb 90-247-2877-0
2. S. Zahorski: *Mechanics of Viscoelastic Fluids*. 1982 ISBN 90-247-2687-5
3. J.A. Sparenberg: *Elements of Hydrodynamics Propulsion*. 1984 ISBN 90-247-2871-1
4. B.K. Shivamoggi: *Theoretical Fluid Dynamics*. 1984 ISBN 90-247-2999-8
5. R. Timman, A.J. Hermans and G.C. Hsiao: *Water Waves and Ship Hydrodynamics*. An Introduction. 1985 ISBN 90-247-3218-2
6. M. Lesieur: *Turbulence in Fluids*. Stochastic and Numerical Modelling. 1987 ISBN 90-247-3470-3
7. L.A. Lliboutry: *Very Slow Flows of Solids*. Basics of Modeling in Geodynamics and Glaciology. 1987 ISBN 90-247-3482-7
8. B.K. Shivamoggi: *Introduction to Nonlinear Fluid-Plasma Waves*. 1988 ISBN 90-247-3662-5
9. V. Bojarevičs, Ya. Freibergs, E.I. Shilova and E.V. Shcherbinin: *Electrically Induced Vortical Flows*. 1989 ISBN 90-247-3712-5
10. J. Lielpeteris and R. Moreau (eds.): *Liquid Metal Magnetohydrodynamics*. 1989 ISBN 0-7923-0344-X

MECHANICS OF ELASTIC STABILITY
Editors: H. Leipholz and G.Æ. Oravas

1. H. Leipholz: *Theory of Elasticity*. 1974 ISBN 90-286-0193-7
2. L. Librescu: *Elastostatics and Kinetics of Aniosotropic and Heterogeneous Shell-type Structures*. 1975 ISBN 90-286-0035-3
3. C.L. Dym: *Stability Theory and Its Applications to Structural Mechanics*. 1974 ISBN 90-286-0094-9
4. K. Huseyin: *Nonlinear Theory of Elastic Stability*. 1975 ISBN 90-286-0344-1
5. H. Leipholz: *Direct Variational Methods and Eigenvalue Problems in Engineering*. 1977 ISBN 90-286-0106-6
6. K. Huseyin: *Vibrations and Stability of Multiple Parameter Systems*. 1978 ISBN 90-286-0136-8
7. H. Leipholz: *Stability of Elastic Systems*. 1980 ISBN 90-286-0050-7
8. V.V. Bolotin: *Random Vibrations of Elastic Systems*. 1984 ISBN 90-247-2981-5
9. D. Bushnell: *Computerized Buckling Analysis of Shells*. 1985 ISBN 90-247-3099-6
10. L.M. Kachanov: *Introduction to Continuum Damage Mechanics*. 1986 ISBN 90-247-3319-7
11. H.H.E. Leipholz and M. Abdel-Rohman: *Control of Structures*. 1986 ISBN 90-247-3321-9
12. H.E. Lindberg and A.L. Florence: *Dynamic Pulse Buckling*. Theory and Experiment. 1987 ISBN 90-247-3566-1
13. A. Gajewski and M. Zyczkowski: *Optimal Structural Design under Stability Constraints*. 1988 ISBN 90-247-3612-9

Mechanics

MECHANICS: ANALYSIS
Editors: V.J. Mizel and G.Æ. Oravas

1. M.A. Krasnoselskii, P.P. Zabreiko, E.I. Pustylnik and P.E. Sbolevskii: *Integral Operators in Spaces of Summable Functions*. 1976 ISBN 90-286-0294-1
2. V.V. Ivanov: *The Theory of Approximate Methods and Their Application to the Numerical Solution of Singular Integral Equations*. 1976 ISBN 90-286-0036-1
3. A. Kufner, O. John and S. Pučík: *Function Spaces*. 1977 ISBN 90-286-0015-9
4. S.G. Mikhlin: *Approximation on a Rectangular Grid. With Application to Finite Element Methods and Other Problems*. 1979 ISBN 90-286-0008-6
5. D.G.B. Edelen: *Isovector Methods for Equations of Balance. With Programs for Computer Assistance in Operator Calculations and an Exposition of Practical Topics of the Exterior Calculus*. 1980 ISBN 90-286-0420-0
6. R.S. Anderssen, F.R. de Hoog and M.A. Lukas (eds.): *The Application and Numerical Solution of Integral Equations*. 1980 ISBN 90-286-0450-2
7. R.Z. Has'minskii: *Stochastic Stability of Differential Equations*. 1980 ISBN 90-286-0100-7
8. A.I. Vol'pert and S.I. Hudjaev: *Analysis in Classes of Discontinuous Functions and Equations of Mathematical Physics*. 1985 ISBN 90-247-3109-7
9. A. Georgescu: *Hydrodynamic Stability Theory*. 1985 ISBN 90-247-3120-8
10. W. Noll: *Finite-dimensional Spaces*. Algebra, Geometry and Analysis. Volume I. 1987 ISBN Hb 90-247-3581-5; Pb 90-247-3582-3

MECHANICS: COMPUTATIONAL MECHANICS
Editors: M. Stern and G.Æ. Oravas

1. T.A. Cruse: *Boundary Element Analysis in Computational Fracture Mechanics*. 1988 ISBN 90-247-3614-5

MECHANICS: GENESIS AND METHOD
Editor: G.Æ. Oravas

1. P.-M.-M. Duhem: *The Evolution of Mechanics*. 1980 ISBN 90-286-0688-2

MECHANICS OF CONTINUA
Editors: W.O. Williams and G.Æ. Oravas

1. C.-C. Wang and C. Truesdell: *Introduction to Rational Elasticity*. 1973 ISBN 90-01-93710-1
2. P.J. Chen: *Selected Topics in Wave Propagation*. 1976 ISBN 90-286-0515-0
3. P. Villaggio: *Qualitative Methods in Elasticity*. 1977 ISBN 90-286-0007-8

Mechanics

MECHANICS OF FRACTURE
Editors: G.C. Sih

1. G.C. Sih (ed.): *Methods of Analysis and Solutions of Crack Problems.* 1973
   ISBN 90-01-79860-8
2. M.K. Kassir and G.C. Sih (eds.): *Three-dimensional Crack Problems.* A New Solution of Crack Solutions in Three-dimensional Elasticity. 1975   ISBN 90-286-0414-6
3. G.C. Sih (ed.): *Plates and Shells with Cracks.* 1977   ISBN 90-286-0146-5
4. G.C. Sih (ed.): *Elastodynamic Crack Problems.* 1977   ISBN 90-286-0156-2
5. G.C. Sih (ed.): *Stress Analysis of Notch Problems.* Stress Solutions to a Variety of Notch Geometries used in Engineering Design. 1978   ISBN 90-286-0166-X
6. G.C. Sih and E.P. Chen (eds.): *Cracks in Composite Materials.* A Compilation of Stress Solutions for Composite System with Cracks. 1981   ISBN 90-247-2559-3
7. G.C. Sih (ed.): *Experimental Evaluation of Stress Concentration and Intensity Factors.* Useful Methods and Solutions to Experimentalists in Fracture Mechanics. 1981
   ISBN 90-247-2558-5

MECHANICS OF PLASTIC SOLIDS
Editors: J. Schroeder and G.Æ. Oravas

1. A. Sawczuk (ed.): *Foundations of Plasticity.* 1973   ISBN 90-01-77570-5
2. A. Sawczuk (ed.): *Problems of Plasticity.* 1974   ISBN 90-286-0233-X
3. W. Szczepiński: *Introduction to the Mechanics of Plastic Forming of Metals.* 1979
   ISBN 90-286-0126-0
4. D.A. Gokhfeld and O.F. Cherniavsky: *Limit Analysis of Structures at Thermal Cycling.* 1980   ISBN 90-286-0455-3
5. N. Cristescu and I. Suliciu: *Viscoplasticity.* 1982   ISBN 90-247-2777-4

Kluwer Academic Publishers – Dordrecht / Boston / London